To Neal Sullivan
With Best Wishes.

Jerry Barker
Los Angeles
26 Jan. 2007

Quantitative Analysis of Marine Biological Communities

BICENTENNIAL
1807
WILEY
2007
BICENTENNIAL

THE WILEY BICENTENNIAL—KNOWLEDGE FOR GENERATIONS

*E*ach generation has its unique needs and aspirations. When Charles Wiley first opened his small printing shop in lower Manhattan in 1807, it was a generation of boundless potential searching for an identity. And we were there, helping to define a new American literary tradition. Over half a century later, in the midst of the Second Industrial Revolution, it was a generation focused on building the future. Once again, we were there, supplying the critical scientific, technical, and engineering knowledge that helped frame the world. Throughout the 20th Century, and into the new millennium, nations began to reach out beyond their own borders and a new international community was born. Wiley was there, expanding its operations around the world to enable a global exchange of ideas, opinions, and know-how.

For 200 years, Wiley has been an integral part of each generation's journey, enabling the flow of information and understanding necessary to meet their needs and fulfill their aspirations. Today, bold new technologies are changing the way we live and learn. Wiley will be there, providing you the must-have knowledge you need to imagine new worlds, new possibilities, and new opportunities.

Generations come and go, but you can always count on Wiley to provide you the knowledge you need, when and where you need it!

WILLIAM J. PESCE
PRESIDENT AND CHIEF EXECUTIVE OFFICER

PETER BOOTH WILEY
CHAIRMAN OF THE BOARD

Quantitative Analysis of Marine Biological Communities
Field Biology and Environment

Gerald J. Bakus
Professor of Biology
University of Southern California
Los Angeles, California

WILEY-INTERSCIENCE

A John Wiley & Sons, Inc., Publication

Published by John Wiley & Sons, Inc., Hoboken, New Jersey
Published simultaneously in Canada

For general information on our other products and services or for technical support, please
contact our Customer Care Department within the United States at (800) 762-2974, outside
the United States at (317) 572-3993 or fax (317) 572-4002.

Wiley also publishes its books in a variety of electronic formats. Some content that appears
in print may not be available in electronic formats. For more information about Wiley
products, visit our web site at www.wiley.com.

Library of Congress Cataloging-in-Publication Data:

Bakus, Gerald J.
    Quantitative analysis of marine biological communities : field biology and
environment / Gerald J. Bakus.
        p. cm.
    ISBN-13: 978-0-470-04440-7 (cloth/cd)
    ISBN-10: 0-470-04440-3 (cloth/cd)
    1. Marine biology–Mathematical models.   2. Biotic communities–Mathematical
models.   I. Title.
    QH91.57.M38B35 2007
    578.77–dc22   2006019132

Printed in the United States of America

10 9 8 7 6 5 4 3 2 1

Contents

2. Types of Data, Standardizations and Transformations, Introduction to Biometrics, Experimental Design

3. Quantitative Methods in Field Ecology and Other Useful Techniques and Information

Preface

There have been some outstanding books and computer programs published on quantitative methods. Among these are old and new editions of Cochran (1977 and earlier editions), Southwood (1966, 1978), Southwood and Henderson (2000), Krebs (1989, 1999, 2000 and earlier editions), Davis (1973,1986), Sokal and Rohlf (1981,1995), Manly (1997), Zar (1999), Mead (1988), Legendre and Legendre (1998), Jongman et al. (1987), Buckland et al. (2001), Fowler et al. (1998), Borchers et al. (2002) McCune et al. (2002), Thompson (2002) and others. Selected information was condensed and simplified then included in this book. The Internet was very useful in locating information and in obtaining photos of marine science equipment. The many reviewers have been exceptionally helpful in improving the text.

This book is an attempt to combine ordinary quantitative techniques with relatively new advances in quantitative methodology. These quantitative methods are frequently used in many disciplines outside of biology. The idea is to present one or two specific examples (e.g., equations) for each quantitative topic, hopefully the best techniques. The book is an introduction with few exceptions (e.g., environmental impact assessments are discussed in considerable detail). Some topics receive greater emphasis than others because of the popularity of the topics and the interests and knowledge of the author. Emphasis is placed on shoreline and nearshore habitats, especially intertidal (littoral) and scuba-depth regions. Both tropical and nontropical examples are given. Chapter 8 offers information on equipment used offshore and in deeper waters.

This book is designed for advanced undergraduate and graduate students interested in marine biology and field biology, although much of the information can also be used in terrestrial biology. For this reason, limited terrestrial examples are given. Terrestrial examples are also offered to make marine biologists aware of some techniques in ecology that may be of use to them. The book will also be useful as a general introduction for professionals, such as marine biologists in consulting firms, fish and game or fish and wildlife workers, and pollution specialists. **The emphasis is on marine biology and community ecology**, classical population ecology receiving scant coverage. It is suggested that Google or other search engines be used to locate topics. This exposes the reader to many sources of information on the same topic. Most of the chapters are rather straightforward (Chapter 2 – Biometrics) and some complex (Chapter 4 – Community Analyses; Chapter 5 – Multivariate Techniques). Chapter 3 (Quantitative Methods in Field Ecology) is an eclectic mixture of various topics that have been of interest and of help to the author. Many of them are intended to introduce the student to a discipline rather than offering detailed coverage of the topic. Many references are cited for further information. Some reviewers have legitimately stated that I should have offered my personal opinion on certain subjects. Numerous topics are covered in this book. I can claim to have firsthand field experience with perhaps a couple dozen of them and may be considered a specialist in but a few. Consequently, I have relied on the expertise of many others, citing their opinions frequently, especially when they are conflicting.

My biological career began with studies of terrestrial plants, birds and mammals as an undergraduate student. This was followed by population studies on a stream bird (Dipper) in Montana (Bakus, 1959ab). Several years were then devoted to the taxonomy and development of marine sponges in Washington (Bakus, 1966). This led to 30 years of research on the chemical ecology of coral reefs (e.g., Bakus, 1986), all originally due to the fact that I could not initially locate sponges for a NSF-supported field study at Fanning Island, an atoll in the central Pacific (Bakus, 1964). Later, I discovered that the very few exposed sponges present were toxic to fishes, leading to studies in chemical ecology.

I became interested in quantitative techniques in biology because of co-teaching twice with a visiting professor of marine biology from Australia. My first book on quantitative methods (printed in India where I was involved in training programs, but distributed in Europe and elsewhere by Balkema, Rotterdam – see Bakus, 1990) was dedicated to Prof. William Stephenson, Department of Zoology, University of Queensland, Brisbane, Australia, and to my graduate students. Prof. Stephenson frequently referred to his area of research as "the numbers game". Prof. Stephenson died since then. It is through efforts of fine people such as these that we continue on. My dedication here is to the many people interested in field biology, natural history, and quantitative methods.

Acknowledgments

The author is indebted to the following persons for reviews, for contributing short sections to this book, for ideas, and for information. Many others helped in various ways:

Wilfredo Licuanan (entire text + contribution); Gregory Nishiyama (entire text + contribution); Jim Fowler (Chapter 1); Tom Ebert (Chapter 1); Katrin Iken (Chapter 1); Edward Keith (Chapter 1); Anonymous reviewer (Chapter 1); Stuart Hurlbert (Chapters 1–3); Todd Alonzo (Chapter 2); Todd Anderson (Chapter 3); Robert DeWreede (Chapter 3); Dick Neal (Chapter 3 + contribution); Marti Anderson (Chapters 4–5); Michael Fogarty (Chapter 6); David Dow (Chapter 7); Burt Jones (Chapter 8); Leal Mertes (landscape ecology and large scale sampling); Dale Kiefer (landscape ecology and large scale sampling); Domingo Ochavillo (contribution); Augustus Vogel (contribution); Daniel Geiger (contribution); Ernie Iverson; Robert Smith; Anil Jain; John Dawson; Fazi Mofidi; Stephen Spence; Rudolfo Iturriaga; Joe DeVita; John Griffith; Jim Kremer; Minturn Wright; Mohammad Yazdandoust; Bernie Zahuranec; S.K. Dutt; Richard Dugdale; Bruce Schulte; Bill Brostoff; Mary-Frances Thompson; John Morrow; Dick Murphy; Tim Stebbins; Charles H. Peterson; Paul Delaney; George Moore; Rick Pieper; Bruce McCune; Leslie Karr; K. Robert Clarke; Michael Arbib; Jin Yao; Kim Berger; Andrew Clark; Benoit Mandelbrot; Irving Epstein; Detlof von Winterfeldt; Phil Dustan; Mike Risk; Mark Clapham; Stanley Azen; Chona Sister; Chris Chatfield; Douglass Birch; Don Cadian; Suzanne Edmands; Charles Hall; Chris Battershill; Jed Fuhrman; Paul Hudson; Bill Stephenson (who taught me the "numbers game"); Ralph Thompson (who taught me computer programming); Miki Oguri (who taught me oceanographic techniques); Cathy Link (for drawings completed long ago); Former graduate students (who helped in many ways); and John Wiley & Sons editors Jonathan Rose and Danielle Lacourciere.

Contributors

Daniel Geiger, Santa Barbara Museum of Natural History, Santa Barbara, CA 93105, USA

Wilfredo Licuanan, Marine Science Institute, U.P. Campus, Diliman, Quezon City, 1101 Philippines

Dick Neal, Department of Biology, University of Saskatchewan, 112 Science Place, Saskatoon, SK S7N 5E2, Canada

Gregory Nishiyama, Department of Biology, College of the Canyons, Santa Clarita, CA, USA

Domingo Ochavillo, Oriental Gardens Makati, Unit IOK, Pasong Tamo Street, Makati City, 1229 Philippines

Augustine Vogel, Department of Biological Sciences, University of Southern California, Los Angeles, CA 90089-0371, USA

Chapter 1

Biological Sampling Design and Related Topics

1.1 PROFILING METHODS AND UNDERWATER TECHNIQUES

1.1.1 Introduction

Because so many marine studies are conducted in the intertidal or littoral zone, **a review of methods of profiling beaches is now given**.

1.1.2 Profiling a Beach

Profiling a beach involves measurements of changes in elevation from the top of the beach to the water. These changes are then plotted as a figure, appearing as if you were looking at the slope of the beach from the side. This enables one to then **record the zonation of species above mean lower low water so that you know at what tidal level a density study of a species occurred**. There are often two high tides and two low tides each 24 h on the Pacific coast of North America. Thus there is a high low and a low low tide each day. The yearly average of the low low tides is the mean lower low reference point.

There are several methods of obtaining profiles on a beach. Some of these are easier than others; some are more accurate. The method chosen will depend on the availability of equipment and time. The Sight method: Stand at point A (facing the ocean) and ask someone to stand at point B (perpendicular to the shoreline and in line with point A), X m downslope from point A (Fig. 1-1). The distance between points A and B will depend on the slope. The steeper the slope, the shorter the distance. The individual at point B holds a calibrated rod (2–3 m long) in a vertical position with the lower end of the rod resting on the average basal level of the substratum (e.g., between rocks on a rocky shore). The individual at point A then sights the horizon at point B and reads the intercepted height value on the rod. The distance

Quantitative Analysis of Marine Biology Communities: Field Biology and Environment
by Gerald J. Bakus
Copyright © 2007 John Wiley & Sons, Inc.

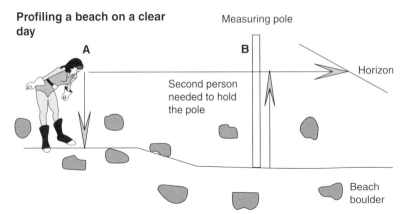

Figure 1-1.   Profiling the beach. Sight from the upper part of the beach to a spot lower down the beach. Measure (1) the height from the eyes to the average beach floor, (2) the height from the average beach floor to the height at the pole where the visual sighting intercepts the horizon, and (3) the distance from the eyes to the pole. Subtract the two height measurements. This is the change in slope over the distance measured. Continue doing this down the beach to the water's edge. Combine the measurements and draw a simple profile figure of the beach. Now that you have a beach profile, you need to determine the tidal level of the profile. Record the position of the water's edge with respect to your profile and the time. Go to a tide table (source: fish tackle store or library). Look for the high and low tidal levels for the day you were profiling the beach (unfortunately not given in metric measurements). Interpolate the tidal level (between high and low tide), based on the time you recorded at the beach, and the times of high and low tides. This gives a reference point tidal level. You can then plot your final beach profile as tidal height (y-axis) vs. distance (x-axis), converting English units (i.e. feet) to metric units (meters) if you wish.

from the average basal level at point A to the individual's eye line is measured and this value is subtracted from the horizon height at point B, giving the change in elevation over a selected distance. A string or twine placed between points A and B (Fig. 1-1) and leveled with a carpenter's level can be used as a substitute for the horizon in foggy weather.

Other methods include using a hand level (with internal bubble for leveling), a Brunton compass (Fig. 1-2), plastic tubing with water, a self-leveling Surveyor's instrument, and a Geographic Positioning System (GPS) or an altimeter with a high degree of accuracy for elevations.

1.1.3   Underwater Profiles

The angle of slopes underwater can be measured with a homemade inclinometer. View the slope sideways, estimating the angle (Fig. 1-3). [An inclinometer can also be used to measure slopes as well as the height of trees on land.]

For oceanographic studies, underwater seafloor profiles are obtained with a precision depth recorder or with sidescan sonar (Fig. 1-4) coupled with the GPS. Sidescan sonar can cover vast areas of the seafloor with a single sweep (the system GLORIA has a two-mile swath).

Figure 1-2. The Brunton compass is used extensively by field geologists. It is tricky to operate as you must peer through two metal holes to the waterline and simultaneously look into the mirror and rotate a knob on the back until the bubble is level, then read off the angle or grade. The advantages are that you can easily measure slopes over long distances with only one person. The distance (d in m) between where you are standing (point A) and the site (Point B or waterline) is measured. Trigonometric functions are applied. $h = d \sin a$ where: h = change in elevation (m), d = distance measured (m) between two points, a = angle measured (degrees). Once h is calculated, the distance from your eye level to the average substratum needs to be subtracted from h. The Brunton compass is accurate to about $1/2°$ for elevations. Example: $d = 50\,m$ $a = 20°$ ($\sin a = 0.0159$) therefore $50 \times 0.0159 = 0.8\,m$ change in the level of the substratum.

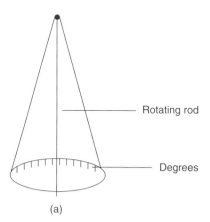

Rotating rod

Degrees

(a)

Figure 1-3a. The plastic clinometer is held up sideways underwater on a reef and the angle of the slope estimated by moving the rotating rod until it follows the average slope line.

1.1.4 Underwater Techniques

Coyer and Whitman (1990) present a comprehensive book on **underwater techniques for temperate and colder waters**. A book by Kingsford and Battershill (2000) is recommended for **techniques of studying temperate marine**

(b)

Figure 1-3b. The metal clinometer reads % slope or angle and is used on land to measure the height of trees. Stand above the base of the tree and measure the % slope to the base of the tree. Measure the % slope to the top of the tree. Measure the distance from the clinometer to the tree. Add the percentages and multiply by the distance measured. For example, the % slope to the base = 30%, the % slope to the treetop = 60%, and the distance = 40 m. Then 30% + 60% = 90% × 40 m = 36 m (the height of the tree).

Figure 1-4. Sidescan sonar. This sonar system (fish) is lowered aft of the ship and towed underwater. It sends out radar and records the topography of the seafloor back on deck. We used it successfully to locate a ship anchor and chain lost offshore from the Port of Los Angeles after two days of operation.

Source: http://www.woodshole.er.usgs.gov/stmapping/images/dataacq/towvehicles/sisi000.jpg

environments by habitat type. An excellent book on **sampling techniques in the tropics** is by English et al. (1997). Hallacher (2004) presents an interesting overview of underwater sampling techniques on coral reefs. See also Fager et al. (1966), UNESCO (1984), and especially Munro (2005). Divers can use a clipboard and waterproof paper (polypaper). The sheets are held down with two large rubber bands (Fig. 1-5). A pencil is tied to the clipboard and the clipboard attached to a brass link on the diver's belt. Alternatively, small polypaper notebooks are available. A very useful tool for measuring distances [e.g., using the Point-Center Quarter (PCQ)

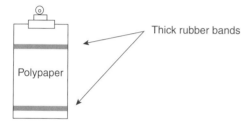

Thick rubber bands

Polypaper

Figure 1-5. A diver's clipboard with polypaper (waterproof paper) and two stout rubber bands to hold the paper down. The clipboard is attached with twine to the diver's belt clip. This mode of operation was designed by Tim Stebbins as a graduate student.

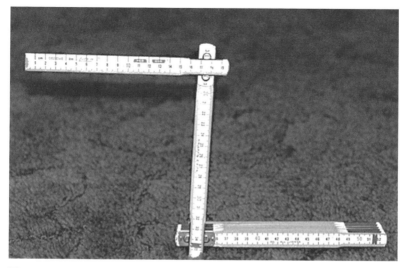

Figure 1-6. Carpenter's collapsible rule. A handy underwater tool for measuring distances. When configured into a square, it forms a $0.25\,m^2$ quadrat.

method] is the collapsible rule. This rule can also easily substitute as a $0.25\,m^2$ quadrat frame (Fig. 1-6). Underwater recording systems are available for divers but they are expensive. WetPC and SeaSlate are recently developed underwater recording systems (see p. 310 in Chapter 8)

1.2   SAMPLING POPULATIONS

1.2.1   Introduction

The procedure by which the sample of units is selected from a population is called the sampling design. Adequate sampling design requires that the correct questions are asked and the study is carried out in a logical, systematic manner. The activities or stages in the study should flow as follows: **purpose → question → hypothesis → sampling design → data collection → statistical analysis → test of hypothesis → interpretation and presentation of the results**. Reasons for sampling populations often involve the need for estimates of densities of organisms and their distribution patterns (e.g., random, clumped, even). These data can then be used to compare community structure or to conduct population studies.

Sampling populations can be accomplished by survey designs (e.g., quadrats, line intercepts) or by model-based inference (Buckland et al., 2001). In a design-based approach to survey sampling, the values of a variable of interest of the population are viewed as fixed quantities. In the model-based approach, the values of the variables of interest in the population are viewed as random variables (Thompson, 2002). Model–based methods use a statistical model of the distribution of organisms based on likelihood methods (e.g., maximum likelihood estimation, Bayes estimation). One area of sampling in which the model-based approach has received considerable attention is with ratio and regression estimation (Thompson, 2002). It has been prevalent in sampling for mining and geological studies. Here we emphasize the use of survey designs. The classical text on sample design is Cochran (1977). An informative book on sampling is Thompson (2002). Murray et al. (2006) recently authored a book on monitoring rocky shores, a valuable source of information on sampling techniques with marine algae and macroinvertebrates.

Krebs (1999), in a leading text on ecological methodology, and Green (1979), in an excellent review of sampling design and statistical methods, **each** present 10 commandments for the field biologist. **They are combined here**. Italic or boldface fonts are explanations, additions, or emphases by the present author.

(1) Find a problem and state concisely what question you are asking.

(2) Not everything that can be measured should be. Use ecological insight to determine what are the important parameters to measure.

(3) Conduct a preliminary survey to evaluate sampling design and statistical analysis options. **Preliminary surveys are critical for well-designed studies**.

(4) Collect data that will achieve your objectives and make a statistician happy. Take replicate samples for each condition (time, space, etc.). See Hessler and Jumars (1974).

(5) **Take an equal number of random replicate samples (at least two) for each condition**. Replicate samples often have 50–90% similarity. **Equal numbers of samples are required for many statistical tests**.

(6) Verify that your sampling device or method is sampling the population with equal and adequate frequency over the entire range of sampling conditions to be encountered.

(7) If the sampling area is large-scale, break it up into relatively homogenous subareas and sample them independently. Allocate samples proportional to the size of the subarea. If an estimate of total abundance is desired, allocate samples proportional to the number of animals in the subarea. *Optimal allotment is to allocate on the basis of within stratum variances (Stuart Hurlbert, pers. comm.).*

(8) Adjust the sample unit size (i.e., number of samples needed) relative to sizes, densities, and spatial distribution of organisms to be sampled. Choose the optimal quadrat size (*see Southwood, 1978 and p. 17 in this chapter*). *Estimate the number of replicates needed to obtain the precision you want*

(Gonor and Kemp, 1978, Krebs, 1999, and see p. 10 in this chapter).
Fractal methods (Chapter 3, p. 168), analysis of variance (Chapter 2,
p. 88), and power analysis (Chapter 2, p. 100) can also be used to deter-
mine the required sample size.

(9) Test the data to determine whether the error variation is homogenous, normally distributed (i.e., has a bell-shaped curve), and independent of the mean. If not, as in most field data, (a) transform the data (Chapter 2, p. 66), (b) use nonparametric analysis (Chapter 2, p. 102), (c) use sequential sampling design *(see p. 27 in this chapter and Krebs, 1999)*, or (d) test against simulated H_0 (null hypothesis) data *(Connor and Simberloff, 1986 and Chapter 3, p. 141)*.

(10) Stick with the result. Do not hunt for a better one.

(11) Some ecological questions are impossible to answer at the present time. For example, historical events that have helped establish future ecological patterns (e.g., *asteroid impacts, rats*).

(12) Decide on the number of significant figures needed in continuous data before an experiment is started.

(13) Never report an ecological estimate without some measure of its possible error.

(14) Always include controls *(in experimental studies)*.

(15) Be skeptical about the results of statistical tests of significance. Cut-off points such as $P = 0.05$ (95% confidence level in your statistical answer) should be considered as shades of gray instead of absolute boundaries.

(16) Never confuse statistical significance with biological significance. Biological characteristics are often much more important than results from a statistical test.

(17) Code all your ecological data and enter it on a computer.

(18) Garbage in, garbage out. Poor data give poor results, no matter what kind of data analysis is used.

Two worthwhile books on terrestrial statistical ecology are those of Ludwig and Reynolds (1988) and Young and Young (1998). Dale (1999) and Fortin and Dale (2005) discuss spatial analysis. Sutherland (1996) discusses basic ecological census techniques then covers specific taxa (plants, invertebrates, fishes, amphibians, reptiles, birds, mammals) and environmental variables. For standard methods in freshwater biology see p. 353 in Chapter 8. See also Elliott (1977) and Gonor and Kemp (1978).

   The most important thing one can do when planning a field study is to make a preliminary survey of the study site. This will indicate whether the organisms are present and provide some information on their density, distribution, and possibly their role in community structure. This preliminary step automatically biases the sampling procedure since further sampling will often take place where the organisms are relatively abundant, but it saves considerable time, effort, and costs for the definitive study.

Four major methods of obtaining population estimates include (1) sampling a unit of habitat and counting organisms in that unit, (2) distance or nearest neighbor techniques, (3) mark-recapture, and (4) removal trapping (Southwood and Henderson, 2000). Removal methods have poor precision and the potential for a large degree of bias (Buckland et al., 2001), thus will not be considered here. Frontier (1983) discusses sampling strategies in ecology.

1.2.2   Sampling Design

Sampling design varies considerably with habitat type and specific taxonomic groups. Kingsford and Battershill (1998) present sampling designs and data analysis based on specific marine habitats. Design analysis in benthic surveys is discussed by Underwood and Chapman (2005). Sampling design begins with a clear statement of the question(s) being asked. This may be the most difficult part of the procedure because the quality of the results is dependent on the nature of the original design. A preliminary survey of the proposed study area is essential as spatial and temporal patterns of selected species can be assessed. If the sampling is for densities of organisms then at least five replicate samples per sampling site are needed because many statistical tests require that minimal number. Better yet, consider 20 replicates per sampling site and in some cases 50 or more. If sample replicates are less than five then bootstrapping techniques can be used to analyze the data (see Chapter 2, p. 113). Some type of random sampling should be attempted (e.g., stratified random sampling) or a line intercept method used to estimate densities (e.g., Strong Method). Measurements of important physical–chemical variables should be made (e.g., temperature, salinity, sediment grain size, etc. – see Chapter 8). Field experiments need to be carried out with carefully designed controls (see Chapter 2, p. 97). The correct spatial scale needs to be considered when planning experiments (Stiling, 2002). Environmental impact assessments ideally attempt to compare before and after studies. For example, a coastline destined to have a new sewage outfall constructed could be studied in detail prior to its initial operation. This study then could be repeated two years after the outfall system begins operation. Because before and after studies are often not feasible, an alternative is to compare impacted areas with nearby unaffected (control) areas.

Peterson et al. (2001) analyzed four major sampling designs in shoreline studies of the impacts of the Exxon Valdez oil spill in the Gulf of Alaska. Two studies employed stratified random sampling techniques and two had fixed (nonrandom) sites. For an explanation of these methods, see pp. 20 and 23 in this chapter. There were differences in sampling sites, sampling dates, effort, replication, taxonomic categories, and recovery data. That the studies came to different conclusions is no surprise (for a similar example of differing interpretations but with the same ecological data see Ferson et al., 1986). The results emphasize how important is sampling design. Gotelli and Ellison (2004) and Odum and Barrett (2005) have informative chapters on sampling design. Diserud and Aagaard (2002) present a method that tests for changes in community structure based on repeated sampling. This may be especially useful in pollution studies and studies on natural catastrophes. See also Cuff and Coleman (1979), Bernstein and Zalinski (1983), Frontier (1983), Andrew and Mapstone (1987), Gilbert (1987), Eberhardt and Thomas (1991), Fairweather (1991), Thompson (1992), Stewart-Oaten and Bence (2001), Peterson et al. (2002), and Lindsey (2003).

1.2.3   Physical–Chemical Factors

Physical and chemical measurements (temperature, salinity, etc.) are frequently carried out when sampling organisms. Techniques for collecting physical–chemical data are discussed in Chapter 8 for marine biology and oceanography. Multivariate analysis of physical–chemical–biological data is discussed in Chapter 5.

1.2.4   Timing of Sampling

The timing of sampling varies with season, age, tides, sex, and other factors. For example, many nocturnal fishes are inactive during the day and seldom observed at that time (Bakus, 1969), thus sampling needs to be done at dawn, dusk, or during nighttime hours for these fishes. Some abundant tropical holothurians move from cryptic habitats and subtidal depths into shallower waters as they mature (Bakus, 1973). There are numerous others changes that occur among species over space and time. These behaviors need to be considered to optimize field studies.

1.2.5   Size of the Sampling Area

The size of the sampling area is highly variable. One must compromise between the overall size of the habitat and the distribution, size, and habits of the organisms, and the statistical measures to be employed before all data have been collected.

1.2.6   Scale

The effects of scale on the interpretation of data have become a very important issue in ecology. The scales commonly encountered in ecology include the individual, patch of individuals, community, and ecosystem (Stiling, 2002). Data based on different spatial scales can yield answers to different questions or even result in different conclusions. One of the earliest discussions on the effects of scale on the interpretation of data from the marine environment is that of Hatcher et al. (1987). For more recent developments see Podani et al. (1993), Schneider (1994), Peterson and Parker (1998), Scott et al. (2002), and Seuront and Strutton (2003). See Fig. 3-1 on p. 124 for examples of how changes in scale can result in different interpretations of the same data. Also see Mann and Lazier (2005).

1.2.7   Modus Operandi

The following sections describe quantitative techniques that give numbers of samples required or densities of organisms. Many of these techniques originated in terrestrial studies and were later employed in aquatic habitats. The examples described herein often center around shorelines or terrestrial sites because most people are familiar with these habitats. Moreover, relatively few students have had shipboard experience to relate to. Nevertheless, these quantitative techniques are often modified and used in seafloor and water column studies as well. For example, plankton sampling can be performed haphazardly, by systematic sampling, or by following a transect line. Infaunal sampling can be carried out with simple random sampling and coordinate lines, stratified random sampling, or line transects. A submersible can perform systematic sampling, belt or strip transects, line intercepts, and so forth. For information on benthic and water column sampling devices see Chapter 8. For information on seafloor sampling techniques see Holme and McIntyre (1984), Mudroch and MacKnight (1994), and Eleftheriou and McIntyre (2005). For information on water column sampling techniques see Hardy (1958), Strickland (1966), Harris et al. (2000), and Paul (2001).

Many of the sampling designs are relatively simple but some (e.g., sequential sampling, mark or tag and recover) can be complex and involve pages of equations and calculations. For those cases, the author refers the reader to references that provide details. A number of special sampling techniques (e.g., coral reef surveys, large scale sampling, etc.) are presented after the discussion of common plot and plotless methods. Collected data can be stored on Microsoft Excel spreadsheets for analysis.

1.2.8   Sample Size or Number of Sample Units Required

Density is the number of individuals per unit area or unit volume. The number of sample units required for a density study is dependent on the variation in population density and the degree of precision required. There are numerous methods for estimating the sample size (i.e., number of samples) needed in any study. The traditional methods have emphasized the variance to mean ratio, such as in the following example for a **normal distribution** (Cochran, 1977):

$$n = \frac{t^2 \, SD^2}{(E \, \bar{X})^2}$$

where

n = number of sample units

t = t value

SD = standard deviation

E = allowable error (e.g., $10\% = 0.1$)

$\bar{X}$ = mean

First conduct a preliminary sampling then calculate the sample mean and the sample variance (σ^2 – see Chapter 2, pp. 76 and 77). Look up the critical t value at $P = 0.05$ and the degrees of freedom (number of samples – 1). Enter the table t value in the equation and the allowable error, say 10% (use 0.1).

Example: The density of brown giant kelp (*Macrocystis pyrifera*) or trees per 100 m^2 is: 17, 7, 8, 5, 3, 5.

The t value for 5 degrees of freedom at $P = 0.05$ is 2.6.
The mean $= 7.5$ and the variance $= 24.7$. With an allowable error of 0.1 (10% error):

$$\text{No. of samples units needed} = \frac{(2.6)^2\,(24.7)}{(0.1 \times 7.5)^2} = 223$$

This large number is based on limited preliminary sampling. Taking more sample units during preliminary sampling could further reduce the number of sample units (decrease the variance) required for the definitive study. A preliminary survey is essential in obtaining precursory density estimates in order to use a preferred method to estimate how many sample units will be needed for a final or definitive study. If this is not possible then a survey of the literature of similar studies is essential.

For population studies, the approximate **number of sample units needed with a Poisson (random) Distribution** is estimated by Krebs (1999:244) as follows:

$$n \cong \left(\frac{200}{r}\right)^2 \frac{1}{\bar{X}}$$

where

    $n =$ sample units required (e.g., number of quadrats or plots)

    $\cong$ approximately equal to

    $r =$ allowable error (%)

    $\bar{X} =$ mean

Example

For a mean of 10, a 10% allowable error, and $\alpha = 0.05$ (95% confidence level – see Chapter 2, p. 81):

$$n \cong \left(\frac{200}{10}\right)^2 \left(\frac{1}{10}\right)$$

$$n \cong (400)\,(0.1)$$

$$n \cong 40 \text{ samples (e.g., quadrats).}$$

Krebs (2000a) has a computer program for this – listed under "quadrat sampling." See the Appendix.

The approximate **number of sample units needed with a negative binomial (aggregated) distribution** is estimated by Krebs (1999:245) as follows:

$$n \cong \frac{\left(100\, t_\alpha\right)^2}{r^2} \left(\frac{1}{X} + \frac{1}{k} \right)$$

where

n = sample units required (e.g., number of quadrats)

$\cong$ = approximately equal to

t_α = t value for $n-1$ degrees of freedom (= 2 for 95% confidence level)

$\overline{X}$ = mean

k = estimated negative binomial exponent

r = allowable error (%).

Approximate estimation of $k = \dfrac{(\overline{X})^2}{(S)^2 - \overline{X}}$

where

$\overline{X}$ = mean

S = standard deviation.

Krebs (2000a) has a maximum likelihood estimation computer program for this – listed under "quadrat sampling." This produces a more precise estimate of k.

Example

For a mean of 4, error of 10%, and negative binomial exponent of 3.

$$n = \frac{(200)^2}{(10)^2} \left(\frac{1}{4} + \frac{1}{3} \right)$$

$$n = 400\,(0.25 + 0.33)$$

$$n = 232 \text{ samples (e.g., quadrats).}$$

The major problem with many of these equations is that the precision level (i.e., 10% allowable error, an arbitrary value) results in too many sample units being required (i.e., often several hundred in the intertidal zone). Hayek and Buzas (1997) state that a precision level of 25–50% is all that is reasonably attainable in many field studies. The 10% sample error may often be met by terrestrial plant ecologists. They contend neither with the tides nor with slow underwater operations. I call this the 1:5:10 rule of thumb, that is, intertidal density studies may take about five times longer, and subtidal studies 10 times longer to obtain the same amount of density data (using plot sampling) as that of many terrestrial studies (e.g., tree densities). When temporal or spatial variation in a population is large, a small number of sample units provides imprecise estimates of population values, so that models derived from such data may be quite distorted (Houston, 1985).

The best sample unit number is the largest sample unit number (Green, 1979). It is better to sample the same total area or volume by taking many small sample units rather than few large ones, according to Green (1979) and Southwood and Henderson (2000). However, this does not consider edge effects, cost considerations, and so forth. Population density (and variance) is always fluctuating thus too much emphasis should not be placed on a precise determination of the optimum size of the sampling unit (Southwood and Henderson, 2000). See Krebs (1999) and Southwood and Henderson (2000) for a discussion of this topic and Krebs (2000a) for a computer program. If one wishes to sample community structure, another method of determining sample size is to use a species area curve (see Chapter 3, p. 145). A newer method of estimating required sample unit number is **power analysis**, discussed in Chapter 2, p. 100, regarding experimental methods. See also Green (1989).

Bakanov (1984) published a nomogram for estimating the number of sample units needed with an aggregated distribution. Manly (1992) discusses bootstrapping techniques for determining sample unit sizes in biological studies. Keltunen (1992) estimates the number of test replicates required using ANOVA.

A correction factor (fpc or finite population correction factor) is employed when sample unit sizes represent more than about 5% of the population. This can be used to reduce the sampling error or the sample unit size required. The equation is:

$$\text{fpc} = \sqrt{\frac{N-n}{N-1}}$$

where

   fpc = finite population correction

   N = size of the population

   n = size of the sample

Assume $N = 2000$ and $n = 200$

$$\text{fpc} = 0.901$$

For example, if the estimated number of sample units needed is 162 and the fpc = 0.901, then the corrected number of sample units needed is:

$$162 \times 0.901 = 146 \text{ samples}$$

In sampling small populations, the fpc factor may have an appreciable effect in reducing the variance of the estimator (Thompson, 2002). For further information see the Internet for numerous examples.

For pollution studies, if you want to know **how many sample units to take** in order to determine if pollution standards have been exceeded, the following equation has been used:

$$N = Y\,\frac{Zs^2}{D^2\,\bar{X}^2}$$

where

N = no. of sample units required

Y = expected level of change (% expressed as a decimal)

s = standard deviation

D = allowable error (10% or 0.1)

$\bar{X}$ = mean

Z = a function of the distance from the mean in standard deviation units.

2-tailed test: $Z\,(p = 0.05) = 1.96\ (=95\%\ \text{confidence level})$

$Z\,(p = 0.01) = 2.58\ (=99\%\ \text{confidence level})$

Example

Assume a Z of 1.96 (95% confidence level), 20% change, allowable error of 10%, mean of 10, and standard deviation of 4.

$$Y = 0.2\,\frac{(1.96)(4)^2}{(0.01)(10)}$$

$$Y = 63\ \text{samples}$$

1.3   QUANTITATIVE SAMPLING METHODS

1.3.1   Introduction

Major methods of sampling marine benthic organisms for abundance can be conveniently categorized as plot and plotless. This section will give only a brief introduction as to how these sampling programs are carried out. The reader is referred to Southwood (1978), Seber (1982), Hayek and Buzas (1997), Krebs (1999), and Thompson (2002) for detailed information. Eleftheriou and McIntyre (2005) discuss methods for the study of marine benthos. The seasonal timing of sampling is determined by the life cycle (Southwood and Henderson, 2000). **Plot methods incorporate the use of rigid boundaries**, that is, squares (quadrats), rectangles, or circles (circlets, unfortunately also called quadrats by some investigators), and circumscribe a given area in which organisms are counted or collected. **They are used to save time, instead of conducting total counts or a census of organisms, and to remove bias in sampling.** Bias is a systematic, directional error (McCune et al., 2002).

Some traditionally plotless sampling techniques become plot techniques when boundaries are added for convenience (e.g., PCQ – see below), and coordinate lines in simple random sampling create sample points rather than fixed boundaries or plots. Establishing transect lines or cluster sampling can be followed by either plot or plotless sampling techniques. Thus plot and plotless are somewhat flexible terms yet are convenient to use.

The plot method of sampling generally consists of three major types: (1) simple random or random sampling without replacement, (2) stratified random, and (3) systematic (Cochran, 1977). Simple random sampling with replacement is inherently less efficient than simple random sampling without replacement (Thompson, 2002). It is important not to have to determine whether any unit in the data is included more than once. Simple random sampling consists of using a grid or a series of coordinate lines (transects) and a table of random numbers to select several plots (quadrats), the size depending on the dimensions and densities of the organisms present (Fig. 1-7 and see p. 19). The advantage of using these standardized sizes is that comparisons can be easily made between the densities of species in different regions and with data collected from the past. Some divers have used circular frames (e.g., using 3 lb. metal coffee cans [approximately 8 inches (20 cm) high by 6 inches (15 cm) in diameter] to core surface sediments in the shallow waters of the coastal Arctic Ocean because this is a convenient way to collect infauna in that region).

The basal area of trees or forest stands has more functional significance than most descriptors of forest structure. Density measurements are of relatively little value with plants unless applied to restricted size classes (McCune et al., 2002).

See Arvantis and Portier (2005) for information on natural resource sampling methodology.

1.3.2   Table of Random Numbers

In the past, few texts had tables of random numbers in columns of two digits, which gave numbers from 1 to 99, convenient for ecologists. The tables were typically columns of four digits. A random number generator starts with an initial number then uses a deterministic algorithm to create pseudorandom numbers (Michael Arbib, pers. comm.). A table of random numbers is shown in Table 1-1. Tables of random numbers are used to take samples randomly. Samples are taken randomly to remove bias.

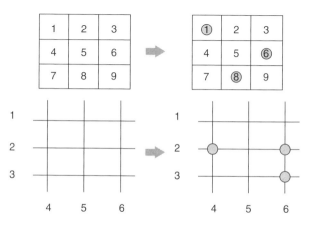

Figure 1-7. Simple random sampling. Random numbers from a table of random numbers give 1,6,8 for the squares and 2-4, 2-6, 3-6 for the coordinate lines, indicating the areas or points to be sampled (e.g., to count animals).

Table 1-1.   A table of random numbers.

20 17	42 28	23 17	59 66	38 61	02 10	86 10	51 55	92 52	44 25
74 49	04 19	03 04	10 33	53 70	11 54	48 63	94 60	94 49	57 38
94 70	49 31	38 67	23 42	29 65	40 88	78 71	37 18	48 64	06 57
22 15	78 15	69 84	32 52	32 54	15 12	54 02	01 37	38 37	12 93
93 29	12 18	27 30	30 55	91 87	50 57	58 51	49 36	12 53	96 40
45 04	77 97	36 14	99 45	52 95	69 85	03 83	51 87	85 56	22 37
44 91	99 49	89 39	94 60	48 49	06 77	64 72	59 26	08 51	25 57
16 23	91 02	19 96	47 59	89 65	27 84	30 92	63 37	26 24	23 66
04 50	65 04	65 65	82 42	70 51	55 04	61 47	88 83	99 34	82 37
32 70	17 72	03 61	66 26	24 71	22 77	88 33	17 78	08 92	73 49
03 64	59 07	42 95	81 39	06 41	20 81	92 34	51 90	39 08	21 42
62 49	00 90	67 86	93 48	31 83	19 07	67 68	49 03	27 47	52 03
61 00	95 86	98 36	14 03	48 88	51 07	33 40	06 86	33 76	68 57
89 03	90 49	28 74	21 04	09 96	60 45	22 03	52 80	01 79	33 81
01 72	33 85	52 40	60 07	06 71	89 27	14 29	55 24	85 79	31 96
27 56	49 79	34 34	32 22	60 53	91 17	33 26	44 70	93 14	99 70
49 05	74 48	10 55	35 25	24 28	20 22	35 66	66 34	26 35	91 23
49 74	37 25	97 26	33 94	42 23	01 28	59 58	92 69	03 66	73 82
20 26	22 43	88 08	19 85	08 12	47 65	65 63	56 07	97 85	56 79
48 87	77 96	43 49	76 93	08 79	22 18	54 55	93 75	97 26	90 77
08 72	87 46	75 73	00 11	27 07	05 20	30 85	22 21	04 67	19 13
95 97	98 62	17 27	31 42	64 71	46 22	32 75	19 32	20 99	94 85
37 99	57 31	70 40	46 55	46 12	24 32	36 74	69 20	72 10	95 93
05 79	58 37	85 33	75 18	88 71	23 44	54 28	00 48	96 23	66 45
55 85	63 42	00 79	91 22	29 01	41 39	51 40	36 65	26 11	78 32

The numbers are arranged into columns of two digits, ideal for the field biologist. Other tables of random numbers may have columns of three or four digits. The digits in a two-column random numbers table range from 01 to 99 usable numbers, in a three-column random numbers table from 01 to 999, and in a four-column random numbers table from 01 to 9999. To use the table, one can proceed from top to bottom (e.g., 20 to 55). Begin with the first column and proceed to the bottom then go to the top of the second column and proceed to the bottom, and so forth. You can also start from a haphazard location in the table (Thompson, 2002). Note that some of these numbers are very close to one another (e.g., 32, 35, 33) by chance. This can skew the results of your survey if you are sampling by the simple random sampling method (see Figure 1-9). This is why ecologists use some type of stratified random sampling in plot techniques. If you need more numbers go to the computer and generate more.

1.3.3 Quadrat Shape

Ecologists have used squares, rectangles, and circles (e.g., 3 lb. coffee cans to core sediments by hand; a 1 m long piece of twine tied to a stake and rotated in a circle as one counts benthic organisms; in songbird surveys). The most common shape for sampling benthic marine organisms is a square (67%), followed by circles (19%), and rectangles (14%) (Pringle, 1984). Rectangular frames with a size ratio of 2:1

tend to give slightly better results with population estimates than do square frames in terrestrial studies (Krebs, 1999). Thompson (2002) compared nine types of plots and compared their detectability functions. **Long, thin rectangular plots are more efficient than square or round plots**. Various line transects, variable circular plots (radial transects), and plots with holes in them (i.e., torus or doughnuts) gave intermediate results. However, **if there is a clinal gradient of some type, a rectangular quadrat can be aligned parallel or perpendicular to the cline and the variance in the density can be very different**. Long quadrats cover more patches, whereas narrow rectangles (size ratios higher than 4:1) can create a severe edge effect, in which too many organisms may cross the boundary of the quadrat, resulting in more frequent counting errors. **Typically, animals intercepting the top and left-hand boundaries are counted (Southwood and Henderson, 2000). Edge effects often produce a positive bias or a number greater than the true density** (Krebs, 1999). Edge effects, in theory, are least with circles, intermediate with hexagons, and greatest with squares and rectangles because bias introduced by edge effects are proportional to the ratio between the boundary length and the area within the boundary (Southwood and Henderson, 2000). Circles are the poorest shape for estimation from aggregated distributions, resulting in high variances (McCune et al., 2002). Squares are also poor and rectangles better for aggregated distributions, especially narrow rectangles, but narrow rectangles may exhibit severe edge effects.

1.3.4 Optimal Quadrat Size

The optimal size for a quadrat depends on many factors. Changes in quadrat size (i.e., scale) can result in differences in the interpretation of field data, such as abundance, associations between species, and the degree of aggregation within a species (Fig. 3-1 on p. 124). One rule of thumb is to select a size of quadrat that will not give frequent yields of zero counts of individuals. **Use the smallest quadrat that is practical or easiest to use but will also sample organisms adequately**. The larger the species the larger the quadrat size. The optimal size for aggregated species is the smallest size relative to the size of the species (Green, 1979). For example, when counting small, numerous barnacles, you may use a $0.1\,m^2$ quadrat frame, but then subdivide the frame into 50 or 100 small squares. A smaller size often results in increased precision of estimates with aggregated distributions because the boundary is small, thus one would be less likely to either double-count or undercount individuals. Moreover, smaller sizes often result in a smaller variance around the mean but scaling factors may alter this (Greig-Smith, 1964). Pringle (1984) found that the $0.25\,m^2$ quadrat was the most efficient size for sampling benthic marine macrophytes. Dethier et al. (1993) concluded that $10 \times 10\,cm$ quadrats were effective for visual estimates of the abundance of sessile benthic marine organisms. A compromise in frame size must be made when more than one species is being studied and counted within the same quadrats. Interactions between adjacent organisms (e.g., production of allelochemicals) may result in the species growing only a certain distance from each other. These interactions should also be considered when determining quadrat size, especially on coral reefs (Wilfredo Licuanan, pers. comm.).

Techniques have been developed to determine the most appropriate group frame size (Southwood, 1978) but field experience seems to be the most efficient and effective determinant of frame size. Southwood (1978) suggests that the relative net precision of a unit of a given size is as follows:

$$RNP = \frac{1}{Cu\, S^2\, u}$$

where

RNP = relative net precision

Cu = relative cost of taking a sample (usually time)

S^2u = variance among unit totals.

Example

Cost (Cu) = 4 h

Variance = 25

$$RNP = \frac{1}{4 \times 25} = \frac{1}{100} = 0.01$$

The highest value of RNP is the best unit. For multiple species, sum the relative net precision values for each quadrat size over all species of interest and choose the unit with the highest sum. If certain species were more important than others (i.e., ecologically as numerical dominants or as keystone species), weighting of their relative precision values would be appropriate. Krebs (1999) recommends the Wiegart method (Wiegart, 1962) in which quadrat size (x-axis) is plotted against relative cost (i.e., time, y-axis) (Fig. 1-8). The size of quadrat with the lowest "cost" is preferred. Krebs (2000a) provides computer programs for determining optimal quadrat size. See the Appendix.

In practice, ecologists often use a range in the size of quadrats from 0.1 to $1.0\,m^2$ (but also $0.01\,m^2$ for small organisms such as barnacles and $100\,m^2$ when sampling the distribution and abundance of trees) to cover all of the possibilities in

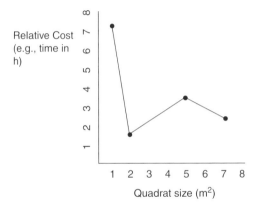

Figure 1-8.   The Weigert method for determining the best quadrat size. It is $2\,m^2$ in this example. Source: modified from Krebs (1999).

a standardized fashion (e.g., number of organisms per 1, 10, or $100\,m^2$). However, one cannot always accurately extrapolate species richness or density in a small area (e.g., $0.1\,m^2$) to species richness or density in a larger area (e.g., $1\,m^2$) because the relationship between the two areal sizes is often nonlinear. Such extrapolations are done frequently for convenience, but must be interpreted carefully. See West (1985) for an interesting discussion on nonlinearity.

When counting organisms in a quadrat, one should examine each quadrat in a similar manner. For example, in looking down on a quadrat from above, you may wish to exclude animals in cracks and crevices (because including cracks and crevices creates numerous complications such as differences in crevice size, shape, depth, etc.). This standardizes the procedure and greatly simplifies the sampling process.

1.3.5 Simple Random Sampling

Simple random sampling consists of using a grid or a series of coordinate lines (transects) and a table of random numbers to select plots (e.g., squares, quadrats). The bottom right side of Fig. 1-9 shows **the main pitfall of the simple random sampling technique, that is, that the random numbers may occur in such a fashion as to concentrate sampling effort mostly in one part of the study area, missing important parts of the study area.** The other major criticism is that the simple random sampling method is unfeasible for large areas (for example, Marsden squares in the ocean or dense forests) since too much time is wasted in moving from one place to a distant site. Marsden squares represent areas on a Mercator chart of the world, each square measuring 10 degrees of latitude by 10 degrees of longitude.

1.3.6 Haphazard (Convenience, Accidental, Arbitrary) Sampling

Haphazard sampling is often carried out in the field to substitute for random sampling. It is sampling without the use of a classical sampling design. Bias is always a problem in haphazard sampling. A diving project in the Maldive Islands required random sampling. Random sampling would have taken an inordinate amount of time and time was limited, thus haphazard sampling was employed. A biologist had

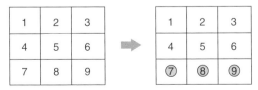

Figure 1-9. Problems with simple random sampling. Three numbers were chosen randomly from a set of number ranging between 1 and 9. By chance they all fell in the lower part of the sampling area. If this were an intertidal site, the study would give an incomplete picture of community structure as it would leave out the middle and upper intertidal zones.

initially and casually swum through the potential site to finally select it as a suitable study area (i.e., it had living hard coral growth rather than continuous sand). He then swam across a flat coral reef area, dropping weights haphazardly every 30 sec, without looking at the seafloor. These weights then became corners of quadrats to be sampled. Some bias was thus removed without random sampling and the effort was highly time efficient. McCune et al. (2002:17) refer to this technique as "arbitrary but without preconceived bias."

1.3.7 Stratified Random Sampling

The sampling design is called stratified random sampling if the design within each stratum (e.g., habitat or elevation) is simple random sampling (Cochran, 1977; Thompson, 2002). In some cases it may be desirable to classify the units of a sample into strata and to use a stratified estimate, even though the sample was selected by simple random sampling, rather than stratified random sampling. Stratified random sampling involves choosing subsamples with a table of random numbers from each of the major plots or quadrats which are arranged in strata in the study area (Fig. 1-10). **This method is frequently used since the sampling is conducted throughout the study area. The advantage of using either simple random or stratified random sampling techniques is that standard statistical procedures can be applied**. Stratified random sampling uses a table of random numbers and is often considered to be the most precise method of estimating population densities other than a direct total count or census, for two reasons. It covers the entire study area and samples randomly from each subdivision of the study area (Southwood and Henderson, 2000). Nevertheless, contrary to assumption, **stratified random sampling is not necessarily the most accurate method of sampling the environment** (because too few samples may be taken and because it may not be as accurate as some line intercept methods with highly aggregated organisms – see p. 43) and **it is often labor intensive for divers and for surveys in dense forests when compared to some plotless methods**.

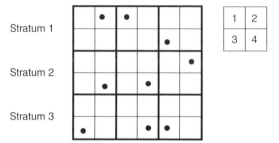

Figure 1-10. Stratified random sampling. The study area is divided into nine large squares (in this example) and each large square into four smaller squares. A table of random numbers is used to select a number (i.e., the dots) between 1 and 4 in each of the larger squares. Thus all strata (3 from top to bottom) are sampled and each large square is sampled randomly.

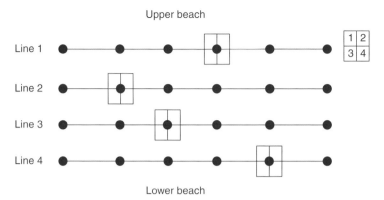

Figure 1-11. Stratified random sampling. A series of transect lines (metric tapes) are lain across the beach. Clothespins are placed at 5 m intervals. A table of random numbers is consulted and one number from 1 to 6 is selected for each transect line. A 0.1 m² quadrat frame is placed in four positions at those random spots and numbered 1–4. A table of random numbers is used to select one number between 1 and 4 for each box on each transect line. The organisms in the selected subunits are then identified and counted, the clothespins removed when the counting is completed. A total of four counts are made in this example.

Stratified random sampling can be carried out in various ways. A grid can be constructed and subdivided into strata, each stratum being subdivided into smaller plots. A table of random numbers is then used to select one of the smaller plots from each of the larger subunits of the stratum (Fig. 1-10). Another method of accomplishing the same goal is to arrange transect lines or coordinates across a study area then mark off every 5 m along each line. A table of random numbers (Table 1-1) is used to select some of the designated points along each line for sampling (Fig. 1-11). A better alternative to this is to mark off the line at each 5 m interval then set up a grid at each point, selecting, for example, one subunit of each set of four subunits per grid using a table of random numbers. (Fig. 1-12). This method covers the entire study area and is sampled randomly.

1.3.8   Systematic Sampling

Systematic sampling is used when a uniform coverage of the area is desired. It can be safely used for convenience when the ordering of the population is essentially random (Cochran, 1977). It is often used in marine studies where the primary interest is to map distributions or monitor sites with respect to environmental gradients or suspected sources of pollution (Southwood and Henderson, 2000; McDonald, 2004). Systematic sampling involves choosing a constant sampling pattern (for example, every other quadrat or every third quadrat, see Fig. 1-13). Note that the systematic pattern may conform with an environmental pattern (e.g., quadrats 3-5-7 in Fig. 1-13) and this biases the overall results. For example, the systematic pattern could follow a ridgeline of serpentine soils or an intrusive ribbon of intertidal rock of a different characteristic

Upper beach

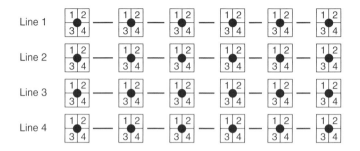

Lower beach

Figure 1-12. Stratified random sampling. A series of transect lines (metric tapes) are lain across the beach. Clothespins are placed at 5 m intervals. A 0.1 m^2 quadrat frame is placed in four positions at each spot and numbered 1–4. A table of random numbers is used to select one number between 1 and 4 for each box on each transect line. The organisms in the selected numbered box are then identified and counted, the clothespins removed when the counting in finished. A total of 24 counts are made.

1	2	3
4	5	6
7	8	9

➡

①	2	③
4	⑤	6
⑦	8	⑨

Figure 1-13. Systematic sampling. Begin with quadrat 1 and select every other quadrat that remains (or every third, fourth, etc.). Note that this has created an artificial diagonal or X pattern. If quadrat Nos. 1, 5, and 9 follow a specific sediment type (e.g., marine clays) then the plants or animals living there may be different than those in other areas and they would be emphasized in the collection data.

than the surroundings (Fig. 1-14). The sampler would thus collect more endemic plants that grow on serpentine soils or a different assemblage of marine invertebrates, thus biasing the overall picture. **Because there is no element of random sampling in this method, standard statistical tests cannot be used** (Southwood and Henderson, 2000). When statistical tests are applied to data from systematic studies, the probability (p) values are not accurate (McCune et al., 2002). One major advantage of the systematic method is that it often simplifies logistics involved in sampling and is useful in fields such as forestry (mensuration) or deep-sea sampling. It may also increase the probability of collecting uncommon species in species-rich areas. A higher density of clams was detected in Prince William Sound, Alaska, in systematically located sites than in preferred clam habitat (McDonald, 2004). One can combine methods, such as using systematic sampling to cover large areas with stratified random sampling within each of the systematic sampling plots. See Buckland et al. (2001), Hayek and Buzas (1996), and Thompson (2002) for general sampling techniques and Keith (1991) and Mueller et al. (1991) for environmental sampling.

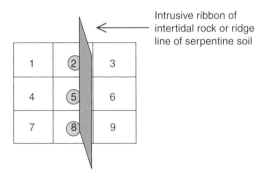

Intrusive ribbon of
intertidal rock or ridge
line of serpentine soil

Figure 1-14. A table of random numbers results in the selection of Nos. 2, 5 and 8 by chance. This happens to follow a vein of serpentine soil. The consequence of this simple random sampling is that sampling will be done where plants are generally sparse and tend to be locally endemic.

1.3.9   Fixed Quadrats

Fixed quadrats can be placed on the reef (e.g., depth 5 or 10 m on the reef flat and 20 or 30 m on the reef slope) to show changes over time. A convenient size is 2 × 2 m divided into four squares for photography. Each 2 × 2 m quadrat can be located some distance apart (e.g., 30 m) for variation in settled species on tropical coral reefs. For example, in temperate latitudes, the rocky intertidal marine biota parallel to the shore often does not change much over a distance of 30 m. However, a study in species rich Fiji showed that the macrofauna (principally hard corals and soft corals) on two pinnacles (only 100 m apart and 5 m below the sea surface) differed in species composition by 95% (Bakus et al., 1989/1990). The quadrats can be visited during wet and dry seasons and photographed from year to year. Joe Connell (pers. comm.) has records on intertidal quadrats that extend over 50 years. For information on coral reefs see Wells (1988), English et al. (1997), ICLARM (2000), and Spalding et al. (2001). Similar techniques can be applied to temperate rocky reefs.

1.3.10   Point Contact (Percentage Cover)

When organisms are modular (e.g., coral colonies), too difficult to distinguish as individuals (e.g., crustose algae, rose bramble), or take too much time to count (e.g., dense population of small barnacles or grass blades), **percentage cover is used in place of direct counts to save time and effort**. A grid with small subdivisions (e.g., small squares measuring 0.01 m^2[k]) is placed over the organisms and the area occupied (as a percentage) is estimated. Another method would be to use 100 points to estimate the percentage cover of species of interest in photographs (see Fig. 1-15 and Rapid Sampling Methods on p. 27). However, Dethier et al. (1993) found that random-point quadrats (RPQ) using 100 points were more accurate and less variable than 50 points, but were still less accurate and much slower to carry out than visual estimates. The RPQ method often missed rare species, that is, those with < 2% cover. Effective visual estimates of sessile benthic organisms were made with 10 × 10 cm quadrats. The advantages of percentage cover estimation are that the

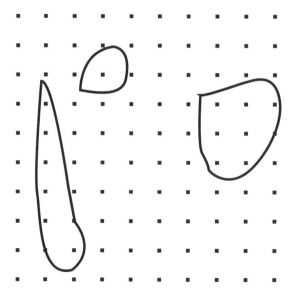

Figure 1-15. A slide is projected on the screen over which a grid of 100 points has been superimposed. The points intercepted by the organism (e.g., encrusting bryozoan) are counted. Alternatively, points are chosen randomly by the computer and those intercepted by the organisms are counted. This procedure was once used with a projector and screen but now can be done on a computer with layering of images.

area covered by each taxon is tabulated, and rare or uncommon species are less frequently overlooked in comparison to point intercept methods (Hallacher, 2004).

Percentage cover is the most commonly used abundance measure for plants, often expressed as cover classes (McCune et al., 2002). Authors use different cut-off points for cover classes. Raw percentage data are often transformed to de-emphasize dominant species whereas percentage cover class data seldom have this problem.

1.3.11   Line and Belt (Strip) Transects

A line transect is characterized by a detectability function giving the probability that an animal or plant at a given location is detected (Thompson, 2002). The probability of detection usually decreases as distance from the transect line increases. Variance estimates based on several transects are preferred over estimates based on a single transect. Many surveyors prefer a systematic selection of transects to avoid the uneven coverage of the study region obtained with random sampling. Transect lines may also be selected with the probability proportional to length by selecting n points independently from a uniform distribution over the entire study area (Thompson, 2002).

Walk or swim along a transect line at a constant speed and record animals observed. This is the mobile analog of the nearest neighbor technique (Southwood, 1978). This technique has some difficulties, such as estimating the velocity of a swimming organism (see the Southwood equation on p. 51). Other estimators of populations with line transects (e.g., Fourier series estimator) are discussed by Krebs (1999). See a discussion of plotless methods below by Bouchon (1981), Heyer et al.

(1994), Sutherland (1996), Boitani and Fuller (2000), Buckland et al. (2001), Elzinga et al. (2001), and Feinsinger (2001).

If studying the densities of several to many species simultaneously then the belt or strip transect method (e.g, a strip 100 m long and 1 m wide) is preferable. It represents an expansion of the quadrat to a long, narrow belt or strip (Buckland et al., 2001). There may be some narrow strip along the line in which detectability is virtually perfect (Thompson, 2002). A wider belt may be needed for fishes and terrestrial plants. One can swim down the belt, recording the numbers of species observed (see below). This technique can also be used in counting small organisms, birds, and so forth. The belt or strip transect method is preferable over many other sampling techniques (e.g., PCQ, line intercept – see pp. 33 and 36 in this chapter) (Steve Buckland, pers. comm.).

Belt transect method for fish surveys: A transect line is lain on the substratum. The length of the line may vary depending on results from a preliminary survey. This may be followed by plotting a species area curve (see Chapter 3, p. 145) if the principal interest is in estimating the species richness of fishes. The line usually follows a depth contour (e.g.,15 m). Fishes are counted on both sides of the line (closer to the line with juvenile fish). The width of the belt transect depends on underwater visibility and the abundance of fishes. A 5-m width for adult fishes and a 1-m width for juvenile fishes work well in clear tropical waters. The diver swims and records counts along the transect. The time of the swim along the line is usually standardized, and replicate transects are traversed. Daily variation occurs in fish activity and underwater light intensity thus transect studies should be done between about 0900 and 1500. However, I have frequently noticed a marked change in coral reef fish faunas in the late afternoon (1600–1800), thus it may be worthwhile to check on this before proceeding to count fishes during this time period. Seasonal variation includes surveys during the wet and dry seasons in tropical regions. McCormick and Choat (1987) compared the precision, accuracy, and cost (time) of five strip-transects. A strip 20×5 m was selected as the best overall size for a single target fish species [the morwong *Cheilodactylus spectabilis* (Hutton)] but the optimal size is likely to be species specific. Problems resulting in sampling error included observer variability (e.g., laying of the tape), edge effect in counting fish, fish characteristics (i.e., crypticity of the fish), and environmental factors (e.g., turbidity).

1.3.12 Adaptive Sampling

Adaptive sampling is a sampling design in which the procedure for selecting sites or units to be included in the sample may depend on values of the variables of interest observed during the survey (Thompson, 2002). Adaptive sampling strategies used with aggregated population units of various locations and shapes may provide a method to increase dramatically the effectiveness of sampling effort. Adaptive sampling is also known as two-stage or even three-stage sampling. Adaptive sampling has been employed with simple random sampling,

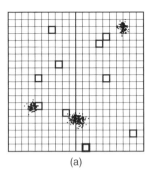

(a)

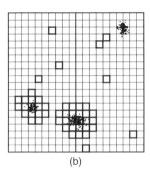

(b)

Figure 1-16. Stratified adaptive cluster sampling. (a) Initial random sampling of five units in each of two strata. (b) Final sampling showing intensive sampling around the clusters indicated in (a). Source: Thompson (2002).

systematic sampling, stratified sampling, strip sampling, and especially with cluster sampling. The simplest adaptive cluster designs are those in which the initial sample is selected by simple random sampling with or without replacement. Once the species of interest is located, further nonrandom sampling is carried out in the same area. Thompson (2002) gives an example of adaptive sampling with an initial sample size of 200. The adaptive strategy was 15 times as efficient as simple random sampling. Stratified adaptive sampling improves the detection of clusters when the locations and shapes of the clusters cannot be predicted prior to the survey (Fig. 1-16). In one example of stratified adaptive sampling, the adaptive strategy was 24% more efficient than the comparable nonadaptive one (Thompson. 2002). See also Thompson (2004).

1.3.13 Sequential Sampling

Sequential sampling is a statistical procedure in which the sample size is not fixed in advance. This may reduce the number of sample units required by up to 50%. Sample units are taken until there is enough information to make a decision (i.e., to stop sampling or to continue sampling). Stopping rules are employed to prevent sampling indefinitely. Sequential sampling is used in ecology, in resource surveys, and in insect pest control. It is rarely used in marine biology. The mathematics are relatively complex. Krebs (1999) discusses in detail sequential sampling involving distributions that are normal (uniform), binomial, or negative binomial (aggregated). The Schnabel method of population estimation is one of several methods suitable for sequential analysis. Krebs (2000a) has computer software programs for sequential sampling.

1.3.14 Rapid Sampling Methods

Some relatively newer techniques of rapid sampling involve photos and videos. These techniques were pioneered by Mark and Diane Littler in the United States who studied marine benthic algae (Littler and Littler, 1985). **Place a quadrat frame on the substratum and take a photo of the quadrat from above**. This technique can be used in both the intertidal and subtidal zones. To sample the images (the slides can be selected randomly), project the slide with the quadrat onto a screen over which has been placed a grid of 100 points (Fig. 1-15). **To determine the areal density of a species, count the points on the screen that intercept the species of interest**. This gives the percentage of the total area intercepted by the species. Repeat the process for the remaining slides and tally the results. One could also use 100 random points to estimate the percentage cover of all species of interest (see above). An easier method of doing this would be to create a grid of 100 points then superimpose or layer the image above a photo of the quadrat image. Alternatively, an image-processing algorithm can be developed that automatically counts or measures the area of interest (see p. 186 in Chapter 3 and Wright et al., 1991).

One can also swim above a transect line or measuring tape and **photograph the organisms with a video camera**. For example, LaPointe et al. (2003) swim slowly along two 50 m long belt transects holding a camcorder 0.4 m off the bottom. A second oblique (45 degrees angle) close-up video is taken along the transects to aid in the identification of the biota. Later, 10 rectangular quadrats are selected from each video transect and quantified for percentage cover of the biota using a randomized point-count method (LaPointe et al., 1997). Phillip Dustan has developed a computer program (PointCount99) that will assign random points to still photos or videotape frames for counting organisms (see the Appendix). Alternatively, each species in the video frames can be assigned a color then counted or the area measured automatically (Whorff & Griffing, 1992). Bezier curves and AutoCAD® have been used to estimate digital cover (Tkachenko, 2005). The automated techniques here and those mentioned above require computers, video cameras, and relatively expensive computer boards and/or software. However, prices have decreased dramatically over the past two decades and a complete, automated system can be purchased for several thousand dollars (U.S.).

The disadvantages of photographic and video techniques are:

(1) Only surface organisms are counted and measured. There may be numerous species that are missed, such as animals living under algae in colder waters and the cryptofauna and infauna of coral reefs. This is a serious deficiency.

(2) One must be able to identify the organisms in the photos or videos, which is not an easy task in many cases.

The major advantages of these rapid assessment techniques are:

(1) rapid data collection
(2) collection of large amounts of data
(3) less tedious data collection.

See Littler and Littler (1987), Littler et al. (1996), English et al. (1997), Kingsford and Battershill (2000), and Smith and Rumohr (2005) for further information.

1.3.15 Introduction to Plotless Sampling

Plotless methods, especially older ones, often assumed that the individuals are randomly distributed. Trees have been successfully estimated in this way. Highly aggregated or rare species are seldom suitable for this type of analysis. Estimates of total density cannot be compared except in terms of relative density. Sample sizes of 40–60 (Krebs, 1999) and 60–80 (Buckland et al., 2001) are recommended for good precision. Buckland et al. (2001) discuss the assumptions, strengths, and weaknesses of many of the plotless sampling techniques. For further information see Southwood (1978) and Southwood and Henderson (2000).

1.3.16 Best Guess or Estimation

The best guess method [waterfowl or sea lion populations estimated from the air – see Krebs (2000) for a computer program] has been frequently used by Fish and Game or Fish and Wildlife agencies. Experienced observers can estimate wildlife populations from the air by eye with a relatively high degree of accuracy. Double sampling or two-phase sampling is used in surveys of the abundance of certain animal or plant species, less accurate counts being made by air and more accurate counts by ground crews (Thompson, 2002).

1.3.17 Catch or Weight Per Unit Effort (CPUE)

There are numerous methods of estimating commercial fish populations. Many of them are quite complex. A relatively simple method used in commercial fisheries to estimate the density of shrimp or fish is to tow a trawl, expressing the results as **CPUE or catch or weight per unit effort** (for example, kg shrimp/h with a 100 ft [32 m] otter trawl). No correction is made for towing with, across, or against a current. Krebs (2000a) has a program for catch per unit effort. These data and fish length data are used in fishery stock assessment (computer program FiSAT II, available from ICLARM, now located in Penang, Malaysia [FAO,

1994]) and in developing fish growth curves (computer program ELEFAN I). Using and interpreting these programs require training. A more recent development is the appearance of freeware computer programs. EcoPath (http://www.ecopath .org) is used to model food webs in marine ecosystems. This information is then incorporated into EcoSim (same Internet address), which predicts changes in fish populations. EcoVal (http://www.fisheries.ubc.ca) shows the economic impacts of information from the former programs. A training tutorial is needed to learn how to use the programs (Christensen et al., 2004). See Ricker (1958), Green (1979), Caddy (1982), FAO (1994), Quinn and Deriso (1999), Krebs (2000a), Southwood and Henderson (2000), Gore et al. (2000), and Walters and Martell (2004) for further information. See Caldrin et al. (2004) for stock identification methods.

1.3.18   Coordinate Lines

Coordinate lines (as opposed to a grid system) are used to save time in sampling under difficult conditions (e.g., random sampling on an intertidal sandy beach in the surf or in a dense forest, see Fig. 1-7). A grid system setup in the intertidal could easily be washed away by the surf. A coordinate system of evenly spaced rocks or poles (spaced 5–10 m apart) placed on the upper sandy beach coupled with poles forced into the sand at spaced intervals down the beach creates a useful sampling pattern for that habitat. In large areas such as an estuary mud flat, one can use a GPS to keep the grid square.

1.3.19   Cluster Sampling

A different type of simple random sampling is cluster sampling. Cluster sampling can use points in space (plotless) or plots. In cluster sampling, a primary unit consists of a cluster of secondary units, usually in close proximity to each other (Thompson, 2002). **Cluster sampling involves sampling a cluster of**

Figure 1-17.   A coastline with villages and artisanal fisherfolk, typical of the west coast of India. The members of each cluster of fisherfolk are numbered then some of the numbers are selected from each cluster using a table of random numbers. This is a special case of simple random sampling with clusters. The fishes are counted and weighed only from the randomly selected boats.

things (e.g., artisanal fishing boats, trees, sampling for bodysize in polychaete worms, capture–recapture of fish) to save time and resources. This technique would be useful in surveying the fish catch of artisanal fisheries along a coast for social and economic information (see Manly, 1986). Cluster sampling is exemplified by having five clusters (groups) of coastal fishing boats. Select from each cluster of boats three boats from a table of random numbers and determine the fish catch per boat in each selected cluster (Fig. 1-17). Cluster sampling is often carried out for reasons of convenience or practicality rather than to obtain the lowest variance for a given number of units observed (Thompson, 2002). The advantage of cluster sampling is that it is usually less costly to sample a collection of units in a cluster than to sample an equal number of units selected at random from the population. Adaptive cluster sampling can be used when organisms are rare and highly clustered (i.e., aggregated). Additional quadrats are sampled near the site of the first occurrence of the species of interest. See p. 25 and Thompson (2002). Conners and Schwager (2002) found that adaptive cluster sampling for spatially patchy and/or rare species gave better results than traditional cluster sampling techniques.

1.3.20 Introduction to Distance Measurements

A number of biological sampling methods are called distance measurements. They include Nearest Neighbor, Point-Center Quarter, and Strip or Belt Transect methods, among others. Distance-based methods are most commonly used for sampling forest structure (McCune et al., 2002). They often perform best when organisms are randomly distributed. For detailed information on distance sampling see Buckland et al. (2001). The computer program "Distance" is available free of charge from the Internet and is based on their book, as follows:

http://www.ruwpa.st-and.ac.uk/distance/

Steve Buckland (pers. comm.) recommends a minimum of 12 lines and preferable 20 or more lines for larger study areas. Differing numbers of lines could give varying results if they are not generally proportional to the size of the study area. Thompson (2002) states that evenly spaced transect lines may be preferable to randomly selected transect lines because randomly selected transect lines may aggregate. Both modes of arranging transect lines have been used in past surveys. Barbour et al. (1999) and Buckland et al. (2001) recommend using random or regular (i.e. evenly spaced) points depending on the type of study. The recommended number of sampling points for distance measurements (e.g., PCQ) ranges from 40 (Krebs, 1999) to 80 (Buckland et al., 2001).

Care must be taken to be certain that relatively stationary organisms (e.g., limpets) have not moved over a period of several tides or corrections for this need to be made (i.e., by recording the positions of marked individuals). Also, arroyos and gullies increase the size of the study's sampling plot resulting in an underestimation of plant density (Barbour et al., 1999).

1.3.21   Nearest Neighbor and Point to Nearest Object

1.3.21.1   Nearest Neighbor Method

The distance from the nearest individual to its nearest neighbor or from a random point to a nearest neighbor is:

$$N = \frac{1}{4\bar{r}^2}$$

where:

N = number of individuals per unit area

$\bar{r}$ = mean distance between nearest neighbors.

This must be measured in the same units as the final density (e.g., meters). Nearest neighbor techniques tend to overemphasize density by a factor of 2 or 3. (Underwood, 1976). **A modification of this method that improves results is to measure the second or especially the third nearest individual** (Fig. 1-18).

1.3.21.2   Point to Third Nearest Object

This technique is almost the same as the third nearest neighbor (3NN). It measures the distance from a **random point** to the third nearest object or individual (Fig. 1-18), whereas the 3NN method measures the distance from a **random individual** to its third nearest neighbor. The equation is the same as the 3NN equation from Krebs (1999).

$$D = \frac{3n - 1}{\pi \sum (d^2)}$$

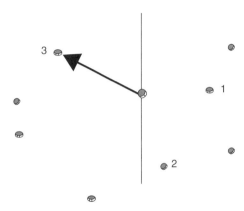

Figure 1-18.   Third nearest object from a random point along a transect line. The dots represent individuals of the same species.

where

n = number of measurements to the 3rd nearest object (3NO)

d = distance (m)

π = 3.14159

D = density (No. of individuals/m^2)

Σ = sum.

The distance to the 3rd nearest object needs to be reported in the same units as the final density (e.g., meters).

Example

Three 3rd nearest neighbor distances measured (20 m, 20 m, 20 m)

$$D = \frac{(3 \times 3) - 1}{3.14159 (400 + 400 + 400)}$$
$$D = 0.00212 \, \text{organisms/m}^2.$$

A computer program for the 3rd nearest neighbor or 3rd nearest object is provided in the Appendix (see Gore et al. (2000), for a mathematical discussion). We found that the 3rd nearest neighbor technique using a laser rangefinder to be the simplest technique for density estimations of trees (Bakus, et al., 2006). However, Buckland et al. (2001) do not recommend nearest neighbor or point to object methods, with the exception of estimating the density of forest stands, because of several logistic and time considerations and the fact that the measurements are bias-prone.

1.3.21.3 General Equation for Nearest Neighbor or Point to Nearest Object Density Data (Gregory Nishiyama)

The following is a general equation for determining densities from any rank nearest neighbor data (1st, 2nd, 3rd, 4th, 5th, 6th, … nearest neighbor). It is not a new equation but a summation of old equations. A regression line (plotting a density coefficient against rank) was constructed from data presented in Krebs (1999:197; see Table 6.1) and used to develop the coefficients 0.316 and 0.068 found in the equation below. It is assumed that the organisms are somewhat randomly distributed and the organisms are allowed to overlap (i.e., corals growing over each other). The user can employ any nearest neighbor data such as the 1st nearest neighbor data or 3rd nearest neighbor data which are the two types of proximity data most frequently used. Insert the rank of the nearest neighbor you are using as well as the average nearest neighbor distance. This will give an estimate of the density of the organism.

$$\text{Density} = \frac{0.316X - 0.068}{(\text{average distance})^2}$$

where:

  X = distance rank (e.g., 3 = 3rd nearest neighbor)

  average distance = average distance between randomly selected points and their nearest neighbors.

Example

$X = 3$ (3rd nearest neighbor)

average distance = 5 m

$$\text{Density} = \frac{0.316(3) - 0.068}{(5)^2}$$

$$\text{Density} = 0.0352/\text{m}^2 \text{ or } 3.52/100 \text{ m}^2$$

The equation above seems to work as well as Kreb's (1999) equation based on simulations. However, it needs to be tested in the field.

1.3.22   Point-Center Quarter or Point Quarter Method

This method involves **measuring the distance from the sampling point to the nearest individual of a species in each of four quadrants** (see Fig. 1-19). One divided by the mean distance calculated[2] = approximate density (e.g., mean distance = 2 m then density = ~1 individual/4 m²). Recommended: At least 13 random points with four measurements each, but preferably 50 random points (=200 measurements total). It is assumed that the distribution of organisms is random (Pollard, 1971).

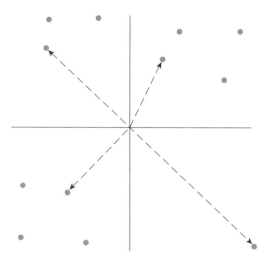

Figure 1-19.   Point-center quarter method. The distance from a random point to the nearest individual is measured in each of the four quadrants. An arbitrary boundary may need to be established. See text for explanation.

Point-center quarter (Source: Krebs, 1999 and Mitchell, 2001).

$$D = \frac{4(4n-1)}{\pi \Sigma(d^2)}$$

where

D = density (No. of individuals/m^2)

n = number of random points

d = distance to the nearest individual (m)

π = 3.14159

Σ = sum.

Mitchell (2001) presents a clear and detailed discussion of the PCQ method.

Note that although the PCQ method was originally designed without boundaries, borders must be established or you may wander out too far from the random point or lose track of the boundaries. This is especially common in diving studies. When the species cannot be found in a quadrant, **you can correct for zero data by using the correction factor table of Warde & Petranka** (*1981*) (Table 1-2). The random points are often located along a transect line. Ideally they need to be spaced so that the same organisms (e.g., sponges or trees) are not counted by two nearby random points. However, this is not always possible (Bakus et al., 2006). Another problem exists with randomly selected transects. They may fall too close to the boundary of the study area so that PCQ measurements occur well beyond the boundary, possibly resulting in density estimation errors. The PCQ method is useful in forestry, particularly practical in moderately dense to dense forests. It has also been used in dense mangrove forests and on coral reefs (see Loya, 1978). Krebs (2000a) has a computer program for the PCQ method. It provides densities but does not include the correction table of Warde and Petranka (1981). Other plotless methods are discussed by Nimis and Crovello (1991) and Krebs (1999).

The **correction for zeros in quadrants** (Warde and Petranka, 1981):

$$\text{Total Approximate Density (No./m}^2) = \frac{1}{(\text{Mean distance})^2} \times \text{CF} \, (n_0/nk)$$

where

CF = correction factor (from Table 1-2)

n_o = number of quadrants or sectors with missing data

nk = total number of quadrants

Example:

Suppose the mean distance between giant kelp (or oak trees) is 10 m and that 5 of the 100 quadrants lacked giant kelp (or oak trees). The approximate density of giant kelp

Table 1-2. Values of correction factors (CF) based on the ratio between the number of samples with no organisms found compared with the total number of samples (n_0/nk) (Source: Warde and Petranka, 1981).

n_0/nk	CF	n_0/nk	CF	n_0/nk	CF	n_0/nk	CF
0.01	0.96670	0.26	0.57018	0.51	0.33103	0.76	0.14685
0.02	0.94012	0.27	0.55896	0.52	0.32284	0.77	0.14023
0.03	0.91630	0.28	0.54791	0.53	0.31473	0.78	0.13367
0.04	0.89431	0.29	0.53704	0.54	0.30670	0.79	0.12715
0.05	0.87368	0.30	0.52633	0.55	0.29874	0.80	0.12068
0.06	0.85416	0.31	0.51579	0.56	0.29086	0.81	0.11425
0.07	0.83554	0.32	0.50539	0.57	0.28306	0.82	0.10787
0.08	0.81771	0.33	0.49514	0.58	0.27532	0.83	0.10153
0.09	0.80056	0.34	0.48504	0.59	0.26766	0.84	0.09524
0.10	0.78401	0.35	0.47507	0.60	0.26006	0.85	0.08899
0.11	0.76800	0.36	0.46523	0.61	0.25254	0.86	0.08278
0.12	0.75248	0.37	0.45552	0.62	0.24508	0.87	0.07662
0.13	0.73741	0.38	0.44594	0.63	0.23768	0.88	0.07050
0.14	0.72275	0.39	0.43647	0.64	0.23035	0.89	0.06441
0.15	0.70845	0.40	0.42712	0.65	0.22308	0.90	0.05837
0.16	0.69451	0.41	0.41789	0.66	0.21586	0.91	0.05236
0.17	0.68090	0.42	0.40876	0.67	0.20871	0.92	0.04640
0.18	0.66759	0.43	0.39974	0.68	0.20162	0.93	0.04047
0.19	0.65456	0.44	0.39082	0.69	0.19458	0.94	0.03458
0.20	0.64182	0.45	0.38200	0.70	0.18761	0.95	0.02873
0.21	0.62931	0.46	0.37327	0.71	0.18068	0.96	0.02291
0.22	0.61705	0.47	0.36465	0.72	0.17381	0.97	0.01713
0.23	0.60502	0.48	0.35611	0.73	0.16699	0.98	0.01139
0.24	0.59320	0.49	0.34766	0.74	0.16023	0.99	0.00568
0.25	0.58159	0.50	0.33930	0.75	0.15351		

(oak trees) using the simple method described above on p. 33 is:

$$\frac{1}{(10\,\text{m})^2}\times\text{CF}\ (0.87368)$$

$$0.01\times\text{CF}\ (0.87368)=0.00873/\text{m}^2$$

$$0.00873/\text{m}^2=87.3/\text{hectare}$$

Using the Krebs equation above:

$$0.0101859\times\text{CF}\ (0.87368)=89.0/\text{hectare}$$

A computer program (PCQ) by the author for PCQ with a built-in correction table is given in the Appendix.

1.3.23 Line Intercepts

Line intercept methods involve laying a transect line (e.g., measuring tape or twine) on the substratum, moving along the line to encounter an organism (e.g., coral), then measuring the length of the organism under or over the line (e.g., tree canopy), and continuing on. Line intercept sampling is an example of an unequal proability design, that is, the larger the patch (e.g., a shrub), the higher the probability of inclusion in the sample (Thompson, 2002). There is no fpc factor (see p. 13) because the selection of positions along the baseline is essentially with replacement.

An example of a line intercept study is for a diver to lay down a length of transect line and to swim the length at a steady known rate while counting fish (see p. 25 or p. 51, and Krebs, 1999 for a computer program). Line intercepts also involve placing a transect line (e.g., a 100-m tape) over a coral reef then measuring the distances intercepted under the line by categories such as hard corals, soft corals, and so forth (Table 1-3). The total length intercepted by a category (e.g., lengths of hard coral directly under the line) is then expressed as a percentage of the total line length. While the investigator is doing this, he or she can make other measurements (see p. 38).

Line intercept reef studies have been conducted in three ways: (1) A taut or straight line; (2) A slack line lying on the reef; and (3) A line that follows the detailed contours of the reef (sometimes called chain length – see Fig. 8-20). Results from the three methods are often variable (3–27% difference between a straight line and a contour line in our marine studies – see Bakus et al., 2006). The easiest and most rapid method is to use the straight line. Straight-line measurements give more accurate results whereas slack line or contour line (chain length) measurements produce underestimations of densities. For greater precision, a plumb line can be hung from a taught intercept line to precisely indicate the start and end points of any organism under the straight line (e.g., hard coral). This increases the accuracy of straight-line measurements slightly but the method takes considerably more time

Table 1-3. Life forms. These categories are used with transect lines or line intercepts on coral reefs. The difficulty in identifying organisms often precludes using generic or specific categories (source: Bakus and Nishiyama, 1999).

Life forms
Live coral (branching, table, massive)
Sand
Coral rubble
Dead coral
Seagrass
Sponges
Turf algae
Pavement algae
Other organisms

than simply sighting the organism below a taut line and measuring the length of tape intercepted by the organism. The ratio between the slack or contour lengths and the straight-line length increases with increasing heterogeneity, resulting in a simple mathematical decrease (i.e., error) in density estimates. Benthic marine populations are typically counted in quadrats as the organisms are observed from above, avoiding vertical surfaces, overhangs, and cracks and crevices in order to standardize and greatly simplify the counting. As substratum heterogeneity increases, the number of species and individuals usually increase. This is not only the result of organisms (e.g., limpets) living on vertical surfaces and overhangs, but sometimes considerable increases in species and individuals living in cracks and crevices and between coral branches (i.e., the cryptofauna) (Bakus, 1969). Thus the straight-line measurements may not reflect either species richness or population increases related to increased physical heterogeneity. Reef heterogeneity can be estimated by comparing straight-line measurements with contour (chain length) measurements to arrive at a ratio between the two.

Line intercepts can also be used to determine if the occurrence of an organism or substratum type is closely associated with or dependent upon another organism or substratum type (e.g., sponges and dead coral; sand and clams). The sequencing of organisms is recorded along transect lines and the frequencies of co-occurring organisms or substratum types are tabulated. A transition frequency matrix is developed (e.g., a matrix showing replacement of substratum types along a transect line) and a Chi-square equation (or G-test nowadays) is applied to the matrix to determine if any of the sequences of associated organisms, categories, or substratum types is significant (see Chapter 3, p. 163 and Davis, 1986 for a detailed explanation of the technique). The line transect can be videotaped and analyzed in the laboratory. Surface animals and plants can be counted directly or indirectly (i.e., by assigning random points to the videotaped frames). A program for this was developed by Phillip Dustan (see PointCount99 in the Appendix).

The line intercept method works well on coral reefs because it can provide considerable information on benthic organisms and substratum types with a minimum of effort (see Marsh et al., 1984). The line intercept method (typically lines arranged perpendicular to the shore) can provide at least the following information:

1. Coral reef profile based on depth measurements (by scuba depth gauge) at 5 or 10 m intercept lengths.

2. The rank dominance of organisms and substratum types expressed as a percentage of the total length of the intercept line.

3. Size-frequency data on hard corals and other organisms. This is an additional measurement made as one moves along a transect line. There is a bias here for large individuals, thus the advantage of using belt or strip transects rather than line intercepts, if there is sufficient time to do so.

4. An estimation of the population densities of selected organisms.

5. Information on whether the occurrence of an organism category (e.g., sponges, turf algae, etc.) is associated with or dependent on another category of organism or a specific substratum type.

6. The incidence of disease (e.g., black band disease in hard corals). This is an additional recording.

The major disadvantage of the line intercept method is that small organisms ($<$3 inches or $<$9 cm) are not often intercepted by the metric tape [measuring 1/2″ or 3/4″ (13 or 19 mm wide)]. For information on distance sampling see Buckland et al. (2001).

1.3.24   Strong Method

The Strong method is a line intercept method, based on the probability of intercepting an object along a transect line. Long ago botanists developed a complex technique using calculus with line intercepts to estimate population densities. This was later simplified and improved upon by Strong (1966) whose technique is still employed today. The equation is based on the concept that the wider an organism is, the greater the probability that a transect line will intercept it. The diver selects a category or taxon of interest (e.g., sponges). When the diver intercepts a sponge under the transect line, he measures the widest width (i.e., orthogonal width) of the sponge and records it (Fig. 1-20). This procedure is continued until the 100-m line has been traversed. The data on sponge widths are then plugged into an equation by Strong (1966) to give an approximation of the sponge density. The **Strong method** (also known as a modified Eberhardt method which appeared later—see Krebs, 1999) can be applied to any organisms that can be distinguished as an individual or as a separate colony (e.g., bryozoan colony). The Strong method is an estimation of density using the harmonic mean.

$$\text{Density (No. of organisms/m}^2) = \sum \frac{1}{M} \frac{\text{unit area}}{\text{total transect length}}$$

where:

$\Sigma = $ sum

$M = $ maximum orthogonal width (m) of an organism intercepting the transect line

unit area ($=1\,\text{m}^2$ for example)

total transect length $= 100\,\text{m}$ (in this example).

Example

Corals (table *Acropora*) or trees intercepting a transect line (measuring the greatest width of the tree canopy)

Intercepted widths (m) are: 2, 3, 1, 4, 3, 1, 3, 2

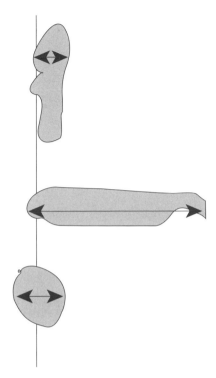

Figure 1-20.   Strong method. Measuring the greatest orthogonal width along a transect or line intercept. For smaller plants – measurements are made below the line (e.g., with grasses, forbs, and small shrubs) and above the line (e.g., with larger plants). For trees, the canopy width is measured.

$$\text{Density} = \frac{1}{2} \quad \frac{1}{3} \quad \frac{1}{1} \quad \frac{1}{4} \quad \frac{1}{3} \quad \frac{1}{1} \quad \frac{1}{3} \quad \frac{1}{2}$$

or

$$0.50 \quad 0.333 \quad 1.0 \quad 0.25 \quad 0.333 \quad 1.0 \quad 0.333 \quad 0.50$$

$$\text{Density} = 4.249 \frac{1}{100} = 0.04249 \text{ organisms}/\text{m}^2$$

1.3.25   Weinberg Method

The Weinberg (1981) method is a line intercept method, based on the probability of an object intercepting the line. It has been buried in the coral reef literature for many years, largely unavailable to terrestrial and freshwater biologists. In the Weinberg method (1981), lengths of organisms intercepted by a transect line are recorded. This method works well with circular organisms (e.g., some hard corals, shrubs) that are larger than about 1.0 m in diameter (Fig. 1-21).

The equation is as follows:

$$\text{Density} = \frac{\dfrac{\text{No. of intercepts}}{\text{transect length}}}{\text{mean organism intercept length} \times 1.156 \, (\text{coral colony size correction factor})}$$

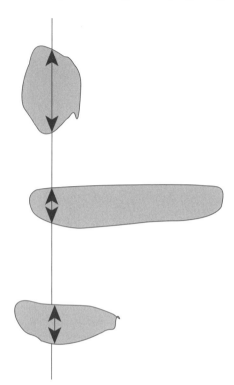

Figure 1-21.   Weinberg method. Measuring the length of an organism intercepted by a line. For terrestrial studies this includes plants below (e.g., shrubs or grasses) and above (trees) the line. For trees the length of the canopy above the line is measured.

Example

Corals intercepting a 50 m transect line

Organism intercept length (m) –2.0, 3.4, 1.2, 3.2, 4.5

$$D = \frac{5/50}{2.86(1.156)} = 0.0303 \text{ coral}/\text{m}^2$$

A computer program (Weinberg) for this is found in the Appendix. See Weinberg (1981) and Nishiyama (1999).

Gregory Nishiyama Correction for the Weinberg Equation:

In the Weinberg equation for the estimation of the density of an organism, he approximates the organism's diameter from the intercept lengths of the organism on the transect line. Thus, he multiplies the average intercept length by the correction factor 1.156. Weinberg assumed that the average intercept length that is found midway between the maximum and minimum possible intercept lengths to be the average intercept length (Fig. 1-22). However, Weinberg's midpoint does not designate the average chord or intercept length. To obtain this, one must add up sample lengths (from maximum to minimum possible lengths) and find the average (Fig. 1-23).

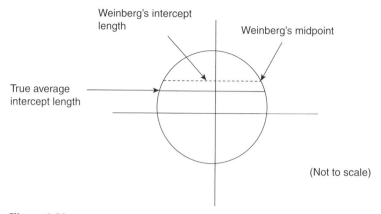

Figure 1-22. Weinberg's intercept length and the true intercept length. See text for explanation.

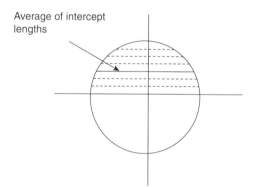

Figure 1-23. Average of intercept lengths. See text for explanation.

The final correction factor comes out to be 1.301. So the Weinberg equation becomes:

$$\frac{\text{No. of intercepts}/\text{transect length}}{\text{mean intercept length} \times 1.301}$$

Although this correction factor may not seem very different from Weinberg's 1.156, there is a 13% difference in these two numbers that translate to a 13% difference in their density estimates. This is a significant difference when ideally an estimation method should have a maximum of a 10% error. This new correction factor was tested by Nishiyama on 30 simulations and shown to produce more accurate estimates than the original Weinberg correction factor. This new correction factor needs to be tested in the field.

1.3.26 Nishiyama Method

The Nishiyama method is a line intercept method. Nishiyama has developed a density estimate technique which takes into consideration the orientation of organisms relative to the transect line. This method was used to estimate the densities of strongly ovate intertidal organisms (e.g., mussels). The transect lines are parallel to the shoreline.

Gregory Nishiyama Density Orientation Method:

Although such methods as Weinberg and Strong can estimate the densities of organisms, they do not take into consideration the effect of the orientation of the organisms. If the orientation of the organism is random, such methods above may work successfully. However, if there is a bias in the orientation of the organism (around 20% or more bias in a particular direction – e.g., organisms oriented more or less parallel to the incoming waves) and it is ovate in shape (especially if the organism's length to width ratio = >3) (e.g., mussels, some chitons) the error in the estimation of density can be considerable. The following equation is an extension of the Weinberg equation, which addresses the problems with orientation and shape.

To use the proposed equation, the investigator must place each organism intercepting the transect line into one of four categories based on its orientation relative to the line. Although each category is assigned an angle designation, the categories actually represent ranges of angles. When an organism has a particular orientation at or close to a particular category, it is placed into that category. For example, if an organism has an orientation of 37 degrees relative to the line, it is placed into the 30-degree category. Alternatively the user can divide the circle into numerous categories with angle ranges, however, this may make the collection of data more difficult. Instead, if it is close to one of four categories, the user should simply put it in its proper angle group. Once these data are gathered, each angle category is analyzed separately and then all the density estimates are summed. This will give the total density.

Each organism intercepting the transect line is placed in one of the orientation categories (Fig. 1-24).

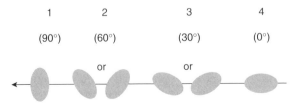

Figure 1-24. The four positions of mussels relative to a tape measure or intercept line, used in the Nishiyama technique for determining densities of intertidal organisms. See text for explanation.

These counts and the organism's dimensions are then placed into the appropriate sections of the following equation:

$$\text{Density (number of individuals/m}^2) = \overset{(90°)}{\dfrac{n}{\dfrac{TL}{L}}} + \overset{(60°)}{\dfrac{n}{\dfrac{TL}{(0.87L)}}} + \overset{(30°)}{\dfrac{n}{\dfrac{TL}{(0.50L)}}} + \overset{(0°)}{\dfrac{n}{\dfrac{TL}{W}}}$$

where

n = number of intercepts (for each orientation category—see Figure 1-24)

TL = transect length (e.g., in meters)

L = mean organism length (e.g., in meters)

W = mean organism width (e.g., in meters)

Note: All measurements must be in the same units

Example

Assume that the numbers of mussels in different orientations = 10, 5, 20, 10; transect length = 100 m; average mussel lengths = 5, 6, and 7 cm; and average mussel width = 3 cm.

$$D = \dfrac{\dfrac{10}{100}}{0.05} + \dfrac{\dfrac{5}{100}}{(0.87)(0.06)} + \dfrac{\dfrac{20}{100}}{(0.5)(0.07)} + \dfrac{\dfrac{10}{100}}{0.03}$$

$$D = 2 + 1 + 6 + 3$$

$$D = 12 \text{ mussels/m}^2$$

In a dozen simulations where there was a bias in the organism's orientation, up to a 33% error between the density estimates of the Weinberg method and Nishiyama method was observed. That is, the Nishiyama method was more accurate than the Weinberg method due to the consideration of the orientation of the organisms. A computer program (Nishiyama) for this is presented in the Appendix. The Nishiyama method needs to be tested in the field.

Some of these plotless methods and others have deficiencies that produce crude estimates and one should consult references or specialists before using them. However, sometimes they are the only practical sampling technique available and therefore of use. **When various plot and plotless estimates are compared with a complete census (i.e., total count), there are often considerable differences in the results** (even using stratified random sampling), an important point to remember. This is especially true when relatively few samples are taken or organisms are highly aggregated, which often produce inaccurate estimates. See Venrick (1971), Weisberg and Bowen (1977), and Yates (1981).

A summary of 12 terrestrial and marine density estimate studies indicated that the Strong method was most accurate among several methods tested for stationary organisms (Bakus et al., 2006). Moreover, it is best used for those organisms measuring more than about 0.3 m (1 ft) in diameter (e.g., corals, shrubs, trees). Stratified random sampling ranked second best and the Third Nearest Object the third best (Bakus et al., 2006). Goedickemeier et al. (1997) found that at a scale of 10–50 km^2, stratified random sampling creates an accurate picture of small-scale vegetation pattern at low sampling effort.

Additional distance sampling methods include point counts, point transect sampling, trapping webs, and cue counting, among others (Buckland et al., 2001).

1.3.27   Mark (or Tag) and Recapture (Mark and Resight) Techniques

Mark-recapture techniques consist of marking or tagging a mobile species one day (Table 1-4), then returning to recapture the species and count the individuals the next or succeeding days. An estimate of the population in the area is thus acquired. Capture–recapture methods are now frequently used for the estimation of birth, death (or survival), and emigration rates (Southwood and Henderson, 2000). Mark-recapture methods have several advantages over line transect sampling for wildlife (see Buckland et al., 2001, for details). However, mark-recapture field costs are substantially higher to achieve the same precision on abundance estimates than line transect sampling (Borchers et al., 2002). For example, mark- and-resight techniques for fishes take 13–15 times more diver time than a visual belt transect estimation (Davis and Anderson, 1989). Although mark-and-resight techniques

Table 1-4.   Methods of marking and tagging organisms.

Molluscs: Quick-drying paint; notches with file, carborundum wheel or dentist's drill; adhesive tape; metal; Peterson discs; Plastic tags

Crustaceans: Spaghetti and dart tags; punching telson, etc. (problem with molting); dyes for shrimp (*Penaeus*).

Echinoderms: Monofilament nylon (sea urchins)

Fishes: Branding (sharks); clipping fins; internal magnetic or radioactive tags; external tags, dart tag best, plastic, inserted just below dorsal fin

Turtles: Magnetic tag injected into muscle

Birds: Aluminum, colored celluloid or plexiglass bands; sonic marker

Mammals:
  a. Seals and sea lions — hot iron branding best, good for over 20 years on fur seals; metal tags in foreflipper good for 8 years.
  b. Cetaceans — stainless steel tube coated with penicillin, shot into whales, recovered in boilers, reward offered.

Newest Methods: Fluorescent paints, injection, tattoos, clipping of wings or fins, rare elements, radioactive isotopes, radioactive labels and incorporation in tissues, transponders and sonic tags, radio-telemetry (radio collars) and micro-videocameras (e.g., attached to elephant seals or great white sharks). For further information see Seber (1982), Sutherland (1996), and especially Southwood and Henderson (2000).

produce more accurate estimations of fish populations than visual belt transects, the method is prohibitively expensive for regular use (Hallacher, 2004). Mark-recapture abundance estimates are more sensitive to failure of assumptions than line transect estimates.

Population marking techniques are often classified as batch (group) or individual tagging techniques (Southwood and Henderson, 2000). Thomson (1962) used 12 different tags and markers on 183,113 shellfish, crustaceans, fishes, and whales. The overall recovery was 5.5%, comparable to other previous large studies on tagging and marking. Because the average recovery rate for many tagged or marked marine organisms is about 5%, hundreds or thousands of animals may need to be marked or tagged to provide meaningful results (Thomson, 1962). For this reason, a sample size estimation needs to be made before starting the marking or tagging program (Krebs, 1999). Dale Kiefer (pers. comm.) reports that tagging recovery for southern sea elephants (*Mirounga leonina*) is as high as 90%. Mark-recapture techniques are only appropriate for use with mobile animals such as shrimp and fishes, but only with fishes that remain in the study area over several days or more.

Some of the recent developments in tagging involve the use of satellite and data storage tags. Seabirds such as the Wandering Albatross (*Diomedia exulans*) are followed with satellite tags (PPT or Platform Terminal Transmitter). Data Storage Tags (DST-CTD) are used for tagging large fishes such as tuna. They measure conductivity (salinity), temperature, and depth. The system consists of a tag, communication box, computer, and software. Pop-up satellite archival tags (PSAT) measure ambient light (for latitude and longitude), temperature, and depth and transmit this information from the sea surface to an Argos satellite system. The tags can be programmed to detach at a preset depth or preset time. Among animals studied in this manner are Bluefin Tuna (*Thunnus thynnus*), Blue Marlin (*Makaira nigricans*), Tiger Shark (*Galeocerdo cuvier*), Blue Shark (*Prionace glauca*), Great White Shark (*Carcharodon carcharias*), Black-footed Albatross (*Phoebastria nigripes*), and the Southern Sea Elephant (*Mirounga leonina*). A fish algorithm developed by Frank O'Brien was incorporated into the computer program Easy (developed by Dale Kiefer, Department of Biological Sciences, University of Southern California) to analyze data collected by satellite tags. This analysis technique can also be applied to seabirds (e.g., Laysan Albatross — *Phoebastria immutabilis*) and marine mammals (e.g., sea lions, elephant seals, whales). One of the latest developments in fish management is the use of Mini GPS fish tags (Gudbjornsson et al., 2004).

Mark-recapture methods used in the past were based on a number of assumptions (Bakus, 1990): (1) The animal will not be affected by the marking or tagging (see Murray and Fuller, 2000); (2) The marks are not lost; (3) The population is sampled randomly and the age groups and sexes are equally available, that is, each individual has an equal probability of capture; (4) The sampling time is small relative to the total time of the study; (5) The population is either closed or immigration and emigration can be measured; (6) No births or deaths occur between sampling periods or some correction for this must be made; and (7) The initial capture does not effect subsequent recapture. It should be emphasized that a number of these assumptions (e.g., random sampling and age groups and sexes are equally available) are no longer present in recent models.

The simplest mark-recapture method is the Simple Lincoln Index (for closed populations). A closed population is one in which there are no gains by births or immigration and no losses by deaths or emigration. This is based on the Petersen method (Krebs, 1999); also see a modified Petersen method (Arnason, 1973). The animals are marked then recaptured on a subsequent day, and marked and unmarked animals are counted. The second count assumes random sampling. This index is seldom used today except perhaps for some student field exercises.

1.3.27.1 Lincoln Index

Example:

Fishes (e.g., territorial damselfish) are collected from the ends and the middle of a large tidepool during low tide. The fishes (64) are tagged then released into the same tidepool. The next day damselfishes are recollected from the ends and middle of the same tidepool. Count the tagged (6) and untagged (56) animals. Use the Simple Lincoln Index to calculate the approximate population size (see below). Note that the answer is not a true density estimate but an estimated total population size, applicable to mobile animals only (see Green, 1979, and Krebs, 1999, for further information). Moreover, the assumption that immigration and emigration has not occurred is likely to be invalid for many species of fishes. Other assumptions are also violated.

$$P = \frac{(a)(n)}{(r)}$$

where

P = total number of individuals (estimated population size)

a = number of animals marked in first sample

n = total number of individuals captured in second sample

r = number of marked individuals captured in second sample

$$P = \frac{(64)(56+6)}{(6)} \quad P = 661 \text{ tidepool damselfish}$$

The 95% confidence limits for a Poisson frequency distribution are 286 tidepool damselfish and 1097 tidepool damselfish (see Krebs, 1999, for a detailed discussion). This is as low or as high as we would expect the damselfish population to be with 95% confidence (i.e., one in 20 chance of being below 286 or above 1097).

1.3.27.2 Sampling at three or more times (for open populations)

Open populations are those in which individuals of a species are subject to gains by birth or immigration and losses by death or emigration. Three methods have been commonly used for this type of sampling, the Seber method, the Schnabel method, and the Manly–Parr method. The Jolly–Seber or Cormack–Jolly–Seber models were often preferred because they account for recruitment and mortality and are fully stochastic (Southwood and Henderson, 2000). The calculations are quite lengthy. The classical Jolly–Seber method for open populations is now presented (Sutherland, 1996). It requires that at least three samples should be taken and animals should be individually marked or have marks that are batch-specific (Tables 1-5 and 1-6).

There are many new mark-recapture techniques for both closed and open populations. Among them are modified Jolly–Seber models such as the Pradel model, Pollock model, Band Recovery model, Joint Live and Dead Encounter model, Known Fate model, Encounter Histories model, and Bayesian models, among others. There are also many computer programs available for these models. One in current favor by many investigators is the program MARK that can handle some 16 variables. The MARK program can be downloaded from the Internet for free:

http://www.cnr.colostate.edu/~gwhite/mark

See also: http://www.phidot.org/software/

Sutherland (1996) discusses several mark-recapture methods with examples. Southwood and Henderson (2000) list computer software for capture-recapture methods and discuss Moran–Ricker curves. Statistical tests used to detect variation in catchability (e.g., Cormack's test) are presented by Southwood and Henderson (2000). A recent handbook on capture-recapture analysis is presented by Amstrup et al. (2006); see also Begon (1979).

Table 1-5. Recording the composition of each collection sample for the Jolly–Seber method when marks or tags are batch-specific (source: Sutherland, 1996).

Days on which previously captured			Number of animals
1	2	3	
0^b	0^b	0^b	40
$+^a$	0^b	0^b	10
0^b	$+^a$	0^b	9
0^b	0^b	$+^a$	11
$+^a$	$+^a$	0^b	2
$+^a$	0^b	$+^a$	2
0^b	$+^a$	$+^a$	1
$+^a$	$+^a$	$+^a$	0^b

[a]Plus signs indicate that an animal was caught on that day.

[b]zeros that it was not. This example refers to sample number 4 in a hypothetical study.

Table 1-6.   An example of the Jolly–Seber method of mark or tag and recapture analysis (source: Sutherland, 1996).

The Jolly–Seber method

Definitions

  n_i = total number of animals caught in the ith sample
  R_i = number of animals that are released after the ith sample
  m_i = number of animals in ith sample that carry marks from previous captures
  m_{hi} = number of animals in the ith sample that were most recently caught in the hth sample

Example

Black-kneed Capsides *Blepharidopterus angulatus* caught at 3- or 4-day intervals in a British apple orchard (Jolly 1965).

The data are best summarized in a table of m_{hi} values, with n_i and R_i values across the top and the m_i values (which are the sums of the m_{hi} values in the column above) across the bottom. Thus in the table below, 169 insects were caught in sample 3 (n_3) of which 164 (R_3) were released; 3 of them bore marks from sample 1 (but not from sample 2) (m_{13}) and 34 bore marks from sample 2 (m_{23}), giving a total of 37 marked insects (m_3).

i	1	2	3	4	5	6	7	8	9	10	11	12	13	r_h	z_i
n_i	54	146	169	209	220	209	250	176	172	127	123	120	142		
R_i	54	143	164	202	214	207	243	175	169	126	120	120	—		
h															
1		10	3	5	2	2	1	0	0	0	1	0	0	24	—
2			34	18	8	4	6	4	2	0	2	1	1	80	14
3				33	13	8	5	0	4	1	3	3	0	70	57
4					30	20	10	3	2	2	1	1	2	71	71
5						43	34	14	11	3	0	1	3	109	89
6							56	19	12	5	4	2	3	101	121
7								46	28	17	8	7	2	108	110
8									51	22	12	4	10	99	132
9										34	16	11	9	70	121
10											30	16	12	58	107
11												26	18	44	88
12													35	35	60
m_i	0	10	37	56	53	77	112	86	110	84	77	72	95		

The table contains two other sets of summations:

  r_h = the number of animals that were released from the hth sample and were subsequently recaptured; these are simply the row sums
  z_i = the number of animals caught both before and after the ith sample but not in the ith sample itself; z_i is the sum of all the m_{hi} that fall in columns to the right of column i and all rows above row i; thus the dashed lines in the table delimit the m_{hi} values that must be summed to obtain z_4, for example

Parameter estimates

  $\hat{M}_i$ = number of marked animals in the population when the ith sample is taken (but not including animals newly marked in the ith sample).
  $= m_i + (R_i + 1)z_i/(r_i + 1)$

Table 1-6. (*continued*)

$\hat{N}_i$ = population size at the time of the ith sample

$\quad = \hat{M}_i\,(n_i + 1)/(m_i + 1)$

Φ_i = proportion of the population surviving (and remaining in the study area) from the ith sampling occasion to the $(i + 1)$th

$\quad = \hat{M}_{i+1}/(\hat{M}_i - m_i + R_i)$

$\hat{B}_i$ = number of animals that enter the population between the ith and $(i + 1)$th samples and which survive until the $(i + 1)$th sampling occasion.

$\quad = \hat{N}_{i+1} - \Phi_i\,(\hat{N}_i - n_i + R_i)$

Note that one cannot calculate $\hat{M}$ for the last sample. $\hat{N}$ for the first or last, Φ for the last two, or $\hat{B}$ for the first or last two. $\hat{M}_1$ is bound to be zero. Calculations are eased if laid out systematically:

$$\hat{M}_2 = 10 + (143 + 1)14/(80 + 1) = 34.89$$

$$\hat{M}_3 = 37 + (164 + 1)57/(70 + 1) = 169.46$$

$$\hat{M}_4 = 56 + (202 + 1)71/(71 + 1) = 256.18$$

etc.

$$\hat{N}_2 = 34.89(146 + 1)/(10 + 1) = 466.12 = 466$$

$$\hat{N}_3 = 169.46(169 + 1)/(37 + 1) = 758.11 = 758$$

$$\hat{N}_4 = 256.18(209 + 1)/(56 + 1) = 943.82 = 944$$

etc.

$$\Phi_1 = 34.89/(0 - 0 + 54) \qquad\qquad = 0.646$$

$$\Phi_2 = 169.46/(34.88 - 10 + 143) \qquad = 1.009$$

$$\Phi_3 = 256.18/(169.46 - 37 + 164) \quad = 0.864$$

etc.

$$\hat{B}_2 = 758.11 - 1.009(466.26 - 146 + 143) = 290.68$$

$$\hat{B}_3 = 943.82 - 0.864(758.11 - 169 + 164) = 293.13$$

etc.

Confidence limits for $\hat{N}_i$

Methods usually presented for calculating confidence limits of Jolly–Seber estimates are inadequate for the commonly encountered sample sizes. The following method, due to Manly, provides better limits.

Calculate a transformation of each $\hat{N}_i$ and the standard error of the transformation.

$$T_i = \log_e \hat{N}_i + 0.5\log_e\left[0.5 - 3n_i/8\hat{N}_i\right]$$

$$s_{T_i} = \sqrt{\left(\frac{\hat{M}_i - m_i + R_i + 1}{\hat{M}_i + 1}\right)\left(\frac{1}{r_i + 1} - \frac{1}{R_i + 1}\right) + \frac{1}{m_i + 1} - \frac{1}{n_i + 1}}$$

For example, the transformation for $\hat{N}_2$ is

$$T_2 = \log_e 466.26 + 0.5\log_e[0.5 - 3(146)/8(466.26)] = 5.6643$$

$$s_{T_2} = \sqrt{\left(\frac{34.89 - 10 + 143 + 1}{34.89 + 1}\right)\left(\frac{1}{80 + 1} - \frac{1}{143 + 1}\right) + \frac{1}{10 + 1} - \frac{1}{146 + 1}} = 0.3309$$

(*continued*)

Table 1-6.   (*continued*)

Calculate confidence limits for T_i, and their exponents:

$$T_{iL} = T_i - 1.65s_{T_i} : \quad W_{iL} = e^T iL$$
$$T_{iU} = T_i + 2.45s_{T_i} : \quad W_{iU} = e^T iU$$

Continuing the same example:

$$T_{2L} = 5.664 - 1.65(0.331) = 5.118; \quad W_{2L} = e^{5.118} = 166.98$$
$$T_{2U} = 5.664 + 2.45(0.331) = 6.475; \quad W_{2U} = e^{6.475} = 648.69$$

95% confidence limits are given by

$$(4W_{iL} + n_i)^2/16W_{iL} \text{ and } (4W_{iU} + n_i)^2/16W_{iU}$$

For the example of $\hat{N}_2$ the limits are

$$[4(166.98) + 146]^2/16(166.98) = 248$$

and

$$[4(648.69) + 146]^2/16(648.69) = 724$$

Goodness-of-fit test

The test is applied to each sample in turn, except the first and the last. All animals caught in the sample are categorized as follows:

f_1 = first captured before this sample, subsequently recaptured
f_2 = first captured before this sample, not subsequently recaptured
f_3 = first captured in this sample, subsequently recaptured
f_4 = first captured in this sample, not subsequently recaptured

Calculate

$$a_1 = f_1 + f_2, \quad a_2 = f_3 + f_4, \quad a_3 = f_1 + f_3, \quad a_4 = f_2 + f_4$$
$$n = f_1 + f_2 = f_3 + f_4$$
$$g_1 = \sum f \log_e f, \quad g_2 = \sum a \log_e a$$
$$G = 2(g_1 - g_2 + n\log_e n)$$

Meadow Voles *Microtus pennsylvanicus* trapped over 6 occasions in Maryland, USA, by J. D. Nichols (Pollock *et al.* 1990) provide an example for this calculation. For the second sample the figures were

$$f_1 = 47, \quad f_2 = 31, \quad f_3 = 29, \quad f_4 = 14$$
$$a_1 = 78, \quad a_2 = 43, \quad a_3 = 76, \quad a_4 = 45$$
$$n = 121$$
$$g_1 = 47\log_e 47 + 31\log_e 31 + 29\log_e 29 + 14\log_e 14 = 422.01$$
$$g_2 = 78\log_e 78 + 43\log_e 43 + 76\log_e 76 + 45\log_e 45 = 1001.99$$
$$n\log_e n = 121\log_e 121 = 580.29$$
$$G = 2(422.01 - 1001.99 + 580.29) = 0.62$$

This G should be compared with χ^2 with 1 degree of freedom.

The G values for samples 2, 3, 4 and 5 of Nichols's study of voles, with their associated probabilities, are

Table 1-6. *(continued)*

sample (i)	G	P
2	0.62	0.43
3	7.69	0.005
4	14.55	0.0001
5	1.83	0.17

Thus samples 3 and 4 give evidence that the assumptions are violated.
The sum of G values provides a test of the overall goodness-of-fit, with a number of degrees of freedom equal to the number of samples providing individual G values. In this example the sum is 24.69, with 4 degrees of freedom, which is highly significant ($P < 0.0001$). Samples for which any of the expected frequencies of the four groups of animals are less than 2 should be left out of this test.

1.3.28 Visual Methods for Fishes

Two major methods of estimating fish populations have been used: (1) **belt transect method** – see pp. 25 and 51, and (2) **rapid visual (roaming or roving) technique** – the biologist swims in a "random" or haphazard direction for 10 min and counts fishes (a plotless method). This is repeated six times and a test of concordance (see Chapter 3, p. 159) is calculated for the rank abundances. The belt transect method is generally more reliable for quantitative estimates of fish populations but the roving method is ideal for a quick and simple assessment. It is important for studies to be time-based or standardized for irregular habitats. Jiangang Luo (University of Miami) has developed a device that projects an array of parallel laser beams through the water onto the body of a fish, helping divers estimate its length. See De Martini and Roberts (1982), Branden et al. (1986), Bohnsack and Bannerot (1986), McCormick and Choat (1987), and Bortone and Kimmel (1991) for detailed information. Example of the estimation of the density of a mobile population (fish):

$$D = \frac{Z}{2rV} \quad \text{(Southwood, 1978)}$$

where:

D = density of the population (No./m^2)

Z = number of encounters per unit time (e.g., h)

r = effective radius (for an encounter with an observer) (e.g., m)

V = average velocity of the organism (the most difficult factor to determine) (e.g., m/h)

Example

Z = 100 fish (1 species) encountered per hour

r = 10 m

V = 500 m/h

$$D = \frac{100}{2(10)(500)} \quad D\text{(single species)} = 0.01 \text{ fish/m}^2 \text{ or } 1 \text{ fish/100 m}^2$$

1.3.29   Narcotizing Agents and Poison Stations

Fishes can be collected alive by using a narcotizing agent such as quinaldene. How-ever, some people are sensitive to the chemical. The quinaldene is fanned among branches of coral or openings in rocks. It works successfully only if there is no cur-rent to disperse the chemical.

Fish populations have been estimated by setting up poison stations. This can be done efficiently using perhaps a dozen divers. Plastic gallon (3.8 l) bags are filled with rotenone or rotenone derivative (Noxfish, Chemfish) and a sol-vent (e.g., alcohol). Each diver carries a plastic bag with chemicals down to the reef. The divers then break the bag in unison and distribute the rotenone by fanning the water over the coral with their swim fins. Fishes affected by the rotenone begin to swim out of the corals around 15 minutes after the applica-tion of rotenone. Typically the last fishes to exit the corals are moray eels, after about 45 min. The diver uses a dip net and goody bag to capture and secure the fishes. The fishes must be captured quickly as they often reenter the reef and probably die. Predators on coral reefs (e.g., large jacks, sharks) may take many of the squirming fishes before the diver has an opportunity to catch them. Thus the resulting fish catch is an underestimation of the population of fishes in the area.

1.4   OTHER METHODS OF ESTIMATING THE ABUNDANCE OF POPULATIONS

1.4.1   Comparison of Estimated Populations with Other Methods

(a) Measure a population by random sampling and compare the results with a relative method (e.g., trap organisms by random sampling and compare them with the number of animals observed per hour).

(b) Calculate the regression of some index with an actual population (e.g., the density of a species of kelp by random sampling and by actual count along a line with depth).

1.4.2   Removal Trapping or Collecting

A known number of animals are removed from a habitat by trapping, and the rate at which the trap catches decrease indicates the size of the total population. Eber-hardt's Removal Method is discussed by Krebs (1999). As mentioned previously, removal methods have poor precision and a large potential for bias (Buckland et al., 2001).

1.4.3   Other Methods

1.4.3.1   Indirect Distance Sampling

The population of elephants in dense forests in Kenya is estimated by counting fresh fecal droppings. These elephants are wary of people and are seldom seen in some biological preserves (e.g., preserve near Malindi). The fresh nests of birds can also be tallied. Point transect sampling can be used for counting birds and whales (Buckland et al., 2001). Typically the lines and points are selected randomly. The area surrounding the point is surveyed over a standardized time period for birds, for example, and the distance from the point to the site where the bird was spotted is measured. Laser binoculars can be used for precise measurements. Spot mapping involves intensive mapping of territories and home ranges of target species (typically birds) at randomly chosen points within the study area (Hallacher, 2004).

1.4.3.2   Rare Species

Thompson (2004) and colleagues discuss the sampling of rare, sparse, or elusive species. **Rare species can be defined as species with a low probability of detection. Sparse species are those that occur in small clumps but over large areas. Elusive species are species with secretive or nocturnal behavior**. Among the many techniques discussed in Thompson (2004) are aerial surveys, distance sampling, intercept sampling, mark-recapture, adaptive sampling, two-phase or stage or double adaptive sampling, adaptive cluster sampling, sequential sampling, point count method of sampling for birds, sighting probability models, patch-occupancy models, occupancy estimation using likelihood methods, resource selection functions for the design of unequal probability sampling, Bayesian estimation, spatial modeling of plants, noninvasive genetic sampling (e.g., mitochondrial DNA or nuclear DNA in hair and feces), photographic sampling of elusive mammals in tropical forests, and banding and PIT tagging in bats.

1.4.3.3   Coral Reef Surveys

Live cover of hard corals has typically been measured in two ways: (1) by eye, and (2) by line transect. Live cover by eye involves: being towed by a rope or Manta board behind a boat (Figure 8-22), or using an underwater scooter (e.g., Farallon scooter – see Fig. 1-25). Towing a person with a boat works well but speed must be reduced because it is difficult to hold on to the rope. A Manta board is preferable. Sometimes the diver may be followed by sharks. The underwater scooter is a wonderful method of surveying reefs although velocities must be reduced until the diver develops sufficient holding strength. There are two dangers in using underwater scooters. The scooters travel so fast that it is easy to reconnoiter down to 60 m or deeper and be unaware of the depth. The battery may die thus it is important to travel in pairs for safety reasons. Although slow, one scooter can tow two people through the water

Figure 1-25. Farallon underwater scooter. An excellent device for the underwater reconnaissance of reefs. Some underwater scooters cost less than $100 but are limited to a depth of 65 ft (20 m). See text for explanation. Source: http://www.faralonusa.com/product.html

back to the ship. In any case, the observer must be competent in coral identification. See p. 24 for information on line transects.

A relatively recent large study of coral reefs in the western Atlantic included the following techniques (Lang, 2003): Benthic organisms – (1) line intercepts (10 m long), (2) measuring hard coral dimensions, (3) recording diseases (e.g., black band), (4) using small quadrats for estimating the percentage of algal cover, and (5) belt transect counts of the sea urchin *Diadema*. Fishes – (1) belt transects (30 m long), (2) a roaming or roving population estimation, (3) counting fish in 7.5 m diameter circles, and (4) fish bites in m^2 plots (total of five plots). See this chapter and Chapter 3 for discussions of these topics.

1.4.3.4   Artificial Reefs

Artificial reefs enhance substrate or habitat heterogeneity and lead to an overall increase in diversity over time (Bortone and Kimmel, 1991). **The primary reason for establishing artificial reefs is to increase lobster and fish populations and to enhance sport diving, especially on sunken ships**. Artificial reefs attract and concentrate commercially viable fish populations at a new site, and then overfishing occurs. The newest approach is to develop artificial reefs to enhance the populations of specific species of fish (e.g., see http://myfwc.com/marine/ar/index.asp). Examples of artificial reefs are shown in Fig. 1-26. Recent innovations include limestone reef restoration using electrodes (cathodic reef stimulation) to induce mineral accretion [$CaCO_3$ and $Mg(OH)_3$] on artificial reef frames. This produces coral reef growth (see http://www.globalcoral.org/third_generation_artificial_reef.htm).

Artificial reef colonization is often very rapid and shows similar and repeatable patterns in temperate waters. When abiotic conditions are stable, biological factors such as competition and predation are usually important for limiting population sizes and controlling assemblage (community) composition.

Bortone and Kimmel (1991) present a list of variables commonly measured on artificial reefs and references to methods of assessing and monitoring artificial habitats. Techniques include collecting data on individual organisms (e.g., body size, feeding habits, territory size), on populations (e.g., natality, mortality, age classes), nonvisual methods for fish and invertebrates (e.g., hook and line, nets, hydroacoutistics), and

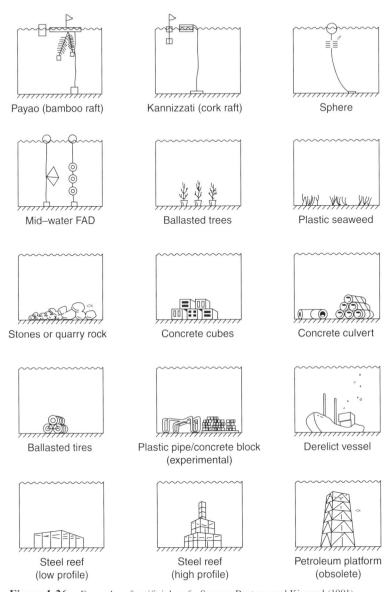

Figure 1-26. Examples of artificial reefs. Source: Bortone and Kimmel (1991).

visual methods (diver transects, species-time counts, predation, still photos, videos, ROVs, and manned submersibles). For further information see Seaman and Sprague (1991), Seaman (2000), and especially the Internet. See Berger (1993 or later) for a **bibliography on artificial reefs**.

Recent reefs in North America available for divers and fisherfolk include: (1) Comanche reef, Charleston, South Carolina [50 subway cars at a depth of up to 29 m (95 ft)]; (2) Tenneco Towers, Fort Lauderdale, Florida [oil and gas exploration platforms at a depth of up to 40 m (130 ft)]; (3) USS Spiegel Grove, Key Largo, Florida [156 m (510 ft)] Navy landing ship at a depth of up to 40 m (130 ft); (3) Snake Island Wrecks (two ships), Vancouver Island, British Colombia, at a depth of up to 40 m (130 ft); and (4) HMCS Yukon, 112 m (366 ft) Canadian destroyer escort ship, San Diego, California, at a depth of up to 31m (101 ft). The U.S. Navy recently sank one of its aircraft carriers (USS Oriskany) off the coast of Pensacola, Florida (Gulf of Mexico).

1.4.3.5   *Further Methods of Small Scale Sampling*

Populations may also be estimated by animal products (e.g., worm tubes, barnacle shells, exoskeleton remains, bird nests, feces) or by effects (e.g., introduction of a known number of animals into a specified habitat or exclusion of grazers from a habitat, amount of plant consumed). Seabird colony populations can be estimated by flushing the birds and whales by cue counting (e.g., whale blows per unit area per unit time). See Borchers et al. (2002) and Gore et al. (2000) for further information. Krebs (1999) discusses methods of estimating abundances of populations with radio transmitters, and a series of enumeration methods used with small mammals. Southwood and Henderson (2000) discuss observation of birds by radar, hydroacoustic methods, automatic fish counters (e.g., for salmon moving upstream), oral detection, trapping including aquatic light traps, baits and lures including kairomones and pheromones, sound, and so forth. Change in ratio methods has poor precision (Borchers et al., 2002). See also Buckland et al. (2001). Other methods of sampling include link-tracing designs (e.g., network sampling), multistage designs, variable circular plots or point transects, and spatial sampling or kriging (Thompson, 2002). Resource or habitat selection sampling is discussed by Manly et al. (2002).

1.4.4   Large Scale Sampling

Techniques that sample large-scale areas include: (1) aerial surveys including photography; (2) satellite imagery; and (3) hyperspectral imaging. Aerial surveys include counts or population estimates of marine birds (e.g., auks, Family Alcidae) and marine mammals (e.g., whales, walruses (*Odobenus rosmarus*) – see Kauwling and Bakus, 1979). Photography from aircraft has been used for many years. Among the methods are infrared photos of forests or of kelp beds along the coast. Other types of aerial photography include photos from cameras mounted on kites and pigeons, unmanned aerial vehicles (UAVs), and balloon airships. Aircrafts (airplanes) generally operate between 600 ft (183 m) and 100,000 ft (30,488 m). Radar offers the greatest potential for foul weather remote sensing, but is relatively expensive.

The main types of instrumentation used in large-scale sampling include the spectrometer (optical), radiometer (thermal), and radar or microwave scanning (weather). Multispectral scanning began in the 1970s with the digital remote sensing Landsat satellite. Two primary instruments are used in multispectral scanning: (1) MSS (multispectral scanner system) provides from orbit visible and infrared images as high as an 80-m resolution in four bandwidths. MSS systems extend the light reception range into the infrared with much higher spatial resolution than photographic systems; and (2) TM (Thematic Mapper) has a 30-m resolution from orbit in seven bandwidths. The panchromatic band has a resolution of over 10 m (see below). The SPOT satellite, with a resolution of 20 m, has been used on coral reefs in the Indo-Pacific (Bour et al., 1986). Space Imaging of Thornton, Colorado (see www.space-image.com) in 1999 produced the first commercial 1-m resolution satellite scanner.

Among the earliest satellites that have been used are Landsat and Seasat. Landsat satellites orbit about 900 km above the earth's surface and scan images of an area of approximately 185 km². Data from Landsat images are used in agriculture, water resources, mineral resources, forestry, land use, marine resources, and environmental science. Landsat images are available to the public for most areas of the world. Seven spectral bands are measured from the TM including red (sensitive to vegetation), green (sensitive to barren areas), and spectral bands in the infrared. The Enhanced Thematic Mapper (ETM+) has a 15-m resolution panchromatic band, a long wave infrared 60-m resolution band, and six 30-m resolution multispectral bands (see Raytheon on the Internet – http://www.raytheon.com/products/etm_plus/). Thermal radiation is measured to determine sea surface temperatures. The results of the color spectra can be combined into one image and each spectrum assigned a false color or pseudocolor for viewing it on the monitor screen. Structural features (i.e., mountains, valleys, rivers), mineral resources, vegetational types, and many other things are observed. The resolution for Landsat today is around 80 m per pixel. A satellite image is shown in Fig. 1-27 and a pseudocolor image is shown in Fig. 1-28. See

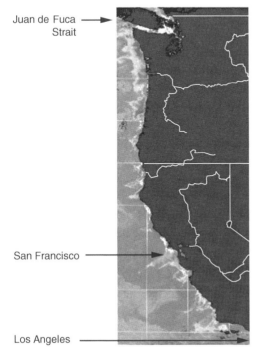

Juan de Fuca Strait

San Francisco

Los Angeles

Figure 1-27. A satellite photo of the west coast of the United States. The colors show chlorophyll a patterns in surface coastal waters. Source: modified from Brink and Cowles (1991).

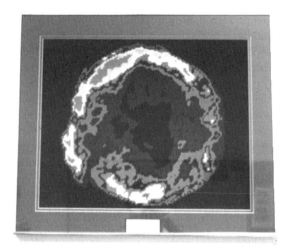

Figure 1-28. False colors or pseudocolors have been used to emphasize certain regions of the earth. A color is assigned to a specific entity (e.g., temperature belt).

Hudson and Colditz (2003) for some of the latest techniques in remote sensing. Two new U.S. Navy satellites include Windsat and the F-16 (Tomaszeski, 2004). Windsat measures ocean surface wind speed and direction. The F-16 measures temperatures, humidity, ocean wind speeds, cloud water content, surface terrain temperatures, soil moisture content, and other variables.

Satellite remote sensing has not yet replaced airborne remote sensing for oil spills (Fingas and Brown, 2000). Laser fluoro-sensors (using the UV spectra) may be the only means of discriminating between oiled and unoiled weeds and to detect oil on different types of beaches. It is the only reliable method of detecting oil in certain ice and snow conditions. The instrument is large (400 kg) and expensive.

Geographical features are commonly georeferenced (**ground-truthed** in industry jargon), that is, observers measure the position of features (e.g., edge of a grassland, edge of a water mass) by using a small, **hand-held or back-packed GPS** (Geographical Positioning System), and often make scientific measurements defining the properties of interest. Garmin, Magellan, and Trimble make a variety of small GPS devices for public use (Fig. 1- 29).

One of the newest methods for large-scale sampling is an advanced *multispectral or hyperspectral sensing system* (Fig. 1-30). Hyperspectral imaging was developed in 1983 by NASAs Jet Propulsion Laboratory. An imaging radiometer scans the land or ocean surface from an airplane or satellite, measuring radiation from 10s to 100s of wavelengths (i.e., color variations in the energy reflected from the ground). Hyperspectral data can also be obtained from a hand-held instrument by holding the scanner above an object. For example, hand-held instruments have been used to discriminate between healthy coral, bleached coral, sand, and algae, even on a species level (Holden and LeDrew, 2001). The 22° FOV (field of view) sensor is hand-held 10 cm above the substratum or organism by a scuba diver. It can measure 175 spectra using a bandwidth of 1.4 nm. Often unrotated PCA (Principal Component Analysis—see Chapter 5, p. 247) is used to reduce the reflectance data

Figure 1-29. The Geographical Positioning System or GPS is used throughout the world to determine exact localities of ocean diving sites, forests, rivers, etc. The accuracy is now about ±3 m for a $300–500 GPS. Source: http://www.garmin.com/products/html.

Figure 1-30. A full-frame video multispectral sensor. Source: Mueksch (2001).

to representative spectra, using the first principal component (Table 1-7), then cluster analysis (see Chapter 4, p. 218) is employed to separate different classes of spectral resolutions. Sometimes there is considerable spectral similarity between certain taxa

Table 1-7. Hyperspectral measurements on organisms and substrates. Source: Holden and LeDrew (2001). [Analysis of Principal Components for Each Bottom Type (Hyperspectral Measurements via Scuba Diving)] See text for details.

Class	Number of Spectra	Variance (%)	Loading
Grass	24	97.89	0.99
Sand	4	99.56	0.99
Rock	6	99.13	0.96
Brown algae	7	97.37	0.99
Green algae	7	97.37	0.99
Rubble	27	85.47	0.99
M. annularis	11	98.99	0.99
D. strigosa	41	97.10	0.99
P. astreoides	7	99.92	0.99
P. porites	6	99.06	0.99
A. palmata	6	99.61	0.99
M. flavida	6	98.99	0.99
Anemone	6	99.83	0.99
Bleached	17	99.62	0.99

(e.g., macroalgae and healthy coral—see brown algae and the hard coral *P. astreoides* in Fig. 1-31). Hyperspectral imaging is used to map invasive plant species and diseased plants with 90% accuracy. For example, it was used to track the spread of alien tree species that are altering the rainforest in Hawaii Volcanoes National Park. Weather conditions can obscure images and species cannot always be distinguished.

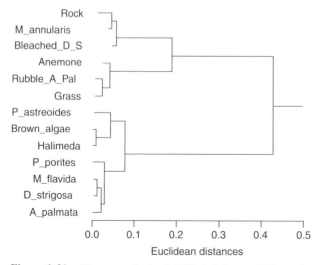

Figure 1-31. Cluster tree for data in Table 1-7. Source: Holden and LeDrew (2001).

See Mumby et al. (1999), Jensen (2000), Bunkin and Voliak (2001), Halpern (2001), Hunsaker et al. (2001), Millington et al. (2001), Lunetta and Elvidge (2002), Avery and Berlin (2004), and the Manual for Remote Sensing (Henderson and Lewis, 1998) for further information on remote sensing. Stein and Ettema (2003) discuss sampling procedures and experimental design for land use studies and comparing ecosystems.

A relatively new discipline involving large-scale sampling is **spatial analysis**. It typically consists of using remote sensing, GPS, expert systems, and GIS to analyze information. In ecology, it involves sampling by distance measurements, patch analysis, spatial autocorrelation, and chaos, among others (Tilman and Kareiva 1997, Hunsaker et al., 2001; Fortin and Dale, 2005). It covers the spatial dynamics of ecological communities on a larger scale (i.e., metacommunities) (Holyoak et al., 2005). A number of these topics are found in various chapters in this book. An example of spatial analysis in biological oceanography is that of Cowen et al. (2006). Flow trajectories from a high resolution Caribbean ocean circulation model in a Lagrangian stochastic scheme were used to generate an individual-based IBM model for coral reef fish larval dispersal. The biological parameters of the model included pelagic larval duration, larval swimming behavior, and adult spawning by season and frequency. They concluded that typical fish larval dispersal distances in the Caribbean are on a scale of 10–100 km.

An example of spatial analysis in operation is that of the University of Washington Fisheries Sciences Spatial Analysis Laboratory (http://sal.ocean. washington.edu/research/research.html). See also the topic of ocean spatial analysis on the Internet.

SUMMARY

For numbers of sampling units required, use the equations most appropriate for the spatial distribution of the species under study. Density estimates are obtained most accurately by belt or strip transects, stratified random sampling, or using the line intercept method of Strong. Rapid sampling techniques involve taking photos and videos of the benthos followed by a random selection of photographic frames for automatic counting. Visual counts of fishes can be easily made by haphazard roaming or more accurately, but more time-consuming, by belt transects. Line intercept methods provide an abundance of data for reef studies. Large-scale sampling by satellite combined with hyperspectral imaging provide an enormous amount of valuable information. Spatial analysis is the latest development using these techniques.

Chapter 2

Types of Data, Standardizations and Transformations, Introduction to Biometrics, Experimental Design

2.1 INTRODUCTION

Statistics can be defined as the study of numerical data (Zar, 1999). Biometrics is statistics applied to biological problems. The major statistician of the twentieth century was Ronald A. Fisher (1890–1962), who developed analysis of variance among many other contributions (see Fisher, 1932) .

Acquisition of Numerical Data

Measurements should be (1) **logical** (do not measure juveniles with adults for data on adults), (2) **comparable** (of the same type; do not mix lengths with weights), (3) **standardized** (measured by the same method; do not combine a mm rule with vernier calipers), (4) **adequate** (sample size or frequency of occurrence should be based on the type of data collected and the allowable error), and (5) **random** (by random sampling methods).

Numbers may be rounded in two ways but be consistent in your choice: (1) numbers 1–4 down and 5–9 up or (2) even numbers up and odd numbers down.

Equipment

Microcomputers are changing so rapidly that practically anything mentioned today will be nearly obsolete within a year or two. On the scene are 64-bit computers and huge memory banks that will soon make microcomputers into what today are called minicomputers. The most rapid advances in the field are occurring in Japan, France, England, and the United States. IBM, Digital, and Cray mainframe computers are three of the biggest names in the computer world. They are used extensively in computer work at many universities, and C++ or Visual C++, Basic or Visual Basic, HTML and Java (HTML and Java for the Internet) are the major computer languages for biologists. FORTRAN is still used by many scientists and engineers outside the United States. The newest developments in computers are the use of wireless techniques, including Bluetooth, Wi-Fi, wireless LAN (WLAN), and Flash memory sticks. MacIntosh computers now contain Intel chips and software that enables them to use PC programs.

CD-ROM

Basic statistics is presented in condensed and animated fashion on a CD-ROM that accompanies this book. (The file is Stats4BN. Be sure to turn the sound on; see the Appendix.)

References

There are numerous texts on statistics but few relate to problems in biology in a clear and concise manner. One of the clearest presentations is that of Fowler et al. (1998); it is recommended for the beginner. Excellent but more detailed presentations are found in Mead (1988), Sokal and Rohlf (1995), Zar (1999), Glover and Mitchell (2002), and Gotelli and Ellison (2004). See also Campbell (1979), Bailey (1981), Hampton (1994), and Townend (2002). A statistics text (without equations) for biologists by Dytham (1999) uses SPSS, Minitab, and Excel software programs. Magnusson and Mourã (2004) present statistics without math. A recommended online (Internet) biostatistical program is that of Gould and Gould (2002).

2.2 TYPES OF DATA

The following section presents two ways of classifying data. The categories for the two presentations are not mutually exclusive.

Classification of Data

(1) **Meristic:** Consists of data represented by integers or whole numbers (counting) (Clifford and Stephenson, 1975). Also called discrete numbers. Problems arise when parts of animals are encountered (e.g., edge effect in sampling with quadrats).

(2) **Metric (continuous):** Consists of data represented by partial numbers (measuring: e.g., 1.8 mm). Rounding off is occasionally a problem.

(3) **Binary:** Consists of two contrasting states. For example, presence/absence in ecology and color red/color blue in taxonomy. Rarer species are overemphasized in binary data. Biogeography employs binary data; for this reason, biogeographers should not overemphasize the extremes of biogeographical ranges. Most ecological data are not binary.

(4) **Ordinal or Ranked (ordered multistate):** Consists of states arranged in a hierarchy (e.g., abundant-common-rare in ecology and crab chela or bird beak long-medium-short in taxonomy). Ranked data often (but not always) apply to one sample, and ordinal data to many.

(5) **Disordered Multistate:** Three or more contrasting states of equal rank (e.g., flowers red, white, blue).

Types of Biological Data

(1) **Data on a ratio scale:** Measurement scales with a constant interval size and true zero point, including lengths, weights, volumes, rates, etc. See Stevens (in Houston, 1985) and Zar (1999).

(2) **Data on an interval scale:** Measurement scales with a constant interval size but not a true zero, including temperature scales such as Celsius and Fahrenheit, and circular scales such as time of day and time of the year.

(3) **Data on an ordinal scale:** Measurements that are ordered or ranked, such as numbers 1–5 or the letters A–E.

(4) **Data on a nominal scale:** attributes defined by some quality, such as hair color blonde, brown, red or black.

(5) **Data that are continuous** (e.g., height of a tree in meters and decimal fractions) *or discrete* (e.g., the number of cleaner fish visiting a host).

Precision for Data Sets

Some biologists use the 30–300 rule to determine precision for data sets (Mitchell, 2001). For example, if you were collecting limpets on a beach and wanted to record their lengths using this rule, you might have measured in 0.1 mm units if your animals ranged in size from 4 to 20 mm. $20 - 4\,mm = 16\,mm$ range. $16\,mm \times 10 - 0.1\,mm$ units $= 160$ units of 0.1 mm. That would fall between the 30 and 300 limits. If you measured the limpets in mm then $20 - 4\,mm = 16\,mm$, which falls below the 30 limit and thus is inappropriate for a measurement scale.

2.3 DATA STANDARDIZATION OR NORMALIZATION (RELATIVIZATION)

Data may be standardized by any of several methods to reduce the values to a common scale for comparison, for groups of unequal size or degree. Many ecological analyses intentionally include standardization (e.g., Bray-Curtis dissimilarity measure – see Chapter 4, p. 212). Normalization allows you to compare apples with oranges!

(1) By Standard Deviation or Variance (by row, column, or both)

Divide raw data by σ or σ^2 (see p. 77). Note here that standardization is built into some coefficients (such as the correlation coefficient). This is the most commonly used method of standardization. However, the classical method is to subtract the mean from each matrix element then divide them by the standard deviation (Fortin and Gurevitch, 2001).

(2) By Totals

Convert raw data to percentages (e.g., the raw values 2000, 1000, 1500, 500 (total = 5000) can be reduced to the percentages 40, 20, 30, 10, representing their proportional equivalent of the total). In community analysis (Chapter 4, p. 217), conversion of raw data can be done with species, sites, species first then sites or sites first then species (the latter two categories are known as double standardization).

(3) By the Norm or Geometric Mean

This is an estimate of the median of a lognormal distribution (Jongman et al., 1995). See p. 66 below, Noy-Meir (1970), and Sokal and Rohlf (1995) for this method and for double standardization (column and row norm).

Example

Sites

		1	2	3	4	5
Species	1	7	2	3	9	1
	2					
	3		etc.			

\# of individuals/site

$$GM = \sqrt[n]{Y_1\, Y_2\, Y_3 \ldots Y_n}$$

Where:

GM = geometric mean (for a row in this example)

Y_i = number of individuals of species 1 at site j

n = root

$$GM\, y = \sqrt[5]{378} \qquad GM\, y = 3.3$$

2.4   DATA TRANSFORMATION

Data transformation stabilizes the variance. **Data transformation is a frequent attempt to normalize data to better resemble a bell-shaped curve**. It is often used to convert somewhat heteroscedastic data to homoscedastic data. It is used to straighten out regression line plots from a slightly curvilinear form to a linear form. It has been used to avoid the excessive importance given to more abundant species and to correct for data gaps (Clifford and Stephenson, 1975). In community analysis (Chapter 4. p. 217), the need for data transformation varies with the similarity or dissimilarity measure used. For example, it is most necessary with Euclidean distance and least with Canberra metric. For further information see Sokal and Rohlf (1995). Not all experimental data sets can be satisfactorily handled by a single transformation (Mead, 1988). Transformations can affect the conclusions of an analysis (Clarke and Warwick, 2001). Transformations should not be taken routinely but only when conditions for statistical tests are grossly violated (Southwood and Henderson, 2000). There are many different types of transformations. It is usually found adequate to transform the data from a regular (i.e., uniform distribution) population by using squares, for a slightly aggregated one by using square roots, and for a highly aggregated one by a logarithmic transformation (Southwood and Henderson, 2000). Those most commonly used in ecology are given below.

(1)   Logarithmic Transformation

Log transformations can make positively skewed data distributions more normal in form, that is, skewed distributions become more symmetrical. It is appropriate when the variance of a sample of count data is larger than the mean (Fowler et al., 1998). It may cause the variance of a sample to become independent of its mean; it may also remove heteroscedasticity (see p. 79). Log transforms are frequently used in relation to growth of organisms (e.g., lengths, weights). Examples: $\log_{10} x$; $\log (x + 1)$ to make zeros count. Hayek and Buzas (1997) discuss log transformation problems. According to Green (1979), logarithmic transformation of data has the effect of making the observations scale-independent. Data for continuous variables (i.e., fractions or decimals) should be transformed to a log scale unless there is good reason to believe that this is unnecessary (Mead, 1988). See also Barnes (1952), Clarke and Warwick (2001), and Gotelli and Ellison (2004) for a discussion on transformations.

Example: $\log_{10}$ of $16 = 1.2$

(2)   Square Root Transformation

Square root transformations are recommended for use with data in the form of counts (Mead, 1988). Square root transforms cause the variance of a sample to become independent of its mean. It makes the data more bell-shaped or normal. It gives less weight to numerically dominant species (van Tongeren, in Jongman et al., 1995). The square root transformation is similar in effect to but less drastic than the log transform (McCune et al., 2002). Roots at a higher power than three transform the data to nearly a presence-absence state.

Examples

$\sqrt{16} = 4$

$\sqrt{x} + 0.5$ is commonly used to avoid $\sqrt{0}$

in community analyses – see p. 213 in Chapter 4.

$\sqrt[n]{x}$ is used with data showing severe discontinuities or spikes.

(3)   Arcsine Transformation

This inverse sine transform is appropriate for binomial distributions such as percentages and proportions (ratios). It contracts the center of the distribution and stretches out the tails. Example: For percentages, use:

Example

$$arcsin \frac{\sqrt{p}}{100}$$

where: $p = $ percentage

when the data are $<20\%$ or $>80\%$. If these percentages occur then **all the data are transformed by arcsine** (not just the $<20\%$ and $>80\%$ data). It is necessary to transform percentages only when several of the percentage points lie outside the 20–80% range (Southwood and Henderson, 2000).

Example

Arcsin of 16% (entered as the decimal 0.16) = 9.2

For proportions use:

Arcsin $\sqrt{p}$, where $p = $ proportion (expressed as a decimal)

One can also use arcsine squareroot transformations for proportion data (McCune et al., 2002). The data are entered in decimal form (between 0 and 1).

Arcsine squareroot $= 2/\pi \ x$ arcsine $(\sqrt{X})$

(4)   Arcsinh Transformation

When there are many zeros ($\geq 25\%$) in the data base, the arcsinh transformation can be used on all the data.

Example: Arcsinh of $0 = 0$, of $58 = 4.754$, and so forth.

(5)   Reciprocal Transformation

Used to change hyperbolic curves (Fig. 2-1) to straight lines, such as in the number of eggs laid per female.

Hyperbola

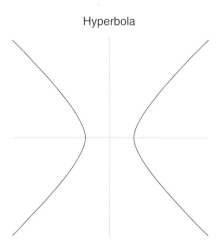

Figure 2-1.   Example of hyperbolic curves.

Examples

$$\frac{1}{x} \qquad 16\,\text{eggs} = \frac{1}{16} = 0.06$$

or

$$\frac{1}{x+1}$$

(6) Box-Cox Transformation

An iterative method for estimating the best transformation to normality within a family of power transformations (see Sokal and Rohf, 1995). Krebs (2000a) has a computer program for this. Taylor (1961, 1971) uses a straightforward power series (examples: $n^{1/2}$; $n^{1/3}$).

Taylor's power law offers an approach to finding a general transformation that makes the variance independent of the mean (Southwood and Henderson, 2000). Krebs (1999) also discusses Taylor's power series. For a recent text on mathematical and statistical ecology (see Gore et al., 2000).

2.5   STATISTICAL DISTRIBUTIONS AND PROCEDURES

Frequency Distributions

Commonly used frequency distributions include scatter diagrams, frequency polygons, histograms, and pie diagrams (Fig. 2-2 a–d). More complex frequency distributions include 3D scatterplots and 3D surface plots (Fig. 2-2 e–f). Mead

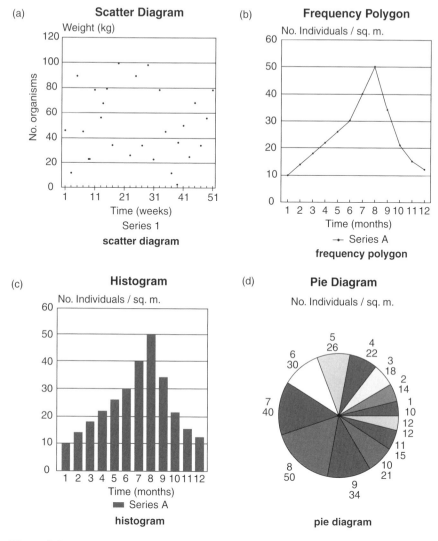

Figure 2-2. Commonly displayed frequency distributions: (a) scatter diagram, (b) frequency polygon, (c) histogram, (d) pie diagram,

(1988) contends that data in tabular form is more desirable than figures because tables contain more detailed information.

Procedures in Statistics

(1) First set up your **null hypothesis**, (see p. 80), (2) determine the **confidence limit or probability level** desired. A confidence limit of 95% ($\alpha = 0.05$) is often used by statisticians. (3) **Work out the statistical problem**. A recent tendency has been to

(e)   3D scatterpict for three items

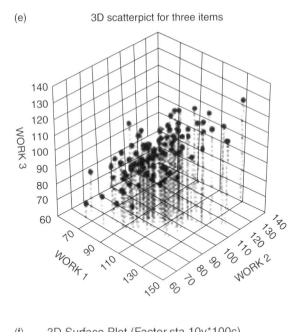

(f)   3D Surface Plot (Factor.sta 10v*100c)
        WORK 3 = Distance Weighted Least Squares

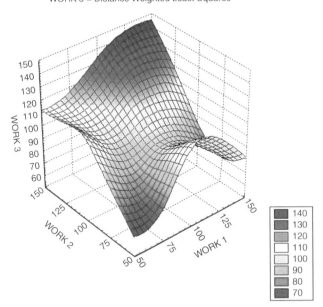

Figure 2-2. *(Continued)* (e) 3D scatterplot (source - http://www.statsoft.com/textbook/popups/
popup14.gif) (f) 3D surface plot (source - Statistica 6.0).

list the exact probability level calculated (if the statistical program can produce this value). (4) **Draw conclusions about your hypothesis without changing the confidence level**. These steps will be explained in this chapter.

Published results should include conclusions and the test of your hypothesis and abbreviated information on (a) design of the study, (b) type of statistical test used, (c) number of samples obtained (or degrees of freedom), (d) frequency number of class intervals, (e) confidence level chosen, and (f) error bars. Three kinds of error bars are common in ecological literature, that is, standard deviations, standard errors, and confidence intervals (Ellison, 2001).

Language and Equations

Statistics is a new language to many persons. In fact, this is one of the most difficult aspects to overcome when you are first becoming acquainted with the discipline. Applied statistics is not difficult to understand; **it is the language barrier or jargon that creates many of the problems**. Unfortunately, few statisticians explain the terminology clearly and concisely. The theory behind statistics may be complex and require a considerable mathematical background.

Equations used to perform statistical problems often differ somewhat from text to text. More confusing is the fact that there may be considerable differences between authors in the symbolism used for statistical terms. For example, σ, **S or SD for standard deviation**. This contributes to a semantic problem in statistics. Statistical tables also differ considerably in how the numbers are arranged and how much explanation is given.

Statistics can be conveniently classified into three major categories: (1) **descriptive statistics**, which characterize one or more sets of data (e.g., mean, standard deviation, skewness, etc.); (2) **statistics** in which two or more sets of data and one or two variables are compared (e.g., using 1-way or 2-way ANOVA); and 3) **multivariate** in which multiple sets of data and multiple variables are analyzed (e.g., PCA or Principal Component Analysis).

Statistical Distributions

There are **four major types of statistical (probability) distributions found in many biological situations**.

(1) Normal or Gaussian

Symmetrical or bell-shaped distribution (for example, length, width) with the following characteristics (Fig. 2-3):

68.3% of the variates (observations or data points) fall between ± 1 σ (standard deviation) from the mean [$\overline{X}$] or average

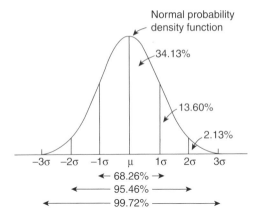

Figure 2-3. Normal distribution (bell-shaped curve).

μ = mean

σ = standard deviation

Source: modified from Sokal and Rohlf (1995)

95.5% of the variates fall between $\pm 2\ \sigma$ from the mean

99.7% of the variates fall between $\pm 3\ \sigma$ from the mean

95% of the variates fall between $\pm 1.96\ \sigma$ from the mean.

This last value is most important to remember since it will appear many times in discussions about statistics. A normal distribution has a skewness of 0 and a kurtosis (peakedness) of 3. Note that approximate normal distributions can be peaked or flattened (see Fig. 2-4, no. 2). Normal distributions are defined by the z statistic. The Central Limit Theorem states that convergence to an approximate normal distribution (i.e., bell-shaped curve) will usually occur with increasing sample size of independent, random variables (Smith, 1988). A special version of the Central Limit Theorem applies to random sampling without replacement from a finite population (Thompson, 2002). Testing data for normality is available in several statistical programs (e.g., see Gould and Gould, 2002). One of the best tests for a normal distribution is the Kolmogorov-Smirnoff goodness of fit test (Legendre and Legendre, 1998). Another test is the Anderson-Darling test (Dytham, 1999).

Count (meristic) data are likely to conform to a binomial, Poisson, or negative binomial distribution (Fowler et al., 1998).

(2)   Binomial

The probability distribution of events that can occur in two classes (e.g., tossing a coin [head/tail], sex [male/female], homozygous recessive or not homozygous recessive). A sign test (with $\pm$ values) or a Wilcoxon paired-sample test (a more powerful test – Zar, 1999) can be used as a binomial test. A binomial distribution is a Bernoulli distribution (Hayek and Buzas, 1997). *In the binomial distribution, the variance is less than the mean* (Southwood and Henderson, 2000).

1. *Binomial* :   Two states (male-female, heads-tails). Involves the binomial expansion $(p + q)^k$. Use binomial coefficients from Pascal's triangle to give expected frequencies.

2. *Normal* :

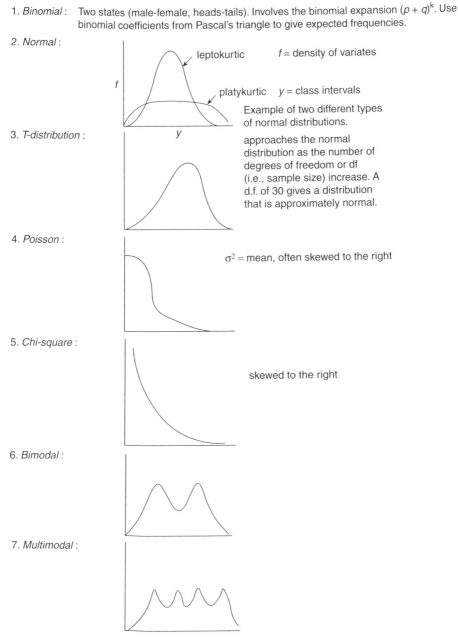

leptokurtic f = density of variates

platykurtic y = class intervals

Example of two different types of normal distributions.

3. *T-distribution* :   approaches the normal distribution as the number of degrees of freedom or df (i.e., sample size) increase. A d.f. of 30 gives a distribution that is approximately normal.

4. *Poisson* : σ^2 = mean, often skewed to the right

5. *Chi-square* :   skewed to the right

6. *Bimodal* :

7. *Multimodal* :

Figure 2-4. Examples of statistical distributions.

(3) Poisson

Poisson is the probability distribution or the count of independent or random events (e.g., distribution of diatoms in a plankton counting chamber or the number of cells in a haemocytometer square). Another way of describing a Poisson distribution – the probability of finding an organism in any quadrat in a grid sampling system is the same for all quadrats. **The variance = the mean in a Poisson distribution.** The goodness of fit of the Poisson distribution to a set of observed data may be tested by χ^2 or the G-test (Zar, 1999). Krebs (1999) has a computer program for determining if the data fit a Poisson distribution. Taylor's power law can also be used with this distribution (Green, 1979).

(4) Negative Binomial Distribution

A distribution that is clumped, characteristic of most species (Southwood and Henderson, 2000). **The variance is greater than the mean** (Southwood and Henderson, 2000). However, the Negative Binomial Distribution does not adequately include many types of aggregated distributions. The statistics involved are somewhat complex. A test for the distribution is the U-statistic Goodness-of-Fit Test. See Krebs (1999) for further information. Krebs (2000a) has a computer program for this test.

Among other statistical distributions are the **hypergeometric** (Zar, 1999), **uniform** (Krebs, 1999), **logarithmic or lognormal** (Sokal and Rohlf, 1995), **t distribution and χ^2 distribution**. Most statistical tests performed by ecologists are based on the z, t, F, or χ^2 distributions (Steidl and Thomas, 2001).

Parametric Statistics

Those based on measures of a specified (often normal) distribution of data (e.g., mean, standard deviation; those with a bell-shaped or Gaussian curve) (Fig. 2-3)

Nonparametric Statistics

Those in which there are no assumptions as to the form of the data distributions. See Segal (1988), Tate and Clelland (1959), Schefler (1969), Daniel (1978), and Hollander and Wolfe (1998). Nonparametric statistics are commonly employed in behavioral and population studies since many of the distributions are not normal (Fig. 2-5 and Southwood and Henderson, 2000). However, some biologists (e.g., Stuart Hurlbert) believe that nonparametric statistics should not be employed in population studies, that log transforms of the data are usually appropriate for allowing parametric statistics to be employed.

General Attributes

Sample: a subset of the population of interest

Variate: each separate datum in a sample (Sokal and Rohlf, 1995).

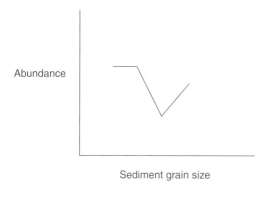

Figure 2-5. A plot of the abundance of organisms versus sediment grain size. This plot indicates a non-normal distribution.

Beware that some science journalists have used the term "variate" to refer to subsets of data assumed to be representative.

Normal or Gaussian curve: discussed above.

Frequency Distributions

(1) Frequency distributions can be **graphically** displayed using various figures (Figure 2-2)

(2) are characterized by a **range** (largest and smallest number)

(3) have **class intervals**.

These are selected for convenience and saving of time. Intervals must be small enough so that unlike data will not be lumped (e.g., length of amphipods should be expressed in class intervals of mm or 0.1 mm rather than class intervals of cm).

2.6 DESCRIPTIVE STATISTICS (SAMPLE STATISTICS)

Descriptive statistics commonly analyze or handle one set of data at a time. A program such as Ezstat (see Appendix) is recommended for biologists (because it is simple to use) although most statistical programs handle descriptive statistics.

Measurements of Central Tendency

The selection of one figure to represent the whole array of data.

(1) **Median:** middle member of a series (or average of two variates at the middle if the total number of variates or class intervals is even); used in nonparametric statistics.

Example: Given the numbers 5, 2, 7, 9, 8, the median is 7

(2) **Mode:** class interval most frequently represented. Used in geology in sediment grain size analysis.

Example: The mode is 8 months in the histogram shown in Fig. 2-2c.

(3) **Mean:** arithmetic average of the measurements

Example: The mean or average of $8+4+6=6$

$$\bar{X} = \frac{\sum X}{N}$$

where

$\bar{X}$ = mean

Σ = sum

X = variate

N = total number of variates

See Welsh et al. (1988) who discuss the fallacy of averages.

(4) **Geometric Mean:** The geometric mean is appropriate only when all the data are positive and the data are of the ratio scale type (Zar, 1999). See data standardizations above (p. 64) for further details.

(5) **Harmonic Mean:** The harmonic mean is the reciprocal of the arithmetic means of reciprocals of data (Zar, 1999). The geometric mean is less than the arithmetic mean and the harmonic mean less than the geometric mean if the data are positive but not identical. If the data are identical then the means are the same (Zar, 1999). The harmonic mean is used in density estimates by the Strong technique (see Chapter 1).

$$\text{Harmonic mean} = \frac{n}{\sum 1/x_i}$$

where

n = total number of data entries

Σ = sum

X_i = datum

Measurements of Dispersion

Tendency to deviate from the mean.

Standard Deviation:

σ = concentration of frequency around the mean or standard deviation

$$\sigma = \sqrt{\frac{\sum d^2}{N-1}}$$

where

    Σ = sum

    d = each deviation from the mean

    N = total number of variates

    -1 = correction factor for underestimation of population variation in samples
        of 20 or less

Example:

    $n = 6, 7, 4, 3$ $\bar{X} = 5$ $d = 1, 2, 1, 1$ $d^2 = 1, 4, 1, 4$

$$\sqrt{\frac{10}{3}} \qquad \sigma = 1.82$$

Variance

Variance = the standard deviation squared (σ^2). This is a very important term to know since it will be used repeatedly in more advanced statistical problems.

$$\text{variance} = \frac{\Sigma d^2}{N-1}$$

where

    σ = standard deviation

    Σ = sum

    d = each deviation from the mean

    N = total number of variates

    -1 = correction factor

Example

Data from Standard Deviation above

$$\sigma^2 = \frac{10}{3} = 3.32$$

or

$$\sigma^2 \text{ or SD}^2 = (1.82)^2 = 3.32$$

Standard Error

This measurement **indicates if the sample mean is a good estimate of the population mean**.

$$S\bar{X} = \bar{X} \pm \frac{\sigma}{\sqrt{N}}$$

where

$S\bar{X}$ = standard error range of standard deviation from the mean range

$\bar{X}$ = mean

σ = standard deviation

N = total number of variates

$\dfrac{\sigma}{\sqrt{N}}$ = standard error

Example

Data from Standard Deviation above

$$S\bar{X} = 5 \pm \frac{1.82}{\sqrt{4}} \quad S\bar{X} = 5 \pm 0.91$$

There are **effects of scale** on the standard error. The standard error (S.E.) varies with sample size (Sokal and Rohlf, 1995). The SE of 3,000 units is 0.1 of the SE of 30 units. Increase in sample size by 10^2 is equal to decrease in standard error by 10^1.

Confidence Limits

Confidence limits are used for estimating confidence in the position of a mean in a normal population. This measurement of dispersion is commonly used in ecology.

$$CL = \bar{X} \pm \frac{d\sigma}{\sqrt{N}}$$

where

CL = confidence limits

$\bar{X}$ = mean

d = 1.96 at P of 0.05

σ = standard deviation

N = total number of variates

Example

Data from Standard Deviation above.

$$d = 1.96 \text{ at } P \text{ of } 0.05$$

$$CL = 5 \pm \frac{1.96(1.82)}{\sqrt{4}} \quad CL = 5 \pm 1.78$$

Coefficient of Variation (CV)

The Coefficient of Variation is used to express quantitatively the variation in a variable (for example, crab carapace length). CV values of 5–6% indicate average variability in animals for morphology. Coefficients of variation for ecological data (pH, population density, etc.) often range from 20 to 60% (Joe DeVita, pers. comm.) but can be as high as 200% (see Eberhardt, 1978).

$$CV = \frac{100\sigma}{\bar{X}}$$

where

CV = coefficient of variation (in %)

σ = standard deviation

$\bar{X}$ = mean

Example

Data from Standard Deviation above

$$CV = \frac{100(1.82)}{5} \qquad CV = 36.4\%$$

Homoscedasticity

Homoscedasticity is a synonym for homogeneity of variances. Tests for homoscedasticity are used to determine if parametric or nonparametric statistics should be used on a data set (see Fig. 2-7 on p. 96). If variances are homogenous (homoscedastic), an approximate test or parametric statistics that require or assume homogeneous variances can be used. Alternatively, one can transform heteroscedastic data and create more homogenous variances (see p. 96).
    Equality of variances are tested for by (see Sokal and Rohlf, 1995, for details):

(a) F-test for two means or samples

(b) Bartlett's test for more than two groups (beware that this test is overly sensitive to non-normality).

(c) Levine's test

(d) Student-Newman-Keuls (SNK) test.

(e) **Hartley's F_{max} test**

Hartley's F_{max} test is a simple test for homogeneity of variances (you need but two data sets to use this test). Suppose the number of sites is six and 10 samples (replicates) are taken at each site. Select the sites with the largest and smallest variances then calculate the maximum/minimum variance ratio.

Example (modified from Sokal and Rohlf, 1995):

Maximum variance = 0.2331   Minimum variance = 0.0237

Maximum variance ratio = 0.2331/0.0237 = 9.84

Examine the F_{max} table (see Appendix) and locate six sites across the top (X-axis) and $n - 1$ replicates $(10 - 1 = 9)$ on the left side (Y-axis). The F_{max} value = 7.80 (using a probability cutoff of $P = 0.05$). Since the calculated F_{max} (9.84) is larger than the F_{max} table value (7.80), the variances are heterogeneous. If the number of replicate samples differs between the sites then select the smallest number for the determination of $n - 1$. Because the variances are heterogeneous in this example, the original data should be transformed (see p. 96) then retested with Hartley's F_{max}. If the data are still heterogeneous, a nonparametric statistical test can be used. Zar (1999) states that no homogeneity test is especially good when sampled populations are non-normal.

Null Hypothesis

This hypothesis is assumed for many statistical tests. It states that **there are no differences between the means** [e.g., in a statistical test comparing means, such as the *t*-test)].

Errors in Statistical Assumptions

Type I (Alpha or Class A) – Rejecting H_0 (null hypothesis – i.e., the means are equal) when it is true. This is assigned the probability of α or alpha. Therefore, accepting H_0 when it is true is $1 - \alpha$. The significance level of $\alpha = 0.05$ (P = 0.05) is generally used as a cutoff point in various statistical tests.

Type II (Beta or Class B) – Accepting H_0 when some alternative hypothesis (H_1) is actually true. Another definition is the probability of not detecting a difference when one exists. This is assigned the probability of β or beta. Therefore, rejecting H_0 when it is false is $1 - \beta$. **The probability of $1 - \beta$ is called the power of the test** (Glover and Mitchell, 2002). The power of a test is a measure of the likelihood of a test reaching the correct conclusion, that is, accepting the null hypothesis when it should be accepted (Fowler et al., 1998). This power value will be of importance in power analysis, used to determine the number of samples required (see p. 100). The power curves of non-parametric tests (i.e., probability of rejecting the null hypothesis plotted against the raw data – see Sokal and Rohlf, 1995) are less affected by a failure of assumptions than are parametric tests. When all assumptions are met, parametric tests have greater overall power (also described as being more robust). Hypothesis testing is discussed by Krebs (2000b).

Probability

Alpha (α) probabilities most frequently used in biometrics are $P = 0.05$ and $P = 0.01$, indicating significance levels or confidence limits of 95 and 99% (a convention established by statisticians), respectively. For example, suppose we compare antennal lengths from two populations of lobsters by using a *t*-test (discussed later) and find that the populations are not similar in antennal lengths at a confidence limit of 95% ($P = 0.05$). The *odds* that the two populations are similar in antennal lengths are at most five in 100. The probability of deviation from the mean of more than $\pm 1.96\,\sigma = 5\%$ for normal distributions.

In summary[1]

If $P = \,<0.05$ – there is **a statistically acceptable level of evidence** that the null hypothesis (e.g., the means are the same) is wrong.

If $P = \,<0.01$ – there is **strong statistical evidence** that the null hypothesis is wrong.

If $P = \,<0.001$ – there is **very strong statistical evidence** that the null hypothesis is wrong.

One- and Two-tailed Tests (One- and Two-sided Tests)

A two-tailed test is a deviation in both directions (e.g., in a bell-shaped curve) from expectation whereas a one-tailed test is a deviation in one direction from expectation (Fig. 2-3). One-tailed tests are justified only rarely according to Campbell (1989) and Fowler et al. (1998) and are never justified according to Stuart Hurlbert (pers. comm.). In a one-sided test, we have some prior knowledge to incorporate (e.g., we expect the mean of one population to be smaller than the mean of another) and so we need less sample evidence to convince us. **Tables of χ^2 (see Appendix) are frequently presented in one-tailed form**, because their probability distributions are highly skewed to one side (Figure 2-4, No. 5), whereas **the normal distribution (called z), the t distribution, and analysis of variance (F distribution) are usually presented in two-tailed form**. For one-tailed tests, use these tables at half the required probability (2.5% in place of 5%).

2.7 STATISTICS WITH ONE OR TWO VARIABLES

The following type of statistics compares two or more sets of data with one or two variables. The samples are assumed to be independent (i.e., there is no effect of one set of data on the other) and uncorrelated. Ezstat is a simple program for biologists that handles these statistics.

Tests of Reliability (Tests of Significance)

Comparison of attributes in samples

[1]see Fowler et al., 1998.

(I) Student's t-test

Test of difference between means of two samples assuming a *normal distribution* (the name "Student" is a pseudonym for a statistician named Gosset – see Smith, 1988).

   This test is useful for samples where the true variances of the two populations are assumed equal. There are other equations for the assumption that variances differ, and so forth. The Student *t*-test is useful for supporting or rejecting hypotheses of similar attributes only (e.g., short spines versus long spines). There are other types of *t*-tests based on different assumptions made concerning the nature of the data (see Bailey, 1961 for examples). For nonparametric or non-normal distributions you may use the Mann–Whitney *U*-test and other tests mentioned in nonparametric statistics tests – see below).

   Follow the procedure for the *t* test in alphabetical order:

   (**a**) Assume Ho: the means are equal.

   (**b**) Choose the confidence limit (95%).

   (**c**) Plug data into the following equation:

$$t = \frac{(\bar{X}_1 - \bar{X}_2)\sqrt{\dfrac{N_1 \times N_2}{N_1 + N_2}}}{\sqrt{\dfrac{(N_1 - 1)\sigma_1^2 + (N_2 - 1)\sigma_2^2}{N_1 + N_2 - 2}}}$$

where

$t = t$ value (unpaired – see below)

$\bar{X}_1 = $ mean of sample 1

$\bar{X}_2 = $ mean of sample 2

$N_1 = $ total number of variates in sample 1

$N_2 = $ total number of variates in sample 2

$\sigma_1 = $ standard deviation of sample 1

$\sigma_2 = $ standard deviation of sample 2

$-1 = $ correction factor for underestimation of population variation in samples of 20 or less

$-2 = $ a constraint on each sample by assuming the total number of observations as fixed.

   (**d**) Look in a *t*-table (see Appendix for statistical tables) under $N_1 + N_2 - 2$ degrees of freedom and find the value of *t*. Thus, the **degrees of freedom equal the number of samples taken minus a small correction factor**. Another definition of degrees of freedom is: the total number of components (e.g., samples) used in the calculation minus the number of estimated parameters (Legendre and Legendre, 1998). **If the calculated t is greater**

than t at P of 0.05 then the means can be assumed to be unequal with 95% confidence. If the means are unequal, then you may suspect demic differences, sexual differences, differences in age classes, and so forth. **The t-test will not tell you what is responsible for giving rise to such a variation**. There are comparable tests for large samples, for comparing a single specimen with a sample, and so forth.

There are paired and unpaired t-tests. If a group of people are weighed then reweighed a week later, *a* **paired t-test** is appropriate. If two groups of people are weighed and their weights compared then an **unpaired t-test** is needed. Ezstat handles both types of *t*-tests.

Many *t* distributions are generally small sample normal or near normal distributions. A sample size of 30 may give a normal distribution. When $N \geq 30$ the *t* value approximates the *z* value from a table of normal distribution (Hayek and Buzas, 1997). Therefore, if there are more than 30 observations in each sample (50 for badly skewed data), use the *z* test (Fowler, et al., 1998).

The *z*-statistic is:

$$z = \frac{\left(\bar{X}_1 - \bar{X}_2\right)}{\sqrt{\dfrac{S_1^2}{N_1} + \dfrac{S_2^2}{N_2}}}$$

where

$\bar{X}$ = sample mean

S = standard deviation

N = number of data entries

If the calculated *z* is greater than 1.96 (95% confidence or $P = 0.05$) then the means are unequal.

If the calculated *z* is greater than 2.58 (99% confidence or $P = 0.01$) then the means are unequal.

Hotelling's T^2-test is the application of the *t*-test to multivariate data (Gotelli and Ellison, 2004).

Example

(Modified from Fowler et al., 1998):

Two samples of intertidal turban snails (*Tegula funebralis*) are collected from different sites and their maximum shell lengths measured.

Site X	Site Y
$N_1 = 40$	$N_2 = 30$
$X_1 = 25$ mm	$X_2 = 20$ mm
$S_1 = 6$	$S_2 = 5$
$S_1^2 = 36$	$S_2^2 = 25$

$$z = \frac{(25-20)}{\sqrt{\frac{36}{40} + \frac{25}{30}}} = \frac{5}{\sqrt{0.9+0.8}} = \frac{5}{1.3}$$

$z = 3.9$

z at $P = 0.05$ (95% confidence) $= 1.96$

Conclusion: The means are unequal.

Analysis of Frequencies

Chi-square (χ^2) test

Test for association. χ^2 is a nonparametric test. **Note that the G test (see below) is now preferred over χ^2 based on theoretical reasons (i.e., in large scale tests) and the fact that it is more simple computationally** (Sokal & Rohlf, 1995). However, Zar (1999) states that while G may result in a more powerful test in some cases, χ^2 tends to provide a test that operates closer to the stated level of α or significance level. G tests are likelihood ratio or maximum likelihood statistical significance tests. Results from G tests are interpreted using χ^2 tables.
    Standard Chi-square equation:

$$\chi^2 = \sum \frac{(0-E)^2}{E}$$

where

$\chi^2 = $ Chi-square

$\Sigma = $ sum

$0 = $ the observed number of a particular association

$E = $ the expected number of a particular association

    This is also known as a "goodness of fit" test where the calculated expected results are compared with the observed results, frequently used in classical genetics. Chi-square should not be used where (1) the expected frequency (E) is < 5 or (2) with percentage values since these give no idea of the absolute numbers involved.
    Included here is an example of using Chi-square with two or more variables.

(a) Assume the Null hypothesis (that there is *no* definite association between a species and a tidal level or zone).

(b) Choose a confidence limit (95% or 99%).

(c) Construct a contingency table by using presence (+) or absence (−) data for the species *versus* the location (assume that they were collected from a number of quadrats (i.e., samples) in an area (e.g., lower and upper intertidal). Plug data into the equation below because it is a short-cut calculation with a correction factor for number of samples < 400.

Contingency table (2 × 2 type):

	No. of samples of species X			
	+	−	Total	
lower intertidal	a (2)	b (4)	a + b (6)	+ = present
upper intertidal	c (5)	d (1)	c + d (6)	− = absent
Total	a + c (7)	b + d (5)	a + b + c + d (12)	

The four letters placed in the boxes are the observed associations. Calculate the total number of observations by adding the numerals in the rows and columns and deriving the total number of observations in the extreme lower right corner *(a + b + c + d)*. Calculate the expected number of observations for the extreme upper left corner *(a)*, by multiplying the total in that column by the total in that row and dividing the resultant figure by the total number of observations (extreme lower right corner) – see Campbell (1989). Repeat the procedure for the remaining three letters *(b, c, d)*. You are now ready for Chi-square calculation.

To speed things up we will use the following shortcut equation with a correction factor for number of samples <400.

$$\chi^2 = \frac{[(a+0.5)(d+0.5)-(b+0.5)(c+0.5)]^2\, N}{(a+b)(c+d)(a+c)(b+d)}$$

$$\chi^2 = \frac{[(2+0.5)(1+0.5)-(4+0.5)(5+0.5)]^2 \times 12}{(6)\quad(6)\quad(7)\quad(9)}$$

$$\chi^2 = 2.3$$

Determine the number of degrees of freedom (d.f.) by the following equation: $(r - 1)(c - 1)$,

where

r = number of rows in the contingency table

c = number of columns in the contingency table.

The −1 in this equation is a constraint by assuming the total number of observations as fixed.

(d) Consult a Chi-square table (see Appendix). If the calculated χ^2 (2.3) is less than the table χ^2 at P of 0.05 (3.84 for the original choice of 95% confidence and 1 d.f.) then you can assume that there is *no* association between the species and a specific part of the intertidal or littoral zone (i.e., no specific preference for one or the other type of substratum or zone). In order **to validate the results** we would be required to construct at least several contingency tables (e.g., one table for every 20 or 30 quadrats, each set of 20–30 quadrats taken from a different part of the beach) and compare the results. In place of numerous individual tables we might obtain better results by pooling the data into one table. See Legendre and Legendre (1998) and Gotelli and Ellison (2004) for a discussion of contingency tables.

Goodness of Fit Test: Observed Versus Expected Values

When data scattering is large, a goodness of fit test is needed to determine if this is acceptable.

Use the G-test (see equation below) recommended by Sokal and Rohlf (1995). Sample size must be at least 5, as in the χ^2 test (use Ezstat). Use the Kolmogorov-Smirnov test for continuous frequency distributions (i.e., metric numbers, such as 6.2) (good for small sample sizes).

The *G* Test

$$G = 2\sum O \ln\left(\frac{O}{E}\right)$$

where

Σ = sum

O = observed frequency (24)

ln = log normal or natural log

E = expected frequency (22)

Example

$$G = 2\left[24 \ln\left(\frac{24}{22}\right)\right]$$

$$G = 12.2$$

For more chi-square data $G = 2\left[\sum O \ln\left(\frac{O}{E}\right) + \cdots + \cdots, \text{etc.}\right]$

Assume the χ^2 table value (d.f. = 8) = 15.51

If the calculated *G* is < the value of χ^2 in a χ^2 table ($n-1$ samples and $P = 0.05$), as indicated above, then the observed and expected frequencies are similar (not statistically different), supporting the null hypothesis (i.e., there are no differences between the observed and expected results).

Circular Statistics

Circular distributions are a type of distribution seldom used in marine biology (e.g., pie-shaped figures that show the percentage of time the wind blows from each direction in a 360° circle – a wind rose; see Fig. 2-2d) but more commonly found in terrestrial studies. Zar (1999) discusses statistics dealing with circular distributions. The distribution of abundances of a species around a circle is an example of an ecological use. The null hypothesis assumes no differences around the circle, that is, the populations are distributed evenly around a circle. This can be tested by the non-parametric Rayleigh z test. The angles (in degrees) and their sine and cosine values are placed in an equation that gives a u value. The u value is compared with a table u value at $P = 0.05$. The null hypothesis is then accepted or rejected.

One of my graduate students (Domingo Ochavillo, 2002) used this technique to test if released Philippine Islands rabbitfish larvae (*Siganus fuscescens*) swam in random directions or not. He found that their swimming directions were very highly significantly different from random (Rayleigh's z test, $P < 0.001$). The fish swam toward the reef. Other types of circular tests are discussed by Zar (2001). Skalski (1987) discusses random sampling in circular plots. Two distributions of circular data can be compared with the Watson-Williams test (Gould and Gould, 2002). A detailed analysis of circular statistics is found in Fisher (1996). See also Batschelet (1981).

Rayleigh z Test

The null hypothesis states that fishes are uniformly distributed around in a circle. The following table shows the location of fish (modified from Zar, 1999):

Angle (degrees)	sin Angle	cosine Angle
45	0.70711	0.70711
55	0.81915	0.57358
81	0.98769	0.15643
96	0.99452	−0.10453
110	0.93969	−0.34202
117	0.89101	−0.45399
132	0.74315	−0.66913
134	0.43837	−0.89879

Σsin Angle $= 6.52069$ Σcos Angle $= -1.03134$

$n = 8$

$$Y = \frac{\Sigma \sin \text{angle}}{n} \qquad X = \frac{\Sigma \cos \text{angle}}{n}$$

$Y = 0.81509$ $X = -0.12892$

$$r = \sqrt{X^2 + Y^2} \qquad r = \sqrt{(-0.12892)^2 + (0.81509)^2}$$

$r = 0.82522$

$R = nr$ $R = (8)(0.82522)$ $R = 6.60176$

$$z = \frac{R^2}{n} \qquad z = \frac{(6.60176)^2}{8} \qquad z = 5.448$$

where

Σ = sum

sin = sine

cos = cosine

n = number of entries (angles)

$\sqrt{\ }$ = square root

z = z value

Using the Rayleigh z table:

table z (P = 0.05 and n = 8) = 2.899

Since 2.899 < 5.448 the null hypothesis is rejected, that is, the fishes are not uniformly distributed around the circle.

Analysis of Variance

Analysis of variance (ANOVA) assumes random sampling, a normal distribution, and similar variances (i.e., with two or more sets of data). Analysis of variance is the most popular category of parametric tests. Its purpose is to test for differences in means by between-group variability. It is superior to the t-test because each factor is tested while controlling for all others and interactions between variables are detected (Statsoft, Inc., 1995). Parametric ANOVA tests are preferred over nonparametric tests because statisticians consider them to be more powerful. **They are more sensitive to borderline probability cutoffs** (e.g., $P = 0.05$ or $P = 0.01$), which result in rejecting the null hypothesis. You need fewer observations to find a significant effect (Statsoft, Inc., 1995). According to Green (1979), the biologist who resorts to nonparametric versions of ANOVA because of slight non-normality of within-group error distributions is being unnecessarily picky. The price paid is loss of power in the test (see discussion of power analysis on p. 100) and loss of information in descriptive presentation of results. Two-tailed tests with F (ANOVA) and t-statistics will generally be valid, even on extremely non-normal populations (Green, 1979). Zar (1999) concludes that often the difference in power between parametric and nonparametric tests is not great, and can be compensated by a small increase in sample size for non-parametric tests.

ANOVA is a statistical test that makes simultaneous comparisons of the effects of a factor(s) on various sample groups (in comparison, the t-test deals with only two sample groups), often used in conjunction with experimental design and sampling design. Types of factors often can be classified in the following major categories: (1) time, (2) locality, (3) environment, (4) physiological factors, (5) taxonomic categories, and (6) techniques.

Analysis of variance can test simultaneously the differences among several populations or test plots, it can determine how much of the overall variation is due to variation in each factor influencing the character being studied, and it saves much time. **Nonparametric or non-normal distributions can be analyzed by this test if no more than moderate deviations from normality occur**, according to some statisticians.

The simplest type of analysis of variance is the single classification analysis of variance or **one-way (one-factor)** *ANOVA*. A more complicated analysis is the **two-way (two-factor)** *ANOVA* (see p. 91). A **three-way (three-factor)** *ANOVA* is given by Bernstein (1966). Often preferred is the **randomized block design** when using ANOVA (see below and Table 2-1) because it is more sensitive to the occurrence of a gradient. If you were interested in weights of male and female euphausiid shrimps at two depths during two seasons this would be called a $2 \times 2 \times 2$ factorial and a *three-way ANOVA* would be used (3 variables are involved – see Fig. 2-6 and p. 90). The number of treatment combinations in such a factorial design would be 8. A three factor ANOVA could be classified as a multivariate ANOVA.

Table 2-1.   A randomized block design corrects for the effects of a single gradient. For example, ABCD represent four different light intensities in an experiment growing plants.

Block	1	C	A	D	B
	2	B	D	A	C
	3	A	B	C	D
	4	B	D	C	A
	5	A	B	D	C
	6	A	D	B	C

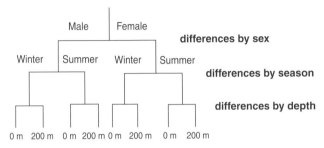

Figure 2-6.   Nested ANOVA. A three-level nested ANOVA (3-way ANOVA) that compares differences in a species of zooplankton by sex, season, and depth.

Experimental Design When Using ANOVA

Randomized Block Design

The objective of the randomized block design is to obtain minimum variation within blocks and maximum variation between blocks (Table 2-1). In other words, it is a method of reducing within-block heterogeneity and exaggerating between-block heterogeneity. **The randomized block design corrects for the effects of a single gradient** (e.g., changes in ambient light intensity in a laboratory where experiments on primary production of phytoplankton or tree saplings are being carried out). In the randomized block design, partitioning the variance into three terms (treatment, block, and error) rather than two (treatment and error) reduces the error degrees of freedom (see Potvin, 2001).

Latin Square Design

The **Latin Square design (Table 2-2) corrects for the effects of two gradients** (e.g., change in moisture and change in N content in soil). See Odeh (1991) and Zar (1999) for further information.

The one-way ANOVA has the following variations: (1) for unequal sample sizes, (2) for equal sample sizes (see below), and (3) other comparisons.

Other types of comparisons include:

(1) Comparison of one sample with a population: use a modified t-test.

(2) Multiple comparisons of pairs of means with equal sample sizes: use the T-method or Tukey method.

(3) Comparison of variation of two populations independent of their means: use a coefficient of variation.

(4) Nested 1-way ANOVA: This is a **hierarchic ANOVA**; one can have two-level nested, three-level nested, on up. **Nested ANOVAs** deal with subgroups (see

Table 2-2.   A Latin-square design corrects for the effects of two gradients. For example, T_1 through T_4 represent four chemical treatments in a field with known moisture and nitrogen (N) gradients (source: Sokal and Rohlf, 1995).

	Drier ⟶ Wetter			
Low N content	T_1	T_4	T_3	T_2
	T_3	T_2	T_1	T_4
	T_2	T_3	T_4	T_1
High N content	T_4	T_1	T_2	T_3

Fig. 2-6) and only more complex statistical programs (e.g., Statistica, SPSS) can analyze such data. Nested ANOVAs are commonly used with laboratory experiments.

(5) Two-way ANOVA significance test: used to test whether or not the results of the two-way ANOVA test are significant at a given probability level (see Sokal and Rohlf, 1995). See Johnson (1999) for comments on significance testing.

(6) Two-way ANOVA without replication: for example, used to test single temperature data at different depths on different days (see Sokal and Rohlf, 1995).

(7) Factorial two-way ANOVA: used to detect the effects of two different factors and their interactions on the measured response. The interactions can be positive (synergistic) or negative (interference) (see Glover and Mitchell, 2002).

Assumptions Using ANOVA

There are two types of ANOVA based on different assumptions, as follows:

Model 1 or fixed-effects Model: treatments are fixed and determined by the experimenter (for example, if you are comparing weight versus age, the treatments dealing with ages are fixed if you specify the ages). This model is commonly presented in statistical texts.

Model 2 or random-effects Model: treatments are not fixed (the ages are chosen at random).

Within-group error may be due to natural variations (genetics, environmental factors, etc.). This is distinguished from between-group differences, which you are attempting to detect.

Analysis of Variance One-way ANOVA (Model 1)

An indication of how an ANOVA can be used in biology is as follows:

Problem: Is there a significant difference in the mean size of species *X* from two plankton tows?

Assumption: There is *no* difference (null hypothesis) at the 95% level of significance.

Table 2-3 gives sizes of species X from sub-samples taken from two plankton tows.

A *t*-test can also be used to compare these two data sets.

Analysis of Variance – Two-way ANOVA (Model 1)

Problem: Are there significant differences in the distribution by biomass of species X in the littoral zone? That is, down the beach and across the beach.

Assumption: There are no differences (Null hypothesis) at the 95% level of significance.

Table 2-3.   Plankton tow. Measurements of the bodylength (mm) of a crustacean from the mesopelagic zone (source: modified from Sokal and Rohlf, 1981).

	I	*II*
	7.2	8.8
	7.1	7.5
	9.1	7.7
	7.2	7.6
	7.3	7.4
	7.2	6.7
	7.5	7.2
Σ	52.6	52.9

Σ = sum
Y_1 = plankton tow 1
$\bar{Y}_1$ = mean length (mm) of species X from plankton tow 1
σ = standard deviation
n = number of subsamples counted from each tow
SS = sums of squares
d.f. = degrees of freedom
F = variance ratio
MS = mean squares

$$\sum Y_1 = 52.6 \qquad \sum Y_2 = 52.9$$
$$\bar{Y}_1 = 7.5 \qquad \bar{Y}_2 = 7.6$$
$$\sum Y_1^2 = 398.3 \qquad \sum Y_2^2 = 402.2$$
$$\sigma^2 = 0.51 \qquad \sigma^2 = 0.41$$
$$\sum Y_1 + \sum Y_2 = 52.6 + 52.9 = 105.5$$
$$Y_1^2 + Y_2^2 = 398.3 + 402.2 = 800.5$$
$$\frac{(Y_1)^2 + (Y_2)^2}{n} = 795.0$$

$$\frac{(Y_1 + Y_2)^2}{n+n} = \frac{(105.5)^2}{14} = 795.0$$

$$SS_{total} = (Y_1^2 + Y_2^2) - \frac{(Y_1)^2 + (Y_2)^2}{n}$$
$$= 800.5 - 795.0$$
$$= 5.5 \text{(within groups)}$$

$$SS_{groups} = \frac{\left(\sum Y_1 - \sum Y_2\right)^2}{n+n} = \frac{(52.6 - 52.9)^2}{14} = 0.006 \text{ (between groups)}$$

$$SS_{within} = SS_{total} - SS_{groups}$$
$$= 5.5 - 0.006$$
$$= 5.5 \text{ (rounded-off)}$$

Table 2-3. (*Continued*)

d.f. (between groups) $=1$
d.f. (within groups) $= n + n - 2 = 12$
MS (between groups) $= 0.006/1 = 0.006$
MS (within groups) $= 5.5/12 = 0.458$
$F_s = $ MS between groups/MS within groups
$F_s = 0.006/0.458$
$F_s = 0.013$
F_{05} (examine the F table for 1 and 12 degrees of freedom at $P = 0.05$)
$F_{05} = 4.75$
$F_s < F_{05}$ therefore the two populations of species X do not differ in mean size at a 95% confidence level.

Table 2-4 gives values of biomass (standing crop) for species X in different parts of the littoral zone as well as in the sublittoral zone. Most statistical programs can perform ANOVAs. **Note that the terminology used in experiments include treatments and blocks.** For this field biology example we use zones (substitute for treatments) and quadrats or plots (substitute for blocks)

Table 2-4. Biomass or standing crop of organisms on a beach (source: modified from Bailey, 1961).

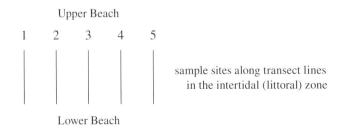

Upper Beach

1   2   3   4   5

sample sites along transect lines
in the intertidal (littoral) zone

Lower Beach

Treatments (zones)	Blocks (plots along transect lines)					Zone totals	Zone means
	1	2	3	4	5		
High littoral	33.4	34.1	35.4	36.7	37.7	177.3	35.46 ± 0.32
Mid littoral	38.2	38.0	40.6	40.3	43.7	200.8	40.16 ± 0.32
Low littoral	31.8	30.5	33.4	32.6	36.2	164.5	32.90 ± 0.32
Sublittoral	33.2	35.2	35.7	37.1	38.9	180.1	36.02 ± 0.32
Plot totals	136.6	137.8	145.1	146.7	156.5	722.7	36.13

(*Continued*)

Table 2-4. (*Continued*)

Procedure:

(1) Total number of observations: $bt = 20$

where:

b (or i) = blocks (quadrats)

t (or y) = treatments (zones)

(2) Grand total: $C = \Sigma X_{ij} = 722.7$

where:

X_{ij} = observation in ith block and jth treatment

(3) Correction factor: $C = \dfrac{G^2}{bt} \dfrac{(722.7)^2}{20} = 26{,}115$

(4) Total sum of squares about the mean:

$$X_{ij}^2 - C = (33.4)^2 + (34.1)^2 + \cdot + (38.9)^2 - 26{,}115 = 206$$

(5) Sum of squares for treatments:

$$\frac{Ti^2}{b} - C = \frac{1}{5}\left[(177.3)^2 + \ldots + (180.1)^2\right] - 26{,}115 = 136$$

(6) Sum of squares for blocks:

$$\frac{Bj^2}{t} - C = \frac{1}{4}\left[(136.6)^2 + \ldots + (156.5)^2\right] - 26{,}115 = 64$$

(7) Estimated $\sigma^2 = \dfrac{\text{residual sum of squares}}{\text{residual degrees of freedom}} = 0.5(6/12)$

(8) M = mean square

e.g., M of treatments = 45 (136/3) etc.

(9) F = variance-ratio**

$$F = \frac{M}{\text{estimated } \sigma^2}$$

e.g., $F = \dfrac{16}{0.5}$ $F = 90$ (for treatments) - down the beach

$F = \dfrac{16}{0.5}$ $F = 32$ (for blocks) - across the beach

Summary

Source of variation	Sum of squares	Degrees of Freedom (d.f.)	Mean squares	Variance ratio
Treatments (zones)	136	3 ($t-1$)	45 (136/3)	90
Blocks (plots)	64	4 ($b-1$)	16 (64/4)	32
Residual (No. 4, No. 5 + No 6)*	6	12 ($t-1$)($b-1$)	0.5 (6/12)	
Total	206	19	–	

*see p. 95

**F is named after Fisher. The F distribution is a theoretical probability distribution like t or χ^2.

For the Two-way ANOVA on the beach data, consult variance-ratio (F) distribution tables (see Appendix). The 5% point (F) corresponding to 3 (f_1 or treatments) and 12 (f_2 or residual) degrees of freedom = 3.49. Since $F = 90$ (for zones) is much greater than 3.49, the differences between zones (treatments) are highly significant at the confidence limit originally chosen. The 5% point ($F*$) corresponding to 4 (f_1 or quadrats, plots) and 12 (f_2 or residual) degrees of freedom = 3.26. Since $F = 32$ (for quadrats) is much greater than 3.26, the differences between quadrats (blocks) are highly significant at the confidence level originally chosen.

Three-way ANOVAs are complex and laborious without a computer. One virtually never sees a four-way ANOVA displayed in a text. One can use MANOVA or other multivariate analyses for dealing with a complex set of variables.

*F is named after the statistician Fisher. The F distribution is a theoretical probability distribution like t or χ^2 (chi-square).

Missing observations (i.e., zeros) such as treatments in a quantitative analysis with no response **should be omitted** (Mead, 1988).

Residual error originates from variations in experimental materials (e.g., different foods), errors in technique (e.g., different technicians), and so forth. Residual errors are used to detect data outliers (e.g., Dixon test). Outliers can be meaningful or be a mistake. The detection and interpretation of outliers is highly subjective (Mead, 1988). The program PC-ORD has an outlier detection algorithm (see the Appendix).

Note that the values in Table 2-4 do not tell anything about the cause of the significant differences, and multivariate analysis (or field experiments) would then be the next step in determining which variables in the intertidal or littoral zone are most important in possibly controlling the biomass of species X (see Chapter 5). Results of multivariate analysis of measurements of these major variables carried out over a year or two would suggest but not prove the relative importance (in quantitative terms) of each factor in controlling the biomass of species X. Experiments would have to be conducted to determine proof.

Other items of importance include:

(**1**) 2-way factorial ANOVA – an ANOVA designed to detect interference interactions between two sets of data.

(**2**) 2-way ANOVA significance test – an example of this type of test is a measurement of oxygen consumption by two species of limpets at three concentrations of seawater and eight replicates for each combination of species and salinity (Sokal and Rohlf, 1995).

(**3**) 2-way ANOVA without replication – an example of this type of test is a measurement of the temperature of lake water on four afternoons during summer at 10 depths (Sokal and Rohlf, 1995).

For these see Sokal and Rohlf (1995), Sund (1981), and Collyer and Enns (1986), Also see Goldberg and Scheiner, 2001). Graham and Edwards (2001) present methods to enhance analyses in a variety of ANOVA models used in ecological studies.

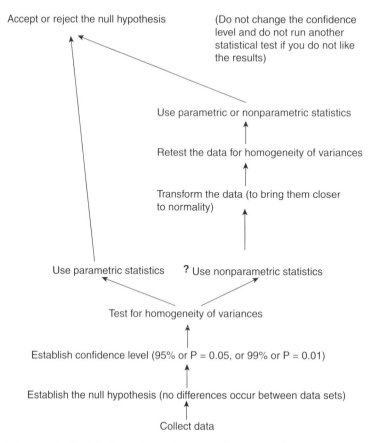

Figure 2-7. Statistical procedure used in comparing two sets of data.

The procedure of comparing two sets of data, whether parametric or non-parametric, is summarized in Fig. 2-7.

2.8 EXPERIMENTAL DESIGN AND ANALYSIS

There are three major difficulties associated with conducting field experiments: (1) field experiments are expensive, (2) true replication may be unattainable, and (3) there is often considerable noise (i.e., natural variation) in the data (ver Hoef and Cressie, 2001). Mead (1988) presents a detailed discussion on experimental design and analysis. This is especially valuable for persons planning to conduct laboratory experiments or working with experimental agricultural plots in the field. The most important process in experimental design is deciding on the purpose of the experiment and asking the proper questions. Each form of a question implies a different form of experimental treatments. The experiment leads to a design appropriate to a particular set of circumstances. The fact that about 85% of experiments employ the standard randomized complete block design suggests that most investigators are following the "cookbook" route in experimentation (Mead, 1988).

One needs to assess the total amount of resources available for the experiment then conduct the experiment. The important decisions about treatments concern the number of factors and the number of levels. Experiments will be predominantly concerned with two to four factors, at least initially. Replication is needed and a good estimate of error requires 10 to 15 d.f. (Mead, 1988), as follows:

Blocks: $b - 1$ d.f.

Treatments: $t - 1$ d.f.

Error: $(b - 1)(t - 1)$ d.f.

Gotelli and Ellison (2004) recommend a minimum of 10 replicates for each category or treatment level. If it is not possible to conduct preliminary sampling (e.g., you have only one low tide to conduct a study) then three replicates per treatment combination is a minimal acceptable number (Green, 1979). Gotelli and Ellison (2004) present an informative discussion of experimental design and analysis. See also Scheiner and Gurevitch (2001), Grafen and Hails (2002), Lindsey (2003), and Ruxton and Colgrave (2003).

Exploratory Data Analysis

Exploratory Data Analysis (EDA) was established largely by the efforts of Tukey (1977). A common use of EDA is to determine whether raw data satisfy the assumptions of the statistical tests suggested by the experimental design. EDA illuminates underlying patterns in noisy data (see Ellison, 2001).

Field Experiments

A preliminary or pilot study needs to be undertaken before commencing field experiments. This study should include natural history observations, and data collection on the size of the organisms, density, sediment transport, predators and other factors. Experiments can be of at least three types: (1) caging experiments, (2) design artifacts, and (3) experimental mimics (Woodin, 1986).

Caging experiments should include an experimental cage (enclosed) and control cages (topless cage, sideless cage, cage with holes on the sides, etc.) to test for the effects of caging, the behavior of predators, and other factors (Figure 2-8). The basic control is a cageless or natural area. **Problems (cage artifacts) that can occur because of caging** include the accretion or erosion of sediment inside or outside the cage, changes in sediment grain size, larval entrapment, and predator enhancement. When conducting caging experiments it is important to make physical measurements (e.g., sediment grain sizes), monitor predators, and monitor animals that disturb sediments by their burrowing activities (e.g., holothurians). Caging experiments are not

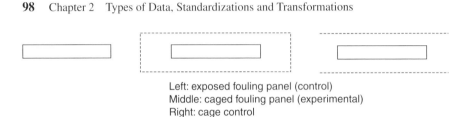

Left: exposed fouling panel (control)
Middle: caged fouling panel (experimental)
Right: cage control

Figure 2-8. Caging experiments in the field (source: Bakus, 1988).

recommended for certain types of studies with fishes because it alters their behavior (Doherty and Sale, 1985; Steele, 1996). Caging also provides shade, protection from wave action or subtidal surge, a surface for attachment, a perch for animals, prevents herbivores, competitors and predators from entering, and also prevents monitored and other species from emigrating. Statistics used to analyze the results by comparing experimentals with controls include the following: one cage-one time, use nested replicates (Green, 1979); one cage over time, use repeated measure design or sequential sampling (Winer, 1971; Krebs 1999 and see p. 98 and Chapter 1, p. 27); different cages-one time and different cages over time, use standard ANOVA statistics (see p. 88 above). See also Hurlbert (1984).

One can simulate what organisms are doing (**design artifacts**) and learn more about how they are changing the sediment and/or the biota (Woodin, 1986). For example, holes can be dug with a spoon or a trowel to simulate the sediment effect of clams digging into mud. One might want to separate the sediment effect from other effects during the experiments. For example, digging holes but not removing the clams versus controls compared with digging holes and removing the clams versus controls.

In order to determine if polychaete tubes are producing only a physical effect by interfering with current movement over the sediment surface or also a biological effect, both natural and artificial tubes (**experimental mimics**) can be added to sediments that lack polychaete tubes. This allows a direct cause and effect to be ascertained. It is a better approach than studying an area with multivariate analysis, the results of which produces a correlation but no proof of cause and effect. Of course, these direct experimental methods are applicable only to areas that are accessible (i.e., intertidal or scuba-diving depths). Such experimentation may be difficult or impossible in deep waters without the aid of a submersible. For further information on experimental ecology see Resetarits (2001) and Scheiner (2001). For a broader summary of methods in ecosystem science see Sala et al. (2000).

For information on experimental design and analysis, see above and Eberhardt and Thomas (1991) and Sala et al. (2000).

Repeated Measures Analysis

Repeated Measures Analysis is used to compare repeated measurements on the same individual animal, experimental unit, or sampling site. For example, comparing changes in the density of organisms in fixed quadrats during tropical wet and dry seasons over five years. The data can be analyzed by parametric methods such as ANOVA or MANOVA (see von End, 2001).

Pseudoreplication

Pseudoreplication is an important problem, a serious category of statistical error (Fig. 2-9). Pseudoreplication can be defined as the statistical treatment of experimental units as independent when they are not (Scheiner, 2001). Hurlbert (1984) and Hurlbert and White (1993) discuss several types of pseudoreplication in experiments, including simple pseudoreplication, temporal pseudoreplication, and sacrificial pseudoreplication. The usual consequence of pseudoreplication is underestimation of the P value, often by several orders of magnitude (Lombardi and Hurlbert, 1996). However, Oksanen (2001) believes that the concept of pseudoreplication is an unwarranted stigmatization of a reasonable way to test predictions referring to large-scale systems. Kozlov and Hurlbert (2006) clarify the difference between experimental units and evaluation units in experimental studies.

Major references that cover the design of experiments include Cox (1958), Fisher (1971), Winer (1971, 1988), Zolman (1993), Underwood (1997), Mead (1988), Morris (1999), Kuehe (2000), Scheiner and Gurevitch (2001), and Quinn and Keough (2002). See Hurlbert (1990b, 1994) for reviews. Mead (1988) has been called the best book on experimental design that has ever been written (Hurlbert, 1990b).

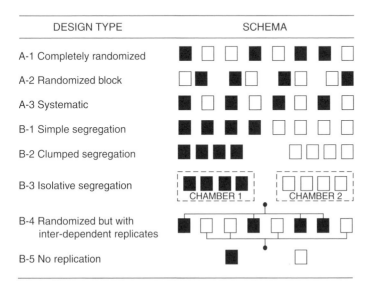

Figure 2-9. Schematic representation of various acceptable modes (A) of interspersing the replicates (boxes) of two treatments (shaded, unshaded) and various ways (B) in which the principle of interspersion can be violated (source: Hurlbert, 1984).

A newer method of determining how many samples are required for laboratory experiments as well as for experimental field studies is power analysis (see Section 2.9). The advantage of power analysis over more conventional schemes of sample requirements is that the procedure produces a few different sampling intensities per calculation rather than just one, which is typical of many other sampling procedures (see Hintze, 1996). **One can simply specify the power with which they wish to detect a particular difference among the population means and then ask how large a sample from each population must be taken** (Zar, 1999).

2.9 POWER ANALYSIS

Power analysis (statistical power analysis) is useful for planning a study in determining how many samples are needed. Power analysis is used to obtain the number of samples that will be detectable within the given level of statistical power. Power is defined as the probability of exceeding F in the central F-distribution (i.e., power $= 1-\beta$; a small beta is very powerful) (Underwood, 1997) or the probability of a Type II error (see p. 80). Beta (β) = Type II error or the failing to reject the null hypothesis when it is false (see Fig. 2-10).

The power of an experiment is based on the following characteristics: (1) type of test, (2) power increases with increasing sample size, (3) power increases with increasing effect size or the difference between two means, that of the experimental versus the control, (4) power increases with a higher α (alpha) level, and (5) power increases with decreasing sample variance. Power analysis consists of fixing three of four variables (i.e., sample size, alpha, beta) and calculating the fourth (i.e., effect size). An example of how power analysis is used to determine the number of samples needed is presented in Fig 2-10. The recommended program available for power analysis is PASS (Hintze, 1996 – see the Appendix). **Although commonly used for laboratory experiments, power analysis is now widely used for field studies**. Sheppard (1999) offers quick guides to sample size and the power of tests. Mumby (2002) discusses the statistical power of non-parametric tests using ordinal (ranked) data. Underwood and Chapman (2003) discuss power and sampling design in assessment of environmental impacts. For further information see Green (1989), Hayek and Buzas (1997), Steidl and Thomas (2001), Bausell and Li (2002), and the Appendix.

2.10 MULTIPLE COMPARISONS TESTS

Multiple comparisons tests are used to compare three or more sets of data simultaneously. The number of data sets is limited by the available computer memory. For example, Zar (1999) compares strontium concentrations in 5 bodies of water (i.e., pond, 3 lakes, and a river). The test compares all the possible combinations (i.e., 1 with 2, 2 with 3, etc.), presenting the results as supporting or rejecting the null hypothesis (i.e., no differences in strontium concentration in the 5 bodies of water).

Find .... n
Sigma m - S.D. of Means 0.10, 0.25, 0.40
Sigma - S.D. of Data 1
Number of Groups (k) 4
Group Sample Size - n .... *Ignored*
Type-I Error Level - Alpha 0.05
Type-II Error Level - Beta 0.20
Show numeric Report .... *Checked*
Show plots .... *Checked*

Plot Setup Tab
Horizontal Axis .... n
Legend (Group) .... Beta

Annotated Output

Click the Run button to perform the calculations and generate the following output.

Numeric Results for One-Way Analysis of Variance

Power	n	k	Alpha	Beta	Sigma m	Sigma	Effect Size
0.80073	274.00	4	0.05000	0.19927	0.10	1.00	0.10000
0.80399	45.00	4	0.05000	0.19601	0.25	1.00	0.25000
0.82340	19.00	4	0.05000	0.17660	0.40	1.00	0.40000

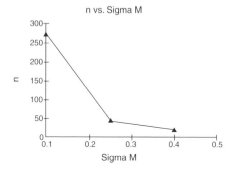

Note the wide range of sample sizes that are necessary to achieve the required alpha and beta values. If our researcher wants to detect only 'large' differences, he can get by with only 19 per group. However, if he wants to detect small differeces, he will need 274 per group.

Figure 2-10.   Power analysis. Determining sample size using PASS (source: Hintze, 1996).

The most popular multiple comparison tests are the Newman–Keuls (SNK or Student–Newman–Keuls) test and the Tukey test, which are very similar statistical tests. **Different variations are used depending on whether equal or unequal sample sizes are being compared**. Multiple comparisons tests assume random sampling, a normal distribution, and homogeneity of variances among the data sets. If the data are strongly heterogeneous or heteroscedastic, use the Kruskall–Wallis or the modified Tukey nonparametric multiple comparisons tests (Zar, 1999). Larger statistical programs contain at least some of these tests. Mead (1988) concludes that multiple comparison tests should be avoided in most cases on the basis of his analysis of the Duncan test. Stuart Hurlbert (pers. comm.) believes that multiple comparison tests should never be used.

2.11 NONPARAMETRIC TESTS, COVARIANCE, CORRELATION, AND REGRESSION

Nonparametric Tests

The following statistical tests do not assume a normal distribution. **Note that emphasis is placed on medians and ranges rather than means and variances**. The distribution of the data around the median cannot be precisely described, as with parametric data (Gould and Gould, 2002). Many nonparametric tests are ranking tests (Legendre and Legendre, 1998).

(1) **Mann–Whitney U-test (or Wilcoxon two-sample test):** differences between medians of two samples. This test is used frequently in biology.

(2) **Kolmogorov–Smirnov two-sample test:** general test of differences between two samples. It is sensitive to *most* differences between two populations (Campbell, 1989). Sample sizes should range between 4 and 40 and be identical. See Table 2-5. Determine the largest difference value between the cumulative frequency distribution (i.e., 9). Examine a K-S two-sample test table (see Appendix) under n sample size. If your calculated largest difference is larger than the K-S table value at the 5% level, the medians are unequal. Stuart Hurlbert (pers. comm.) states that this test can give a value of $P < 0.001$ for two samples with identical medians. He discourages the use of this test and nonparametric tests in general.

(3) **Wilcoxon signed rank test:** difference between medians or means; observations taken in pairs, one from each of two samples. For randomized blocks. This is a binomial test.

(4) **Kruskal–Wallis analysis of variance of ranks or Friedman's test** (nonparametric ANOVA)*:* difference between medians of more than two samples. For randomized blocks (see Zar, 1999).

(5) **Sign test** (see Sokal and Rohlf, 1995). This is another binomial test (also see Zar, 1999).

Outliers will have little or no influence on analyses employing nonparametric two-sample tests (e.g., Mann–Whitney U test) or multi-sample (multi-comparisons) tests (Zar, 1999).

Table 2-5. An example of the data for a Kolmogorov-Smirnov two-sample test (source: modified from Campbell, 1967).

					Sample values of proportions of fishes infected						
Group A	43	62	81	69	73	55	64	18	89	67	59	61
Group B	24	31	19	47	35	29	18	13	43	17	65	28

	Rearranged values		Cumulative frequencies		
Interval	*A*	*B*	*A*	*B*	(Difference)
0–4			0	0	0
5–9			0	0	0
10–14		13	0	1	1
15–19	18	19, 18, 17	1	4	3
20–24		24	1	5	4
25–29		29, 28	1	7	6
30–34		31	1	8	7
35–39		35	1	9	8
40–44	43	43	2	10	8
45–49		47	2	11	9
50–54			2	11	9
55–59	55, 59		4	11	7
60–64	62, 64, 61		7	11	4
65–69	69, 67	65	9	12	3
70–74	73		10	12	2
75–79			10	12	2
80–84	81		11	12	1
85–89	89		12	12	0
90–94			12	12	0
95–100			12	12	0

Analysis of Covariance or ANCOVA

Analysis of covariance is a measure of the degree of correspondence between variables. **One definition is that it is the joint variation of two variables about their common mean** (Davis, 1986; Legendre and Legendre, 1998). A positive covariance indicates that variables X and Y tend to increase together, whereas a negative covariance shows that when one variable (X) increases, the other (Y) tends to decrease. ANCOVA includes continuous variables in the design or covariates (Statsoft, Inc., 1995). Gotelli and Ellison (2004) present an informative discussion on ANCOVA. They describe ANCOVA as an analysis of variance performed on residuals from the regression of the response variable on the covariate (see p. 104). Covariance analysis is commonly displayed in matrix format (i.e., variance-covariance matrix). Standardizing the covariance (see below) results in the coefficient of correlation.

$$\text{COV} = \frac{\text{SP}}{n-1}$$

SP = corrected sum of products

$$\text{SP} = \sum XY - \frac{\left(\sum X\right)\left(\sum Y\right)}{n}$$

COV = covariance

X and Y = the two variables

n = number of pairs of data compared

$$r = \frac{\text{COV}}{\sigma X \; \sigma Y}$$

r = coefficient of correlation (see below)

If $r = +1$: the two variables are increasing or decreasing together (monotonically) with perfect correlation.

If $r = -1$: one variable is changing inversely with relation to the other, with perfect correlation.

For example, when a high negative value for r ($r = -0.9$) is associated with the size of predator (X) and prey (Y) populations, this indicates that a strong increase in predator populations is correlated with a strong decrease in prey populations. It is not a proof. **Covariance is also used to test a dependent variable Y for homogeneity among group means** (Sokal and Rohlf, 1995). This is carried out through a series of linear regression procedures. The mathematics involved are rather complex and lengthy. The main significance test here is a test of homogeneity of all Y intercepts. **Covariance is used to eliminate variables that are highly correlated in a multivariate analysis. It may also be used to remove the effects of disturbing variables from an analysis of variance**. A discussion of the general theory of covariance analysis is presented by Mead (1988). Fryer and Nicholson (2002) discuss the analysis of covariance that changes over time. See also Goldberg and Scheiner (2001).

Correlation

Coefficient of correlation relates the intensity of relationship between two factors. It is not a measure of dependence between two factors. Correlation is the ratio of the covariance of two variables to the product of their standard deviations (see the preceding and following text).

This coefficient is useful for attributes that are likely to be related (e.g., tail length and body length), not those that are improbable (e.g., number of sponges and number of crabs on a beach). It is intended for normal population distributions (some correlation coefficients are nonparametric). It does not supply much specific information about the association, similar to some species diversity indices.

+1 = perfect positive correlation

0 = no correlation

−1 = perfect negative correlation

Null hypothesis: no linear correlation exists between the two factors

Remember that **a strong correlation occurs when r ≥ 0.7 or when r ≥ −0.7** (equal to or more negative than; Fowler et al., 1998).

The following equation gives the Pearson product-moment coefficient of correlation:

$$r = \frac{\sum (X - \bar{X})(Y - \bar{Y})}{\sqrt{\sum (X - \bar{X})^2 \sum (Y - \bar{Y})^2}}$$

where

r = coefficient of correlation

$\bar{X}$ = mean of X

$\bar{Y}$ = mean of Y

X = variate of X

Y = variate of Y

Σ = sum

Example

What is the correlation between the number of grazing fishes and the standing crop of algae on a coral reef flat?

No. Surgeonfish	g algae/m² on reef flat
23	118
92	25
16	74
106	10
5	236

$r = -0.82$ (using the correlation equation above)

This is a strong negative correlation, indicating that algae are decreasing (being eaten) as the fish population increases.

The value of r^2 is often presented along with r. **r^2 is the Coefficient of Determination, the percentage variability in one factor accounted for by the variability in the second factor**.

Coefficients of correlation are used extensively in numerical taxonomy and sometimes plankton studies (see Cassie, 1963). **Beware – you cannot get a good correlation with data within a narrow range. Beware also of spurious self-correlations** (a ratio correlated against its own denominator – e.g., weight-specific epiphyte assimilation rate divided by shrimp weight – see Kenny, 1982). **Correlations may be strongly effected by outliers** (see Chapter 5 and McCune et al., 2002).

After calculations, consult the table for correlation coefficients (see the Appendix). It will tell you if your *r* value is significant at the 95 or 99% confidence limit (procedure similar to that of the *t*-test and Chi-square test).

Further important tests include the following (see Sokal and Rohlf, 1995):

(a) Tests of significance and confidence limits for correlations. The test of significance indicates if the resulting *r* value differs from zero correlation.

(b) Test of homogeneity among correlation coefficients.

(c) If the distribution is not normal, rank the variates and use Kendall's non-parametric coefficient of rank correlation.

(d) Comparing two or more (Tukey-type) correlation coefficients. See Zar (1999).

Regression

Regression measures the nature of the relationship between two (or more) factors (e.g., how much a change in one factor is reflected in a change in another factor). It provides more information than correlation coefficients and is widely used in biology. **Regression is a measure of dependence between two factors or multiple sets of two factors**.

As with ANOVA, there are two different assumptions regarding the data base for regression:

Model 1: X is fixed (under control of the investigator) and *Y* is a random variable. This is the most commonly used model, an example of which is presented below.

Model 2: Both *X* and *Y* are random variables.

Log transformation is most commonly used to straighten the regression line if the data points form a curved pattern (e.g., $\log_{10}$ of dose concentration (*X*-axis) plotted against survivorship (*Y*-axis), and for other dose-response studies). If a log transformation of the raw data is unsuccessful in straightening a line then polynomial or curvilinear regression must be used. See Juliano (2001) for a discussion of nonlinear curve fitting.

Cade and Noon (2003) present an introduction to quantile regression for ecologists.

By convention the Y-axis is the dependent variable and the X-axis the independent variable. Programs to calculate and plot regression lines are found in Ecostat and in Krebs (1999 – by the least squares method – see the following), and

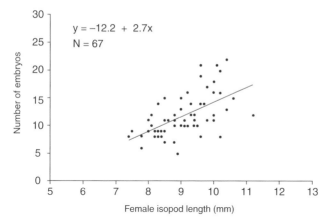

Figure 2-11. *Colidotea rostrata* (isopod crustacean). Relationship between numbers of embryos per brood and female body length (mm). Females carrying embryos at all stages of development are included. Regression line significant at $p < 0.001$ (source: Stebbins, 1988).

other statistical packages. **Least squares regression results in a regression equation with a minimum residual sum of squares. Stepwise regression proceeds in steps in which changes are made to generate smaller residuals.** Data scatter in a regression line can be tested for significance using a *t* test or ANOVA (Fowler et al., 1998).

Procedure: Suppose we wish to determine the quantitative relationship between number of embryos and female body length. Plot the number of embryos (*Y*-axis) against body length (*X*-axis) on a scatter diagram (Fig. 2-11).

If it is obvious that a potentially straight regression line (estimate by eye from the scatter pattern of the data) does not appear likely (it looks like curvilinear regression) then use the square root of *Y* or the $\log_{10}$ of *Y* (square root or log transformation). If the data still curve too much you must use equations dealing with curvilinear regression, usually some type of polynomial expression (least squares fit). See Davis (1986) for *FORTRAN* algorithms to handle this. Sokal and Rohlf (1995) recommend the use of a cubic polynomial or three power states in biology, as follows:

$$Y = a + bX + cX^2 + dX^3$$

where

    Y = dependent variable
        b, c, d include increasing powers of *X* along the curve

Many statistical programs now include the ability to perform curvilinear regression. It is recommended that a computer be used to handle the computations involved in regression analyses. There are, however, rapid formula techniques for carrying out these operations on a statistical calculator. The following represents a "long" calculation for obtaining regression:

Null hypothesis: $b = 0$ (no slope exists)

Constants:

$$b = \frac{\sum (X - \bar{X})(Y - \bar{Y})}{\sum (X - \bar{X})^2}$$ b is a measure of the slope of the regression line

$$a = \bar{Y} - b\bar{X}$$

where

$$\left. \begin{array}{l} \sum (X - \bar{X})^2 \\ \sum (Y - \bar{Y})^2 \end{array} \right\} = \text{sums of squared deviations from the mean}$$

$$\sum (X - \bar{X})(Y - \bar{Y}) = \text{sums of the product of deviations from the mean}$$

$$\bar{X} = \text{mean of } X$$

$$\bar{Y} = \text{mean of } Y$$

Regression line: $Y = a + bX$ (the coefficient b represents all the data)

Calculate the numerical values of b and a and transpose to obtain the value for the regression line. Express Y as equal to one numeral (value of a) plus another numeral (value of b) times X. Determine the numerical value of Y by **choosing a low numerical value for X** (use your lowest body length) and perform the calculation according to the formula $Y = a + bX$. Plot this point on your scatter diagram. Repeat the procedure but this time **choose a high numerical value for X** (use your highest body length). **Draw a line through the two extreme points**. This is your regression line. For accuracy, check that the regression line passes through the mean of X and Y. In the equation given for this example (see Fig. 2-11), the value $-12.2 = a$ or the Y-axis intercept, the value $+ 2.7 = b$ and includes all the data representing the slope, and $x = X$-axis intercept.

To obtain an estimate of the degree of scatter compute the σ of the regression line as follows:

$$\sigma^2 = \frac{1}{N-2} \left(\sum (Y - \bar{Y})^2 - \frac{\left[\sum (X - \bar{X})(Y - \bar{Y}) \right]^2}{\sum (X - \bar{X})^2} \right)$$

where

$\sigma = $ standard deviation

$N = $ total number of points plotted on the scatter diagram

$-2 = $ correction factor

You may also wish to compare the regression coefficients calculated from two small samples (see Sokal and Rohlf, 1995; Guest, 1961; Davis, 1986).

Some important tests are as follows (see Sokal and Rohlf, 1995):

(1) Significance test (*t*-test to test if your data are uncorrelated variables) and confidence limits (for Model 1 only). The significance test indicates if the calculated regression line slope differs from a zero slope.

(2) Comparison of two regression lines: *t*-test or one-way ANOVA.

(3) Nonparametric regression: Kendall's rank correlation of *Y* with *X*.

(4) Comparison of several multiple regressions (see Zar, 1999).

Logistic Regression

Logistic regression is a situation where the ecologist tries to model a categorical (dichotomous) response variable on a continuous predictor variable(s) or on a mixture of continuous and categorical predictor variables. For example, effects of nitrogen in giant kelp (*Macrocystis pyrifera*) fronds on the occurrence of a snail (*Norissia norrisi*) that consumes the plant (see Floyd, 2001 for details). See Addendum for binary regression.

2.12 MULTIVARIATE STATISTICS

Multivariate statistics compare sets of data with two or more variables.

Multiple (Partial) Correlation and Regression

Correlation and regression analyses are used when **two or more independent variables** are plotted against a **dependent variable** (for example, variables *A* and *B* against *C*, *X* variables against *C*). Multiple regression (a multivariate analysis) is used frequently in ecology. It determines simultaneously the influence of independent variables on a dependent variable (for example, predation, sediment size and temperature effects on density of a species; see Chapter 5).

Multiple regression: by least squares method. The least squares method plots the best fit line based on the smallest deviations, determined from the residual sums of squares. A significance test, and standard error can be calculated (see Chapter 5). See Addendum for hierarchical partitioning.

2.13 RANKING ANALYSIS (NONPARAMETRIC CORRELATION)

Rank correlation coefficients (nonparametric tests) are used in ethological studies (see Bernstein, 1966), community analyses, and so forth. They compare sets of ranks, assuming there are no differences in ranks (null hypothesis).

Two types of rank correlations (nonparametric) are in common use. These types are the **Kendall (Tau) Correlation and the Spearman (Rho) Correlations**. The procedure for the use of each will be done on a similar set of data drawn from Kendall (1955) (also see Kendall and Stuart, 1961).

Example

10 species (A-J); Data set 1; Data set 2; ($n = 10$)

	A	B	C	D	E	F	G	H	I	J
Rank of set 1:	7	4	3	10	6	2	9	8	1	5
Rank of set 2:	5	7	3	10	1	9	6	2	8	4

Assumption: Ho: the ranks are similar

Procedure for Kendall (Tau) Rank Correlation

(1) Rank 1 in set 2 has a corresponding 6 in set 1. "6" has four numbers in the natural ranking order to 10. Strike "6" from the order of set 1.

(2) Rank 2 in set 2 has a corresponding 8 in set 1. "8" has two numbers in the natural ranking order to 10. Add this two to the previous four (above). Strike "8" from the order of set 1.

(3) Rank 3 in set 2 has a corresponding 3 in set 1. "3" has five numbers (two have been struck) in the natural order to 10. Add this five to the previous four and two (above). Strike "3" from the order of set 1.

(4) Proceed in a similar way until you have the sum of all remaining numbers in the natural ranking order to 10.

Sum of remaining numbers in the natural ranking order to 10:

$$[4 + 2 + 5] + 3 + 2 + 1 + 1 + 2 + 1 + 0 = 21$$

"21" is your "*P*" value.

Plug this into the Kendall Correlation equation:

$$\tau = \frac{2P}{1/2(n(n-1))} - 1$$

$$\tau = \frac{42}{45} - 1 = -0.070$$

Compare this calculated tau (τ) with tau in a table of values at a probability level of $P = 0.05$ and X number of samples. If the calculated tau is less than the table value of tau then the ranks are similar.

Procedure for Spearman Rank Correlation

Same data set

	A	B	C	D	E	F	G	H	I	J
Rank of set 1:	7	4	3	10	6	2	9	8	1	5
Rank of set 2:	5	7	3	10	1	9	6	2	8	4

Assumption: Ho: the ranks are similar.

(**1**) Subtract the rank of each individual of set 2 from the rank of each individual of set 1.

$$2 - 3\ 0\ 0\ 5 - 7\ 3\ 6 - 7\ 1$$

(**2**) Square this difference:

$$4\ 9\ 0\ 0\ 25\ 49\ 9\ 36\ 49\ 1$$

(**3**) Sum the squared differences $(\Sigma d^2) = 182$

Plug this into the Spearman Correlation equation:

$$\rho = 1 - \frac{6\left(\sum d^2\right)}{n^3 - n}$$

$$\rho = 1 - \frac{6(182)}{990}$$

$$\rho = 1 - 1.103 \qquad\qquad \rho = -0.103$$

Compare the calculated rho (ρ) with the table rho (see the Appendix), following the same procedures as with the Kendall Rank Correlation.

Both Kendall and Spearman rank correlation coefficients range from +1.0 to −1.0, and as can be seen from these two examples, the two coefficients are similar when the two are computed with the same data. There is no fixed rule about using one over the other (Nie et al., 1975). The chief differences between Spearman's and Kendall's correlations seem to be that the Kendall coefficient is more meaningful when the data contain a large number of tied ranks (i.e., ranks that are the same – Legendre and Legendre, 1998). Spearman's coefficient, on the other hand, seems to give a closer approximation to the Pearson's product – moment correlation coefficient when the data are more or less continuous, that is, without a large number of tied ranks (Legendre and Legendre, 1998). The rank correlation coefficient becomes unreliable if more than half of the ranks are tied (Fowler et al., 1998).

As a general rule, it is probably best to use the Kendall correlation when a fairly large number of cases are classified into a relatively small number of categories and use the Spearman correlation when the ratio of cases to categories is small. Spearman may give slightly better results than Kendall when n is large (Zar, 1999). The

Kendall Tau can be generalized to a partial correlation coefficient (Legendre and Legendre, 1998). Each of the procedures has a correction for tied ranks, and these corrections are as follows:

Kendall corrected for tied ranks:

$$\tau = \frac{2P}{\sqrt{1/2\,N\,(N-1)-T_x}\,\sqrt{1/2\,N\,(N-1)-T_y}} - 1$$

where $T_x = 1/2\ t(t-1)$, t is the number of tied observations in each group of ties on the X variable, and T_y is the same quantity for the Y variable.

Spearman (r_s) corrected for tied ranks:

$$\left(r_s\right) = \frac{T_x + T_y - \sum d^2}{2\left(T_x T_y\right)^{1/2}}$$

were d is the difference between ranks of the two variables, T_x and T_y are to be defined by the quantity:

$$\frac{N\left(N^2 - 1\right) - \sum R\left(R^2 - 1\right)}{12}$$

where R is the number of ties at a given rank for X or Y, respectively.

Ezstat conducts the Kendall rank correlation under the category of nonparametric correlation (see the Appendix). If a multiple rank correlation (i.e., correlating more than two sets of data simultaneously) is desired, use Kendall's Coefficient of Concordance (see Chapter 3, p. 159, and Zar, 1999).

Further statistical tests: Statistical methods used in physical oceanography are discussed by Emery and Thomson (1998). There are numerous other statistical measures that have not been mentioned. Consult the Appendix. There are also many computer programs available so be certain to contact statisticians, systems analysts, computer centers, and especially the Internet for information (e.g., using Google or another search engine).

2.14 RANDOMIZATION METHODS

Randomization tests (i.e., randomizing data) are a subset of Monte Carlo methods (i.e., data simulation; Manly, 1997). **Most standard statistical tests can also be conducted using randomization techniques**. In fact, randomization tests may have more power than classical statistics. Randomization tests do not require random sampling although the results are better when having used random sampling. **Randomization methods are good for small groups of data and irregular or missing data. They are especially suitable for analyzing data from non-normal distributions with many tied values and zeros.** Generally 1000 randomizations are sufficient except in borderline cases (i.e., when the calculated P value is close to the cutoff P value of 0.05, use 1500 randomizations). For borderline cases with $P = 0.01$ about 5000 randomizations are needed. Legendre and Legendre (1998) recommend 10,000 permutations (randomizations) for final, published results.

An example of randomization is to compare mandible lengths from museum specimens of Killer Whales (*Orcinus orca*) using 10 mandibles of each sex (a two-sample test). The question might be as follows: Is there a significant difference between mandible lengths of male and female Killer Whales? First the 20 measurements are combined or pooled into a single data group. Using a computer and random number generator algorithm, select 10 measurements (from mandible numbers 1–20) and allocate them to the male group, allocating the remaining 10 measurements to the female group. Repeat this procedure 1000 times. Compare the two groups of data using a t-test. This is but a simple example of dealing with small data sets.

Bootstrap and Jackknife Techniques

Perhaps the most common use of randomization techniques in biology and geology is that of bootstrapping and jackknifing. The purpose of using these techniques is to estimate variability, test hypotheses when the distribution of the test statistic is not known, or take a small, often inadequate data base, and generate reliable statistics from it. In ecology these techniques are used to estimate species richness, population size, and population growth rates (Southwood and Henderson, 2000). Many statistical tests require a minimum sample size of about four or five replicates. Thus, if you have only two, three or four replicates (typical of the majority of marine intertidal experiments), you can quickly generate thousands of replicates from the original data base (Manly, 1997). The key assumption behind bootstrapping is that observations are assumed to be independent samples from some population (Dixon, 2001). However, bootstrapping can accommodate complex dependent data situations also. **Bootstrapping is generally preferred over jackknifing for very small data sets** because the numbers withdrawn from the data pool are replaced in the data pool. Data taken from the jackknife data pool are not replaced. The bootstrap distribution is used to calculate a confidence interval, test a hypothesis, and estimate the standard error. The nonparametric bootstrap is used in most ecological applications because it requires fewer assumptions about the population. However, the parametric bootstrap is often better with very small samples (Dixon, 2001).

For bootstrapping one takes the original data (e.g., 20 numbers), and with a computer program, samples randomly with replacement some of the numbers (e.g., 5 of the original 20 numbers). This sampling is repeated 100 (minimum number for standard error) to 1000 or more times (minimum number for confidence interval). For 95% confidence intervals, at least 1500 bootstrap replicates are needed. With the resulting large pool of data, one can **calculate an estimated mean and standard error** (from the 100 to thousands of runs). The Jackknife technique is basically the same except that the numbers taken from the original data pool are not returned to that data pool, that is, one data point is omitted for each iteration. For example, if you had 100 numbers in the original data set and withdrew three numbers, the next withdrawal of three numbers would be from a data pool of 99 numbers, and so forth until all sets of three numbers are withdrawn. Both techniques give the same answers with very large sample sizes and slightly different answers for small sample sizes. However, the bootstrap method does not always work (Dixon, 2001), nor does the jackknife method always work (Todd Alonzo, pers. comm.). Rohlf (1995) has a computer program that will do bootstrapping and jackknifing. A very nice series of randomization programs are found in the computer program Rundom Projects 2.0 LITE located on the Internet (http://pjadw.tripod.com/). For further information see Manly (1992, 1997), Efron and Tibshirani (1993), Krebs (1999), and Scheiner and Gurevitch (2001).

2.15 GENERAL LINEAR PROGRAMMING

The Generalized Linear Model (GLM) is the mathematical foundation for many statistical analyses, including the t-test, analysis of variance (ANOVA), analysis of covariance (ANCOVA), regression analysis, cluster analysis, and various multivariate analyses (Trochim, 2002). The GLM allows a summary of a wide variety of research outcomes. The general equation for the model takes the following form:

$$Y = b_0 + bx + e$$

where

  Y = a set of outcome variables

  b_0 = a set of intercepts or covariates (value of each Y when each x = 0)

  b = a set of coefficients, one for each x

  x = a set of pre-program variables

  e = error factor

GLM's are math extensions of linear models that do not force data into unnatural scales and allow for nonlinearity. They are more flexible and better suited for ecological studies than are classical Gaussian distributions (Guisan et al., 2002). Generalized additive models (GAMs) deal with highly nonlinear and non-monotonic data.

Example of a GLM

Assume that you have a group of children for whom you are responsible (e.g., a children's camp). You need to provide them with sufficient good quality food to maintain their health. You feed them the following foods:

  Beef

  Baked potatoes

  Popcorn – for snacks

  Chicken

  Skim milk

  Peanut butter

  Raw carrots

You need to obtain and input considerable information on the foods they eat including calories, minerals and vitamins. **The major question being asked: What**

is the cheapest combination of foods that will satisfy all the daily nutritional requirements of the children?

This information was taken from the Internet, based on studies from the Department of Energy (Argonne National Laboratory) and Northwestern University, and modified (Table 2-6).

Table 2-6.   Linear programming using an example of an optimal diet.

Verbal formulation of the diet problem.

Minimize the "cost of the menu"
subject to the nutrition requirements:
 eat enough but not too much of Vitamin A

   .
   .

 eat enough but not too much Vitamin C

Additional constraints can be added to the diet problem. One can put constraints on the amount of which foods are consumed:

 Eat at least a certain minimum number of servings of beef but not more than the maximum number of servings of beef you want.

   .
   .

 Eat at least a certain minimum number of servings of carrots but not more than the maximum number of servings of carrots you want

In order to create a mathematical version of this model, we need to define some variables:

 x(beef) = servings of beef in the menu

   .
   .

x(carrots) = servings of carrots in the menu

and some parameters:

 cost (beef) = cost per serving of beef
 min (beef) = minimum number of servings to eat
 max (beef) = maximum number of servings to eat

   .
   .

 cost (carrots) = cost per serving of carrots
 min (carrots) = minimum number of servings to eat
 max (carrots) = maximum number of servings to eat

 A (beef) = amount of Vitamin A in one serving of beef
 A (carrots) = amount of Vitamin A in one serving of carrrots

   .
   .

 C (beef) = amount of Vitamin C in one serving of beef
 C (carrots) = amount of Vitamin C in one serving of carrots

(Continued)

Table 2-6.   (*Continued*)

and

min(A) = minimum amount of Vitamin A required
max(A) = maximum amount of Vitamin A required

.

.

min(C) = minimum amount of Vitamin C required
max(C) = maximum amount of Vitamin C required

Basic Formulation

Given this notation. The model can be rewritten as follows.

Minimuze:

cost (beef) * x(beef) +. . . + cost(carrots) * x(carrots)

Subject to:

min (A)

Mathematical Formulation

The formulation can be made more mathematical if we define a set of foods and a set of nutrients.

Let Foods = { beef, . . . , carrots}
   Assume there are f foods in Foods.
Let Nutrients = { A, . . . , C}
   Assume there are n nutrients in Nutrients.

and a set of parameters:

$cost[i]$ = costs of the foods for $1 < i < f$.
$x[i]$ = unknown amounts of foods to eat for $1 < i < f$.
$nutr\,[j]\,[i]$ = amount of nutrient j in food i for $1 < i < f$ and $1 < j$
$min\_nutr[i]$ = minimum amount of nutrient i required for $1 < i < n$.
$max\_nutr[i]$ = maximum amount of nutrient i allowed per day for $1 < i$
$min\_food[i]$ = minimum amount of food i desired per day for $1 < i < f$.
$min\_food[i]$ = minimum amount of food i desired per day for $1 < i < f$.

Minimize

$$\cos t = \sum_{i=1}^{f}\left(\cos t[i]\cdot x[i]\right)$$

Subject to:

$$\min_{\text{nutr}}\left[j\right] < \sum_{i=1}^{j} nutr\left[j\right][i]\cdot x[i] < \max_{\text{nutr}}\left[j\right] \quad \text{for } i < j < n, \text{ and}$$

$min\_food[i] < x[i] < max\_food[i]$ for $1 < i < f$.

Table 2-6. (*Continued*)

The AMPL code is provided below.

set NUTR;
set FOOD;

param cost {FOOD} > 0;
param f_min {FOOD}>= 0;
param f_max { f in FOOD} >= f_min[f];

param n_min { NUTR } >= 0;
param n_max {n in NUTR } >= n_min [n];

param amt {NUTR, FOOD} >= 0;

var Buy { f in FOOD} >= f_min[f], <= f_max[f];

minimize total_cost: sum { f in FOOD } cost [f] * Buy[f];

Subject to diet { n in NUTR }:
        n_min[n] <= sum { f in FOOD} amt [n, f] * Buy[f] <= n_max[n];

Your Input

You selected 7 foods to consider for your diet.

The Optimized Menu

The cost of this *optimal* diet is $2.29 per day.

The Solution and Cost Breakdown by Food

Food	Servings	Cost($)
Carrots, Raw	3.11	0.22
Roasted Chicken	0.20	0.17
Frankfurter, Beef	3.90	1.05
Lettuce, Iceberg, Raw	2.87	0.06
Potatoes, Baked	5.99	0.36
Skim Milk	3.31	0.43

Linear programming (LP) software programs can handle thousands of variables and constraints rapidly. Two families of solution techniques are widely used: (1) simplex methods, and (2) interior-point methods. Linear planning has been used to handle diverse types of problems in planning, routing, scheduling, assignment and design in the military and in industries such as energy, telecommunications, manufacturing and transportation. LP software is of two types: (1) algorithmic codes that seek optimal solutions to specific linear planning, and (2) modeling systems. Large-scale modeling systems are expensive commercial products. LP has been employed in statistical courses (engineering) and books. An interesting LP statistical software program for biologists (etc.) is called S-PLUS (Venables and Ripley, 1997). For further information see Odeh (1991), Crawley (1993), and "linear programming" on the Internet.

2.16   MAXIMUM LIKELIHOOD

The Maximum Likelihood (ML) Method has been used in a number of disciplines for many years. It is widely used in biology. It is used where one would use χ^2 or the G-test, for smoothing data, and for population estimates, among other applications. The concept of maximum likelihood was pioneered by Ronald Fisher. It is a type of statistics designed to provide estimates that maximize the probability of zero or negligible error. The likelihood function is a conditional probability distribution that is maximized by calculating the first derivative and solving an equation. It is usually an iterative method by which a parameter estimate is completed. It is often easier to plot or maximize the log of the likelihood function, thus the term log-likelihood. It has many supporters (e.g., Milligan, 2003) but some critics (e.g., Manly, 1997). Maximum likelihood estimators are not in general unbiased but the bias tends to zero as sample size increases (Buckland et al., 2001). Some statisticians prefer bootstrapping because it does not require the assumptions of ML. Clark Labs (developer of the GIS program Idrisi) report that both linear discriminant analysis (Fisher Classifier) and a back propagation neural network classifier are superior to ML in distinguishing classes of data. Hayek and Buzas (1997) consider ML a preferred method of estimating parameters. Buckland et al. (2001) authored a text on estimating animal abundance in closed populations, emphasizing likelihood methods. Nielsen and Lewy (2002) prefer the Bayes estimator over maximum likelihood. Krebs (2000a) has a computer program that calculates the maximum likelihood estimation of the negative binomial k for counts of aggregated or clumped organisms. Krebs (1999) plots estimated population sizes (X-axis) against the Log-likelihood (Y-axis). The plot involves an iteration of an equation in which N or population size is repeatedly changed. The most probable value of N is that which gives the highest likelihood (Fig. 1-8). This is the estimated population size based on radio-tagged animals that have been re-sighted. White (1996) has a computer program for this.

    See Edwards (1972) for an introduction to likelihood methods. Also see Morrison (1976), Davis (1986), Manly (1997), Zar (1999), Buckland et al. (2001), Gelman et al. (2003), and the Internet. Ver Hoef and Cressie (2001) discuss Restricted Maximum-Liklihood Estimation, which gave results superior to classical ANOVA.

2.17   BAYESIAN STATISTICS

Bayesian statistics are used to update probabilities that are sometimes called conditional probabilities, for example, the probability of event A occurring given that event B has occurred. Bayesian statistics allow the incorporation of external information in the interpretation of experimental or observational data, can provide for probability statements, and can minimize sample sizes and study durations (Clark and Lavine, 2001). A Bayesian analysis combines a sampling distribution or likelihood estimate with additional (or prior) information to produce a density estimate (posterior density) or updated probability distribution (Clark and Lavine, 2001).

Example

Fish were censused one year in a tidepool. A total of 132 fish were counted. The census was repeated one month later and 127 fish were counted.

First calculate the likelihood function (θ) as follows:

$$\theta = \frac{n}{k}$$

where

   k = No. fish from the first census

   n = No. fish from the second census

   $k = 132$ $n = 127$

   $\theta = 0.96$ (representing the fraction that survived)

Next combine the sampling distribution or likelihood function with the prior density (e.g. $\theta = 0.9$ or density from a previous study at the same tidepool) to produce a posterior or updated density probability distribution.

$$PD = \frac{\text{prior density} \times \text{likelihood}}{\text{integrated} (\int) \text{prior density} \times \text{likelihood } d\theta}$$

where

   PD = posterior probability density distribution

   $\int$ = integral (normalizing constant)

   θ = likelihood

   d = density function

This second procedure, involving a series of mathematical steps, produces a figure of the updated density distribution (Fig. 2-12).

Clark and Lavine (2001) recommend the use of several different priors, computing a posterior for each, and then comparing the posteriors (i.e., results). The several posteriors may be similar. If they differ then the sample size may need to be increased (or the posteriors may not be robust to the choice of the priors – Todd Alonzo, pers. comm.).

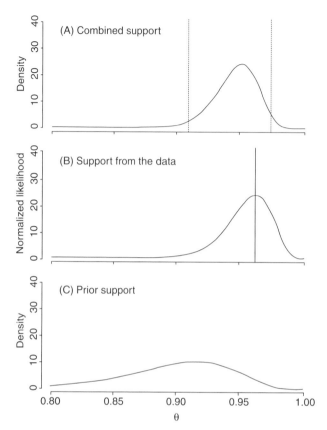

Figure 2-12.   The elements of a Bayesian analysis. The posterior density (A) represents the support of values of the survival rate θ in light of data at hand (B) and prior inputs (C) (source: Clark and Lavine, 2001).

A simple example:

A group of 1000 elderly people are given a screening test for cancer (e.g., PSA for detecting colon cancer). The true cancer status (e.g., as determined by biopsy) was also determined for the 1000 people. The resulting positives and negatives are as follows:

Epidemiology: Bayesian statistics – to update disease estimates

Data		1000		
people with cancer symptoms	$50 = (1/20)$		950	people without cancer symptoms
people tested for cancer	5	45	95	855
	$-$	$+$	$+$	$-$
	false	true	false	true

A = patient has cancer symptoms

B = positive test for cancer

$$P(A/B) = \frac{P(A)P(B/A)}{P(B)} = \text{conditional probability that a patient with symptoms has cancer}$$

$$P(B) = \frac{45 + 95 \text{ all positives}}{1000} = 0.14$$

$$P(A/B) = \frac{1/20(45/50)}{0.14} = 0.32$$

The prior estimate of 0.14 is the initial probability of being diagnosed with cancer. A biopsy changes this value to a posterior estimate of 0.32. Thus, P(A) is the prior probability and P(A/B) is the posterior probability after receiving the additional (updated) information B.

Another example:

Raftery and Zeh (1993) counted bowhead whales seen and heard near Pt. Barrow, Alaska. The likelihood function came from counts during a 1988 census. The prior included location of the observers and sonar arrays, visibility, the physics of sonar location, and bowhead whale migratory behavior. The posterior estimated the most likely population size of bowheads as 7500 with a reasonable range of 6500 to 9000 (Fig. 2-13).

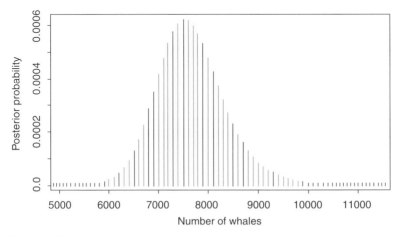

Figure 2-13. Posterior distribution of the number of whales for 1988 (source Raftery and Zeh, 1993).

The mathematics are formidable (for most biologists) and include probabilities and numerous pages of equations.

2.18   HOW TO LIE WITH STATISTICS

The following references tell us how to lie with statistics:

Huff (1954), Wheeler (1976), Hooke (1983), Jaffe and Spirer (1987), and Best (2001). These are worth looking at.

Chapter 3

Quantitative Methods in Field Ecology and Other Useful Techniques and Information

A. INTRODUCTION

3.1 Introduction

Population patterns may vary with (1) sampling method, (2) size of sampling unit (e.g., even-spaced organisms in a smaller area may be aggregates in a larger area), (3) shape of sampling unit (e.g., limpets on a rock may be limited in distribution by sand around the rock), (4) patchiness of the habitat (e.g., differential predation, or snails on rough or smooth surfaces), and (5) behavior of the animals (e.g., aggregation during the mating season).

B. POPULATION PATTERNS

3.2 Distributions (Dispersion)

The three major distributions of organisms are even (uniform), clumped (aggregated, contagious, contagion), and random. The scale or size of a sample area is the prime determinant of aggregation (Fig. 3-1). **A Poisson distribution (a test for randomness) occurs when the variance is equal to the mean**. It means that there is an equal probability of an organism occupying any point in space and assumes that the presence of one individual does not affect the distribution of others. Given a large enough scale, no species are randomly distributed (Hayek and Buzas, 1997). **The random distribution of small populations and the aggregated distributions of larger populations have often been observed** (Southwood and Henderson, 2000).

Quantitative Analysis of Marine Biology Communities: Field Biology and Environment
by Gerald J. Bakus
Copyright © 2007 John Wiley & Sons, Inc.

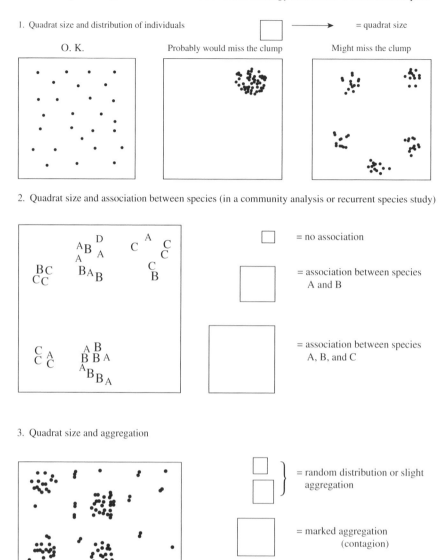

Figure 3-1.   Effects of quadrat size on the interpretation of field data (source: modified from Kershaw, 1973).

Krebs (1999) tests if a group of data fits a Poisson distribution, which would signify a random dispersion. **If the standard deviation equals the mean then considerable aggregation occurs** (Hayek and Buzas, 1997). Most aggregation indices have been based on the variance to mean ratio. Hurlbert (1990a) indicated that the variance to mean ratio is useless as a measure of departure from randomness and uninterpretable as a measure of aggregation (see Hurlbert, 1990a). When he censused populations using a grid of 1 m² quadrats, all populations showed a variance to mean ratio of 1.0 although each population showed a different pattern of aggregation and none had a Poisson distribution.

Paul Delaney, a former student, tested 15 aggregation indices and determined that the best plot method is Morisita's Index (see Grassle et al., 1975, for an example of how this index has been used).

In the equation that follows (i.e., Morisita's index), the sample size should be greater than five.

Morisita's Index (MI):

$$MI = \frac{S\left(\sum n^2 - N\right)}{N(N-1)}$$

where

n = total number of individuals in a quadrat

N = total number of all individuals (or colonies)

S = total number of quadrats

If - MI = < 1: regular or uniform dispersion

MI = 1.0: random dispersion

MI = >1.0 to X: clumped dispersion

The greater the MI value over 1, the greater the degree of aggregation. The advantage of Morisita's index is that it does not vary with population density. It does vary slightly with sample size. To determine the effect of scale on aggregation, one can employ a range of quadrat sizes and calculate an I_{mr} curve (see Hurlbert, 1990a). The Morisita's index value MI can reach 10 or more for dense populations of small sea anemones and small barnacles.

Example:

Quadrat No. (10 m² each)	No. colonies of soft corals
1	10
2	9
3	11
4	9
5	10

$$MI = \frac{5 \times (483 - 49)}{2352}$$

MI $= 0.923$ (indicating $\sim$ random distribution, with a very slight tendency toward an even distribution).

Krebs (1999) suggests using an improved Morisita's index (see below), which converts it to a scale ranging from -1 to $+1$ and is independent of sample size. A minimum of 50 quadrats is recommended, but 200 or more quadrats with a highly aggregated population. Krebs (2000a) has a program for the Morisita's index and a standardized *Morisita's index*, listed under Spatial Pattern – Indices of Dispersion. See also Brown and West (2000).

Standardized *Morisita's Index* (Krebs, 1999)

1. Calculate the Morisita's index of dispersion as discussed above.

2. Calculate two new indices, the Uniform Index and the Clumped Index, as follows:

$$\text{Uniform Index } (M_u) = \frac{\chi^2_{.975} - n + \sum x_i}{\left(\sum x_i\right) - 1}$$

where:

$\chi^2_{.975}$ = value of chi-squared from the table with $(n-1)$ degrees of freedom that has 97.5% of the area to the right

x_i = number of organisms in quadrat i

n = number of quadrats

$$\text{Clumped Index } (M_c) = \frac{\chi^2_{.025} - n + \sum x_i}{\left(\sum x_i\right) - 1}$$

where:

$\chi^2_{.025}$ = value of chi-squared from the table with $(n-1)$ degrees of freedom that has 2.5% of the area to the right

3. Calculate the standardized Morisita's Index (MIS) by one of the following equations:

When MI $\geq M_c > 1.0$ then:

$$\text{MIS} = 0.5 + 0.5\left(\frac{\text{MI} - M_c}{n - M_c}\right)$$

When $M_c > $ MI ≥ 1.0 then:

$$\text{MIS} = 0.5\left(\frac{\text{MI} - 1}{M_u - 1}\right)$$

When $1.0 > \text{MI} > M_u$ then:

$$\text{MIS} = -0.5\left(\frac{\text{MI} - 1}{M_u - 1}\right)$$

When $1.0 > M_u > \text{MI}$ then:

$$\text{MIS} = -0.5 + 0.5\left(\frac{\text{MI} - M_u}{M_u}\right)$$

$\text{MIS} > 0 =$ clumped or aggregated distribution

$\text{MIS} = 0 =$ random distribution

$\text{MIS} < 0 =$ uniform distribution

or

$+1 =$ maximum aggregation

$0 =$ random distribution

$-1 =$ perfectly uniform distribution

Using the data from the Morisita's index example given above, the standardized Morisita's Index $= -0.50$. This suggests a stronger tendency toward an even distribution than suggested by the original Morisita index.

See Dale (1999) for a review of spatial pattern analysis in plant ecology.

3.3 Dispersal

Dispersal of animals as movement away from an aggregation can be measured in several ways:

(**1**) *Heterogeneity of Dispersal*   It is described as Ku or kurtosis and expressed as a plot of numbers against time (normal-flat-steep) – see Southwood (1978). Kurtosis is a measure of the distribution of data between the center and the tails in a unimodal curve (see Fig. 2-4 No. 2).

(**2**) *Distance Traveled*

$$D = \sqrt{\frac{\sum(d^2)}{N}}$$

where

$D =$ distance traveled

$d =$ distance from recapture point to release point of an individual

$N =$ total number of animals recaptured

The value of D is plotted against time for a graphic display. This equation is very similar to the standard deviation (Southwood, 1978).

(**3**) *Regression Equations* Southwood (1978) gives several equations that plot a regression line between the distance traveled and the density.

(**4**) *Marked Animal Index* Determine the marked animals leaving the area (Petersen index – see Chapter 1, p. 46), assuming that the total population has been measured by some other method and that no births or deaths has occurred.

(**5**) *Other Methods* Southwood and Henderson (2000) discuss rather complex techniques of studying dispersion including using a diffusion equation and the rate of population interchange between two areas, among other methods. For random walk methods see below.

3.4 Home Range

A home range is an area regularly traversed by an animal (vs. a territory, which is a defended area). Knowledge of home range is valuable in analyzing competition and density effects. A total of 10 recaptures is considered sufficient to measure the home range. See Powell (2000) for a discussion of home range.

(**1**) *Minimum Area Method* Map the points of animal recapture and join the outermost points. See Fig. 3-2.

(**2**) *Range Length* The distance between the most widely separated animal captures.

(**3**) *Index of Range Activity*

$$I = \sqrt{\frac{\sum D^2}{N-1}}$$

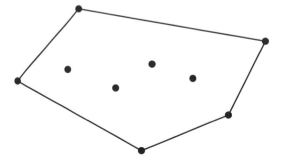

Figure 3-2. Minimum area method, connecting the outermost points.

where:

I = standard diameter

D = animal recapture distance from the geometric center of all points (the Mahalanobis distance)

N = number of observations

Σ = sum

See Southwood (1978) and Southwood and Henderson (2000) for the above methods and for the matrix index technique and Boitani and Fuller (2000) for further information.

3.5 Random Walk

For many years marine biologists conducted behavioral research on marine invertebrates with choice experiments consisting of Y-shaped tubes (Davenport, 1955). Host substance in seawater was released into one arm of the Y-shaped tube but only seawater into the other arm. A symbiont would tend to move into the arm with the host substance, demonstrating **directional walking or swimming behavior**. Another method of determining whether or not the movement of an animal is random is by comparing its motion with random movement. This can be accomplished graphically (see below) or by Monte Carlo simulation or bootstrapping techniques. See Kareiva and Shigesada (1983), Weiss (1983), and Underwood and Chapman (1985) for further information.

Random Walk Equation:

$$(D)^2 = nL^2$$

where

$(D)^2$ = root mean square distance (displacement) from the origin

n = cumulative number of turns

L = average distance between turns

See Feynman et al. (1963) and Hammond (1982).

Example:

Deposit feeders associated with plant cover.

(1) Plot cumulative daily straight line distance traveled per day (dots in Figure 3-3).

(2) Indicate distance (mean, standard deviation) traveled each day from the starting point (stars in Figure 3-3).

(3) Calculate random walk distances (D) from the original starting point each day (crosses in Figure 3-3).

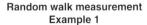

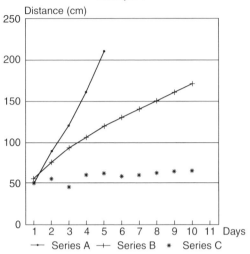

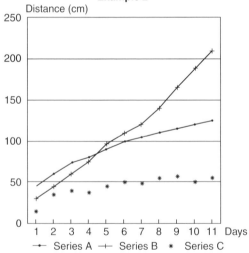

• = cumulative straight line distance traveled per day
+ = distance (cumulative) from original starting point each day as predicted by the random walk theorem.
∗ = distance ($\bar{X}$, σ) traveled each day from the starting point (average of 10 measurements)

Figure 3-3. Examples of random walks. See figure for additional description.

(**4**) Use a *t*-test or one-way ANOVA to compare the lines (slopes), that is, the plotted cumulative line distance with the plotted random walk distance.

3.6 Feeding Ecology

Ratio of Feeding Types

$$I = \frac{1}{S/D + 1}$$

where

I = ratio

S = live plankton feeder (no. of species)

D = detritus feeder (no. of species)

I varies from 0 (only live plankton feeding) to 1.0 (only detritus feeding). This is a simple ratio calculation with no statistical test.

Feeding Ecology Analyses

Vodopich and Hoover (1981) present a summary of equations used in ecological studies on feeding in animals. See also Strauss (1979). Krebs (2000a) presents a computer program on the ranking of foods. A discussion of feeding ecology is presented by Litvaitis (2000). See also van der Meer et al. (2005). Ecopath (http://ecopath.org) is used to model food webs in marine ecosystems.

Food Selectivity or Electivity

Feeding selectivity is a measurement of the degree to which foods are preferred or rejected by animals (Ivlev, 1961).

$$L = r_i - p_i$$

where

L = linear food selection index

r_i = proportion of food item in the gut

p_i = proportion of food item in the environment

L ranges from -1 (complete aversion to food item) to $+1$ (highest level selection). L values can be compared with the *t*-statistic.

Example

What is the overall feeding selectivity of this shark?

	Resource			
	Dead fish	Young turtles	Surgeonfish	Parrotfish
Black-tip reef shark (r_i) (gut contents)	0.100	0.504	0.300	0.096
Resource availability (p_i) (relative abundance in the field)	0.100	0.300	0.502	0.098

L for dead fish $= 0.001 - 0.001 = 0.0$

L for young turtles $= 0.504 - 0.300 = 0.204$

L for surgeonfish $= 0.300 - 0.502 = -0.202$

L for parrotfish $= 0.096 - 0.098 = -0.002$

Mean of L (overall feeding selectivity) $= 0$

Conclusion: The shark is not selective. It eats many things with nearly equal frequency.

This is the simplest of the indices (also see Jacobs, 1974). Krebs (1999) recommends Manly's alpha. However, the index is strongly affected by rare dietary items. The Manly alpha program is available in Krebs (2000a) under Niche Measures. The Rank Preference Index of Johnson (1980) is also preferred because the index is usually not affected by the inclusion or exclusion of foods that are rare in the diet. Program RANK in Johnson (1980) will do the calculations. Walter et al. (1986) describe a serological technique (enzyme-linked immunosorbent assay) with which one can identify marine organisms consumed by predaceous birds. Fatty acid composition in seastars and in the blubber of marine mammals and trace elements in the skeletons of terrestrial mammals (e.g. martens) have been used in feeding ecology analyses (see May, 1986; Southwood and Henderson, 2000; and the internet for further information).

C.   POPULATION GROWTH

The following sections are limited to a brief introduction to some of the highlights of population ecology in animals.

3.7   Size-frequency Distribution

One simple method of **determining age classes** within a population of animals is to plot the size of organisms (ordinate or *X*-axis) against their frequency (abcissa or

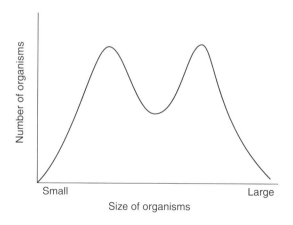

Figure 3-4. Size frequency distribution of a species showing two age classes.

Small   Large
Size of organisms

Y-axis). This gives a visual representation of age classes (Fig. 3-4). In some cases the determination of age classes by sight is difficult and probability paper can be used for the graphical analysis of polymodal frequency distributions (see Harding, 1949). ENORMSEP, a FORTRAN IV program, is used to separate polymodal frequency distributions into their component groups (Yong and Skillman, 1975). For quantitative comparisons of populations with different age structures, see Dapson (1971). See also autocorrelation in Chapter 6, p. 266.

3.8   Growth of Individuals in a Population

The growth of individuals in a population can be determined by various methods, such as **marking and tagging** individuals (see Chapter 1, p. 44). To determine growth and mortality without marking or tagging individuals, one simply **plots a size-frequency distribution** of the species at a site (Fig. 3-4), **return to the site at periodic intervals** (e.g., every 2 months) and construct size-frequency distributions. The **movement of the modal points to the right** will offer some indication of growth rates, total mortality rate, and in some cases, of immigration and emigration.

3.9   Natality

Natality (i.e., birthrate) equals number of births/female/unit time. It is determined in three common ways: (1) **Trap animals** when they move from one habitat to another at a specific stage in the life history (e.g., using emergence traps); (2) **Measure fecundity directly** (e.g., count eggs) and plot fecundity against weight (usually a linear relationship). Fecundity is the number of gametes, fertilized eggs, seeds, or live offspring produced by an individual (Neal, 2004), and (3) **Use mark and recapture data**. In temperate waters, benthic surveys carried out in the fall season will often show a population dominated by recent recruits (Southwood and Henderson, 2000).

3.10   Mortality

Mortality equals number of deaths/unit time. This is determined by: (1) direct observation, (2) recovery of dead or unhealthy organisms, (3) mark and recapture data, (4) experiments in the lab or the field (for example, effects of abrasion on survival of limpets), (5) age frequency distribution, and (6) biotic factors. Biotic factors would include indirect methods of estimating mortality such as a count of parasites; for predation it would include (a) direct observation of predation (rarely observed underwater with sharks and fishes), (b) labeling of prey with isotopes, (c) constructing barriers against the parasite or predator to exclude or eliminate them (d) serological techniques (e.g., Walter et al., 1986), (e) analysis of feces and pellets, (f) analysis of gut contents, (g) gas chromatography (h) electrophoresis, (i) precipitation methods, and (j) agglutination methods (see Southwood, 1978 and Southwood and Henderson, 2000).

3.11   Construction of Life Tables

(1) *Age Specific Life Tables*   Constructed for members of a single generation with the population stationary or fluctuating. The data are plotted as in survivorship curves for humans. Life table data are usually designed for females because they are the sole reproducers. Life table data are used for sexually reproducing species. Life tables usually assume a stable age distribution (i.e., the number of females in each age class or cohort remains constant over time), although the age distribution often shifts over time in natural populations (Krebs, 2001). Life tables for plants are often more complex because there are separate longevities for the entire plant and for its parts (e.g., whole trees, buds, flowers, leaves, cones, etc.).

The following is a brief summary of major steps in the complex process of constructing a life table. Neal (2004) gives instructions on how to construct a life table using a spreadsheet (also see Gotelli, 2001).

(a) Determine the total abundance of a stage: figure 3-5 and see Southwood (1978). See Southwood and Henderson (2000) for more advanced analyses.
(b) Determine the total change in the population.
(c) Determine the number entering the stage.
(d) Determine emigration, immigration, or migration.
(e) Describe these data above by the following:
   (1) survivorship curve
   (2) life table – see pp. 136–137

The growth rate of a population can be described as follows:

$$N_t = N_0 e^{rt}$$

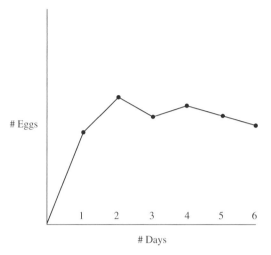

Eggs

Days

Figure 3-5. Methods of estimating the fecundity of females in a population. Place a grid over the figure corresponding to the intervals on the X and Y axes. Count the number of squares under the line (assuming a constant mortality rate) and divide by the mean development time under natural conditions. Alternatively, a random point method (i.e., throwing darts randomly) can be used for calculating the area under the curve. A third method is to use an image analysis program (e.g., Sigmascan) to automatically calculate the area under the curve. The correct scale must be entered before the area is calculated.

where

N_t = population size at time t

N_0 = population size at time 0

e = base of natural logarithm ($\log_n$ or $2.3 \times \log_{10}$)

r = population growth (i.e., biotic potential) in an unlimited environment and with a stable age distribution

t = time

There are two major types of population growth curves: (1) exponential or J-shaped curves (intrinsic rates of population increase or biotic potential and characteristic of population explosions such as locust swarms or red tides), and (2) Logistic or S-shaped curves (characteristic of many species that have been studied [Fig. 3-6]). The S shape is the result of a natural population increase modified by climate, food availability, predation, and competition (i.e., population growth inhibiting factors or environmental resistance). Population growth curves can be simulated using the computer program Ecosim. Enter an initial population size, birth rate, and death rate, and the program plots a population growth curve. See Roughgarden (1998) for ecological theory on populations, Elzinga et al. (2001) for monitoring plant and animal populations, and Elseth and Baumgardner (1981) and Neal (2004) for an introduction to population ecology.

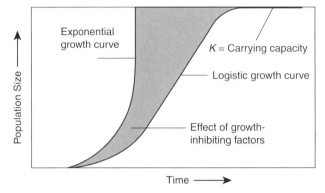

Figure 3-6. Major population growth curves. The exponential growth curve (J-shaped curve) and the logistic growth curve (S-shaped curve). The major growth inhibiting factors are food, climate, predators, and parasites.

(2) *Life and Fertility Tables*   Most animal life tables (Tables 3-1 and 3-2) assume a stable age distribution, that is, the proportion of each age class remains approximately the same from generation to generation. Ro is the number of young produced per female per generation. When Ro > 1 the population is increasing, when

Table 3-1. Example of a life table for Dall mountain sheep (*Ovis dalli*) in Denali National Park, Alaska (source: Neal, 2004).

Age Class (years) (x)	Number dying in age class (D_x)	Number surviving at start of age class (S_x)	Probability of survival from birth (l_x)	Number dying in age class (d_x)	Mortality rate (q_x)	Mean expectation of life (e_x)
0–1	121	608	1.000	0.199	0.199	7.1
1–2	7	487	0.801	0.012	0.014	7.7
2–3	8	480	0.789	0.013	0.017	6.8
3–4	7	472	0.776	0.012	0.015	5.9
4–5	18	465	0.765	0.030	0.039	5.0
5–6	28	447	0.735	0.046	0.063	4.2
6–7	29	419	0.689	0.048	0.069	3.4
7–8	42	390	0.641	0.069	0.108	2.6
8–9	80	348	0.572	0.132	0.230	1.9
9–10	114	268	0.441	0.188	0.425	1.3
10–11	95	154	0.253	0.156	0.617	0.9
11–12	55	59	0.097	0.090	0.932	0.6
12–13	2	4	0.007	0.003	0.500	1.0
13–14	2	2	0.003	0.003	1.000	0.5
14–15	0	0	0.000	0.000	—	—

Table 3-2. Cohort life table and age specific birth rate data to show a population increase (Ro) (source: modified from Gotelli, 2001 and Neal, 2004). See table for further information.

X (age class)	P (population size)	lx (survivorship)	mx (female fertility rate)	lxmx (fertility rate per age class)
0	500	1.0	0	0.0
1	400	0.8	2	1.6
2	200	0.4	3	1.2
3	50	0.1	1	0.1
4	0	0.0	0	0.0

Ro = 2.9 offspring (adjusted for mortality)

$$G \ (\text{generation time}) = \frac{\sum l(x)m(x)X}{\sum l(x)m(x)} = 1.483 \text{ years}$$

r (estimated intrinsic rate of population increase) = ln (Ro)/G = 0.718 individuals/ individual year

r (Euler's equation – true intrinsic rate of population increase) = 0.776 individuals/ individual year

$N_t = N_0 \, e^{rt}$ gives the population increase over time (t)

Euler's equation: $1 = \int_0^\infty e^{-r_m x} l_x m_x$ (calculated by trial and error) - see Neal (2004)

The age class is in years (i.e., the fish lives for 4 years). The population size represents the number of fish at time t, starting with a population of 500 fish. The survivorship is the percentage decrease in the population over time expressed as a decimal (1.0 = 100%). The female fertility rate is the number of young produced per female at a certain age (age class). The reason the rate is so low is because the fish is viviparous (i.e., bears its young live rather than eggs). The average fertility rate per female for an age class is lxmx. Ro is the net reproductive rate or the mean number of female offspring produced per female over her lifetime. The Ro value indicates that the population is increasing slightly each year. The generation time is the average lifetime of the species. The estimated intrinsic rate of increase of the population = r, with a correction (G) for generation time. The actual intrinsic rate of population increase is derived from the Euler equation. Once r is known, the population increase over time can be predicted by using the equation $N_t = N_0 \, e^{rt}$, shown above on p. 135.

Ro = 1 the population is stable, and when Ro < 1 the population is decreasing. See Krebs (2000a) for a computer program, Gore et al. (2000) for a mathematical discussion, and Sibly et al. (2003) for a discourse on wildlife population growth rates. Estimates of population density for life tables require an accuracy of a standard error of 10% of the mean (Southwood and Henderson, 2000). Failure-time statistics can be used to evaluate data in life tables (Fox, 2001).

(3) *Time Specific Life Tables*   The procedure for arriving at these tables is:

(a) age group the organisms (at a single instant of time)

(b) determine time specific survivorship curves by plotting frequencies against age groups

(c) determine if there is a stable age distribution by geometrical progression (i.e., by comparing age class frequencies over time)

(d) calculate the rate of population increase

Analysis of Life Table Data

By multiple regression analysis and graphical analysis, the investigator can incorporate key factors such as predation and parasitism against number of eggs laid or hatched (Southwood, 1978). See Southwood and Henderson (2000) for a detailed discussion of life table data, models, and Lewis-Leslie matrices. Krebs (1999) presents a variety of techniques for estimating survival rates. These include finite survival rates, estimation from life tables, key factor analysis, estimation from age composition, radiotelemetry estimates of survival, Maximum Likelihood method, Kaplan-Meier method, estimation of bird survival rates, and tests for differences in survival rates (also see Manly and Parr, 1968; Borchers et al., 2002; Turchin, 2003; Vandermeer and Goldberg, 2003; and Skalski et al., 2005 for further information).

3.12   Population Dynamics Models

Population dynamics models reveal how the number of individuals in a population change over time. These dynamics models can be separated into analytical models and simulation models (see Chapter 7 on modeling). **Analytical models** include the following types: (1) reaction diffusion models that use partial differential equations for animal populations, (2) patchy environment models that include biogeographics, landscape, habitat-patch models, patches, and barriers (these are similar to the Lotka-Volterra classic predator-prey model), and (3) neighborhood models used to study the population dynamics of sessile organisms (Czaran and Bartha, 1992; also see Kot, 2001). Mathematics used in population ecology is presented by Thieme (2003).

   Simulation models include three types of computer simulations: (1) cellular automata models (see below), (2) distance models (e.g., minimum spanning trees or a network of straight lines connecting all sample points which has the minimum total line length – Green, 1979), and (3) tessellation models (e.g., cluster growth according to a deterministic set of rules). **Cellular automata are the most popular population dynamics models** (see Gutowitz, 1991 and Tilman and Kareiva, 1997). They employ a gridwork of cells or units (on a computer monitor), each cell representing one individual animal or plant, colony or clone. Cellular automata typically use Monte Carlo (randomization) methods, Leslie matrices, and stochastics (i.e., probabilities of death from competition, predation, and environmental disturbances such as fires, typhoons, and vessel anchors).

Wilfredo Licuanan, a former student, developed a probabalistic automata model of a Philippine backreef coral community of two dominant species of hard corals living on unconsolidated sediments (Licuanan, 1995). Transition probabilities (see below) were derived from actual field studies whereas recruitment rates were estimated using projection matrix simulations (Leslie) on the basis of field data and implemented as a Poisson (random) process. Three size classes were developed for the two species of hard corals, one species colored yellow on blue, the other species red on green. The size or intensity of yellow or red indicates the size class (Fig. 3-7). Disturbances are randomly distributed anchor impacts. Fig. 3-8a shows the initial configuration of 100 corals per size class per species (600 total), all randomly distributed. Figure 3-8b shows the changes that have occurred in the populations over a period of 30 years. The executable file called *100persc* on the CD-ROM (see the AUTOMATA folder) shows changes

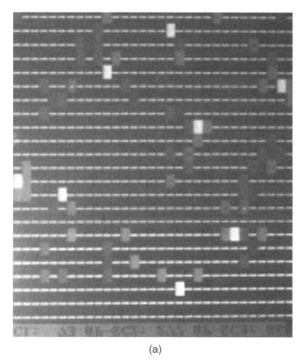

(a)

Figure 3-7.   (a) Cellular automata model showing the initial configuration of corals, all randomly distributed (a). One species of coral is colored yellow on blue, the second competing species of coral is colored red on green. The size or intensity of yellow or red indicates the size class.

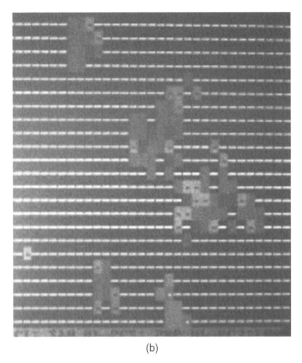

(b)

Figure 3-7.   (b) Configuration of corals after 30 years. See text for explanation.

in color over 30 years with a selected disturbance rate of 12 anchor hits per year (one boat anchorage per month with impact intensity of two). Release 2 and Kilimanjaro (GPS) models of Idrisi incorporate changes based on cellular automata (see Appendix). For information on cellular automata and complexity theory

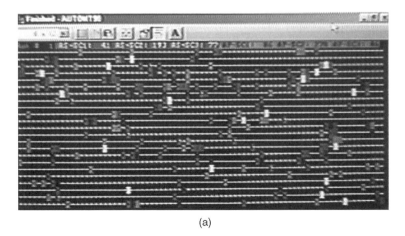

(a)

Figure 3-8.   Cellular automata model showing the initial configuration of competing corals (a).

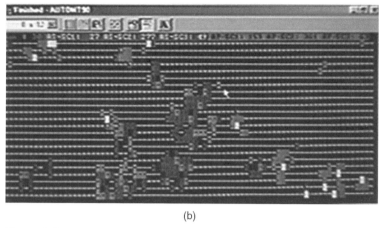

(b)

Figure 3-8.   (b) The configuration after 30 years. See caption for Figure 3-7.

contact Steven E. Phelan on the internet (s.phelan@latrobe.edu.au). Also see the journal Complex Systems.

3.13   Population Growth and Productivity

Southwood and Henderson (2000) and van der Meer et al. (2005) discuss productivity and energy budgets. Ecopath (http://ecopath.org) is used to model food webs in marine ecosystems (also see Bakus, 1969 and Sala et al., 2000).

3.14   Null Models

Null models began to develop in 1979 and became fashionable in ecology in the 1980s (Connor and Simberloff 1979, 1983, 1986). Since that time the field has developed rapidly and is a trendy topic in modern ecology. Null models are recommended for the analysis of non-experimental data (e.g., field data). Total species lists or assemblages (typically animals) are often used (Gotelli and Graves, 1996). Random species occurrences and non-normal or nonparametric data are assumed. Null models begin with actual species lists, the lists are randomized, then the actual species lists are compared with the randomized lists (Erlich and Roughgarden, 1987). The data are often treated by randomization techniques or Monte Carlo simulation and statistics are generated (see Randomization Techniques on p. 117 in Chapter 2). Null models elucidate patterns but not causes, allow for the outcome of "no effect", control for Type 1 statistical errors, and exclude a mechanism in order to gauge its effect. Thus, null models provide powerful tools in analyzing ecological data.

Null models are based on the null hypothesis, such as: (1) nothing has happened, (2) there are no differences, (3) a process has not occurred, and (4) change has not been produced. Uses of null models are shown in Table 3-3. For further information see Gotelli and Graves (1996).

Table 3-3.   Examples of uses of null models in ecology (source: Gotelli and Graves, 1996).

Species richness (Sander's rarefaction curves) and eveness (Hurlbert's PIE)
Species-area curves
Species diversity indices and resource partitioning
Biogeographic patterns
Niche overlap (Hutchinsonian niche)
Co-occurrence (Diamond's assembly rules)
Diel activities
Size ratios of coexisting species and competition
Food web models and concordance

Example:

Grossman (1982) repeatedly depopulated California tidepools of fishes over a period of 29 months and applied Kendall's W statistic of rank concordance (i.e., a measurement of the fidelity or constancy of rank abundance over time – see p. 159). The null hypothesis is that the species rank abundances do not change over the years. The alternative hypothesis is that rank abundances vary over time. Grossman (1982) found that the fish assemblages were stable over time thus supporting the null hypothesis.

Problems with using Kendall's W include that the test is sensitive to both sample size and the underlying species abundance distribution, and the presence of many rare species in samples results in creating concordance (Gotelli and Graves, 1996). Major criticisms of null models include the lack of power, results of using null models are sensitive to the species pool, and what one learns by falsifying a null model is unclear (Erlich and Roughgarden, 1987). Nevertheless, the use of null models for field studies organizes and standardizes a methodology, similar to that of the null hypothesis in statistics, and thus is useful for ecologists.

D.   DIVERSITY AND RELATED INDICES

3.15   Species Richness, Diversity, Evenness, and Dominance

Good discussions of diversity indices are given by Krebs (1999) and the BIODIV software program manual (Baev and Penev, 1993 – see the Appendix). Southwood and Henderson (2000) classify diversity into (1) alpha – diversity of species within a community or habitat, (2) beta – comparison of diversity between different habitats, and (3) gamma – large scale diversity including both alpha and beta diversities. Hughes et al. (2001) discuss four approaches to diversity estimates. They include (1) rarefaction, richness estimators; (2) accumulation curves; (3) parametric estimators (lognormal, Poisson lognormal); and (4) nonparametric estimators.

(1) Species Richness (Species Density): The total number of species (plant or animal) per unit area or unit volume.

Species-Area Curve (for species richness or diversity)

This method indicates how many samples must be taken (i.e., at what point the curve reaches an asymptote or levels off) to obtain the majority of species when examining community assemblage structure (Fig. 3-9). It is preferable to have separate samples, each randomly selected (see table of random numbers in Table 1-1) to construct a species area curve.

Example

How many samples would you need to collect most of the species for estimating species richness?

Collect 1 m^2 samples and add the new species found in each quadrat to give a cumulative series of total species. Species area curves will likely differ with latitude, as in warm temperate latitudes (e.g., S. California – Fig. 3-9a), cold temperate sites (e.g., Prince William Sound, Alaska – Fig. 3-9b), and at a tropical site (Fig. 3-9c). One can also plot the number of species (Y-axis) against the number of individuals (X-axis).

Alternatively, the number of samples needed can be calculated from a program in Krebs (2000a) using a Bootstrap or Jackknife randomization technique. If the samples are of uneven sizes (e.g., grab sample volumes), one can use the rarefaction program in Krebs (2000), based on Sanders (1968). Rarefaction is the process of predicting species richness based on random sub-samples of individuals (see Gotelli and Graves, 1996, for a discussion of this topic).

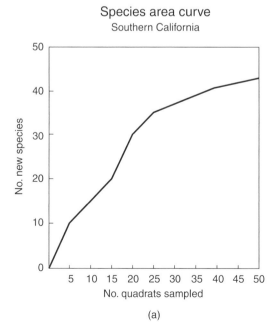

Species area curve
Southern California

Figure 3-9a. Typical rocky intertidal species area curve for southern California (Warm Temperate marine biogeographical region).

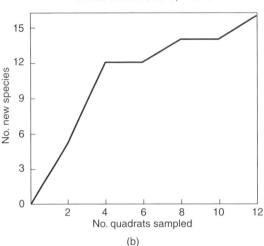

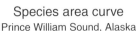

Figure 3-9b. Typical rocky intitial species area curve for Prince William Sound, Alaska (Cold Temperate marine biogeographical region).

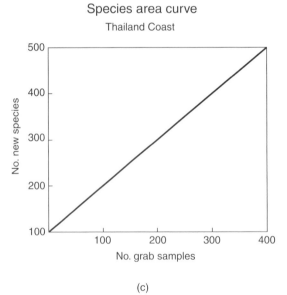

Figure 3-9c. Typical infaunal species area curve for the coast of Thailand (Tropical marine biogeographical region).

(2) Species Diversity: Species diversity differs from species richness in that both the number of species and the numbers of individuals of each species of plant or animal are considered simultaneously. Some biologists use the term "species diversity" or just the term "diversity" to mean species richness. Others use the Brouillon index or the rarefaction technique of Sanders (1968) to express

species diversity (see Clarke and Warwick, 2001). **A high species diversity does not necessarily correlate positively with high environmental quality** (Green, 1979). **Diversity indices are not independent of sample size** (i.e., number of samples taken – see Soetaert and Heip, 2003, and Oksanen, 1997). A plot with species having even abundances is generally more diverse than a plot with the same species that are unevenly abundant (McCune et al., 2002). The following species diversity indices are among those most commonly used by biologists at this time:

(**a**) *Shannon-Wiener index* (on the basis of information theory)

The Shannon-Wiener index (H') was introduced to biology (from communications theory) by MacArthur (1955).

$$H' = -\sum_{i=1}^{s}(p_i)(\log p_i)$$

where

Σ = sum

s = number of species

p_i = proportion of individuals found in the *i*th species and estimated as n_i/N

n_i = number of individuals of species i

N = the total number of individuals of all species

$\log = \log_2$ ($\log_{10}$ and $\log_e$ have also been used)

This index emphasizes rare species. It is used by most aquatic biologists. Higher numbers indicate high species diversity and low numbers low species diversity (see below).

Example

Species No.	1	2	3	4	5
No. individuals	24	2	36	7	19 = 88 total

$$p_i1 = \frac{24}{88} \quad p_i2 = \frac{2}{88} \quad p_i3 = \frac{36}{88} \quad p_i4 = \frac{7}{88} \quad p_i5 = \frac{19}{88}$$

$p_i1 = 0.27 \quad p_i2 = 0.02 \quad p_i3 = 0.41 \quad p_i4 = 0.08 \quad p_i5 = 0.22$

$\log_2 : p_i1 = -1.88 \quad p_i2 = -5.61 \quad p_i3 = -1.28 \quad p_i4 = 3.62 \quad p_i5 = -2.17$

$(0.027)(-1.88) = -0.51$

$(0.02)(-5.61) = -0.11$

$(0.41)(-1.28) = -0.53$

$(0.08)(-3.62) = -0.29$

$(0.22)(-2.17) = \dfrac{-0.48}{-1.92} \qquad H' = 1.92$

The Shannon index varies from 0 to 5 using $\log_2$ and from 0 to 10 using $\log_{10}$.

(b) *Simpson index (Simpson-Yule index)*

$$S = 1 - D$$

$$S = 1 - \sum \frac{n_i(n_i - 1)}{N(N - 1)}$$

where

S = Simpson's Index

D = dominance

n = number of individuals of each species

N = total number of individuals of all the species

Example

Using the data for the Shannon index above -

$$\frac{24(23)}{88(87)} = 0.072$$

$$\frac{2(1)}{88(87)} = 0.00026$$

$$\frac{36(35)}{88(87)} = 0.165$$

$$\frac{7(6)}{88(87)} = 0.006$$

$$\frac{19(18)}{88(87)} = 0.045$$

$$\sum = 0.29 \quad 1 - D = 0.71 \; (\text{Simpson index})$$

The Simpson Index varies from 0 (no diversity) to nearly 1 (maximum diversity). This index emphasizes dominant species. It is used more frequently by terrestrial biologists. The Simpson Index is relatively stable with sample size (McCune et al., 2002). Both the Shannon and Simpson indices are found in the Ecostat computer package.

Be aware that $\log_{10}$, $\log_2$ ($3.3 \times \log_{10}$) or $\log_e$ [$= \log_n$] ($2.3 \times \log_{10}$) have been used in the literature. For $\log_2$, values often range from 0-5 (Margalef, 1972; Frontier, 1985). Comparisons between diversity indices are valid only if there are approximately equal numbers of organisms in the samples from which the indices are calculated (Fowler, et al., 1998). Since the H' value has little ecological meaning, its importance as a tool in biology has greatly diminished (see Hurlbert, 1971). Peet (1974) concluded that ecologists working with diversity indices may have lost a fair amount of time calculating relatively meaningless numbers. The local extinction of species by pollution can be masked by little or no change in the value of H'. H' states nothing about species composition. H' can remain constant while the entire species pool turns over. The Shannon diversity indices of fish communities between mined and control sites in the Maldive Islands were similar yet MDS ordination of the data demonstrated a clear distinction between the two sets of sites (Clarke, 1993). Southwood and Henderson (2000) conclude that the Shannon index should be regarded as a distraction in ecological analysis. Cousins (1991) states that diversity indices should include body size and whether the animal is a predator or parasite. No indices are yet available that include these parameters, so far as is known. **For unequal sample sizes, such as grab samples from soft bottoms, use the Sanders rarefaction technique.** Krebs (2000a) presents a program for this. One problem with rarefaction is that if species are found in clusters (frequently the case), rarefaction will consistently overestimate the expected number of species for small sample sizes (Clarke and Warwick, 2001).

Hughes et al. (2001) recommend nonparametric estimators for microbial studies. The most promising are based on mark-release-capture (MRR) statistics and include the MRR-like ratio techniques of Chao1 (total species richness) and ACE (abundance-based coverage estimators). Both Chao1 and ACE underestimate true species richness at low sample sizes. The Chao estimator appears to be one of the most reliable and cost-effective techniques available (Southwood and Henderson, 2000).

Chao Estimator

Example:

$$S_{Chao1} = S_{obs} + \left(\frac{a^2}{2b}\right)$$

where

S_{Chao1} = total species richness

S_{obs} = number of species in the sample

a = number of species captured once

b = number of species captured twice

Example:

$S_{obs} = 20$

$a = 17$

$b = 3$

$$S_{chao1} = 20 + \frac{(17)^2}{2(3)}$$

$$S_{chao1} = 20 + 48 = 68 \, species$$

See Peet (1974) and Magurran (1988, 2003) for an overview of species diversity indices and Gore and Paranjpe (2001) for a mathematical discussion. Zar (1999) presents a *t*-test for comparing differences between two diversity indices. See also Heip et al. (1988). Clarke and Warwick (2001) recommend replacing species richness and various density indices with an Average Taxonomic Distinctness measure based on species abundances and taxonomic distances through a classification tree. Rosenzweig et al. (2003) provide suggestions on estimating large-scale diversity in biogeographical provinces. Ferson and Burgman (1998) discuss quantitative methods for conservation biology and Anderson et al. (2002) for environmental issues. Feinsinger (2001) discusses the design of field studies for biodiversity conservation. The evolution of species diversity is reviewed by Magurran and May (1999). See also Dean and Connell (1987) and the Addendum.

The Shannon and Simpson species diversity indices are easy to calculate and heavily ingrained in the biologist mentality and the literature. It will take considerable time before some more complex technique of expressing species diversity is widely used.

(3) Evenness: The relationship between the Shannon-Wiener value (H') and the maximum mathematical evenness possible (H_{max}) (see below and Birch, 1981). An example of a high degree of evenness would be a collection of 10 species, each with two individuals.

$$J' = \frac{H'}{H_{max}}$$

and

$$H_{max} = \log_e S$$

where

J' = evenness

H' = diversity index value

S = total number of species

range: 0–1 (greatest evenness)

Use Ecostat to calculate these indices. This index (J') gives a sufficient approximation of evenness (but conveys little ecological information). McCune et al. (2002) contend that the main problem with J' is that it is a ratio between a relatively stable number (H') and an unstable number (S). Krebs (1999) recommends a complex index of evenness developed by Smith and Wilson (1996).

Example: using the 5 species data listed for the Shannon equation above

$$J' = \frac{1.93}{2.32} \quad J' = 0.83 \,(\text{moderately high evenness})$$

(4) *Dominance:* the inverse of evenness. See Birch (1981).

$$D = 1 - J'$$

range: 0–1 (greatest dominance)

Example: Using the results from the evenness data –

$$D = 1 - 0.83 \qquad D = 0.17 \text{ (low dominance)}$$

In Fig. 3-10a, the species richness of community plot A > B > C, the species diversity of plot A > B > C, dominance in plot C > B > A, and evenness in plot A > B > C. However, in Fig. 3-10b the species richness of plot D > E, the species diversity of plot D and E are approximately equal, dominance in plot D = E, and evenness in

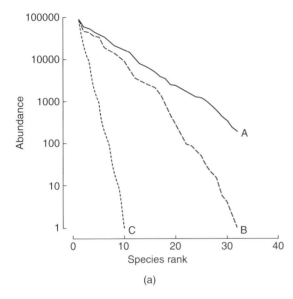

(a)

Figure 3-10a. Species ranked by abundance from three communities and plotted (source: Dick Neal; see text for further information).

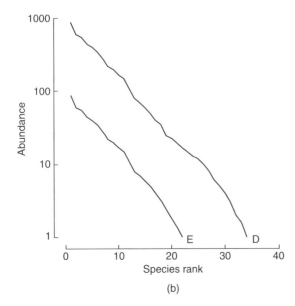

(b)

Figure 3-10b. Species ranked by abundance for two communities and plotted (source: Dick Neal; see text for further information).

Table 3-4.   Community diversity measures that are based on data that were used to plot Figures 3-10ab (source: Dick Neal).

Community	A	B	C	D	E
Diversity measures					
Species richness (S)	32	32	10	33	23
Margalef (D_{MG})	2.37	2.43	0.77	3.77	3.57
Simpson's ($1/D$)	10.88	7.63	1.76	10.87	10.81
Simpson's ($1-D$)	0.91	0.87	0.43	0.91	0.91
Shannon (H')	2.68	2.33	0.85	2.67	2.61
Dominance					
Berger-Parker (d)	0.18	0.25	0.73	0.18	0.18
Simpson's (D)	0.09	0.13	0.57	0.09	0.09
Evenness					
Shannon (E)	0.77	0.67	0.37	0.76	0.83

plot D = E. Certain of the values may not be intuitively apparent in Fig. 3-10b but compare them with the data on which the plots are based (Table 3-4).

Lambshead et al. (1983) and Frontier (1985) pioneered the use of the **K-dominance curve or cumulative dominance curve**, which gives much more information on community structure than any of the species diversity indices (Fig. 3-11). See also Clarke (1990) and Warwick and Clarke (1994). A biomass comparison curve can be added to the cumulative abundance curve to produce the Warwick ABC (Abundance/Biomass Comparison) curves (Fig. 3-12a). Difference curves can be constructed (i.e., the abundance is subtracted from the biomass for each species rank) (Clarke and Warwick, 2001). The curves can be compared using a 1-way ANOVA. Because of difficulty in distinguishing curves in highly dominated communities and the fact that the visual information presented is over-dependent on the single most dominant species, partial dominance curves can be constructed to overcome these difficulties, as follows:

$$p_1 = \frac{100 \, a_1}{\sum_{j=1}^{s} a_j}, \quad p_2 = \frac{100 \, a_2}{\sum_{j=2}^{s} a_j}, \quad \text{and so forth}$$

where

p_1 = partial dominance of species 1

a_1 = absolute (or percentage) abundance of species 1

a_j = abundance of the jth species

Σ = sum

s = total species

j = species 1, and so forth

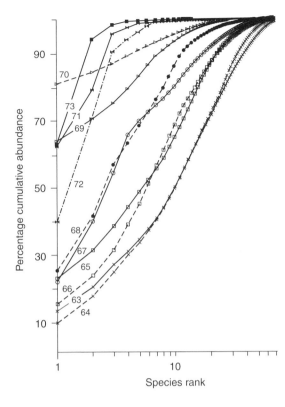

Figure 3-11. K-dominance curves for several sites. Curves that peak rapidly (station 70) generally show dominance and fewer species. Those that peak slowly (station 64) generally have greater numbers of species and more evenness (source: Lambshead et al., 1983).

The first ranked (most abundant or greatest biomass) species is plotted independent of the remaining species. See Fig. 3-12b for an example of a plot.

Botanists use the powerful importance value (powerful because it is based on three types of data) for quantifying dominance, as follows:

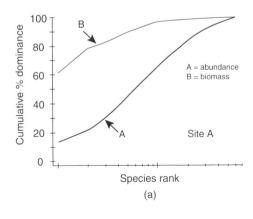

Figure 3-12a. Cumulative dominance curve showing abundance (A) and biomass (B) of benthic marine macrofauna at Site A in the Frierfjord (source: Clarke and Warwick, 2001).

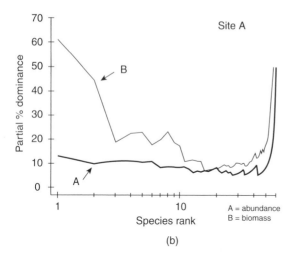

Figure 3-12b. Partial dominance curves (data from Figure 3-12a) showing abundance (A) and biomass (B) of macrofauna (source: Clarke and Warwick, 2001).

Importance Value =

1. Relative Density:

$$\frac{\text{No. individuals of a species}}{\text{Total No. individuals of all the species}} \times 100$$

plus

2. Relative Cover:

$$\frac{\% \text{ cover of a species}}{\text{Total } \% \text{ cover for all species}} \times 100$$

plus

3. Relative Frequency:

$$\frac{\text{Number of points intercepted by a species}}{\text{Total number of points intercepted by all species}} \times 100$$

Example

(modified from Barbour et al., 1999)

Tree Species	Relative Density	Relative Cover	Relative Frequency	Total	Rank Importance
1	45	2	39	86	2
2	20	14	23	57	3
3	5	6	8	19	4
4	30	78	30	138	1

Importance values can be identical for two species that have quite different values for density, cover, and frequency; thus, it may be unclear as to what the importance number represents. This is similar to problems with Shannon-Wiener diversity values.

Another way of conceptualizing the importance of a species in a community is that of general functional importance (Hurlbert, 1997). This is defined as the sum, over all species, of the changes in their individual productivities, which would occur on removal of the particular species from the biocenosis (= sum of all living organisms found in a given area or habitat). This is expressed as:

$$I_i = \sum_{J=1}^{s} \left| P_{jt=1} - P_{j,t} = 0 \right|$$

where

I_i = general functional importance of a species

P_j = productivity of the (jth) species

$t = 0$ = before removal of a species

$t = 1$ = after removal of a species

s = species

Obviously, this treatment is of theoretical interest as measurements of all population changes in a community could not be achieved.

3.16   Keystone Species

Keystone species are typically assumed to be animals that have a large impact on community structure, such as the seastar *Acanthaster planci* in the Indo-Pacific, the seastar *Pisaster ochraceus* on the west coast of North America, and the sea urchin *Diadema antillarum* in the Caribbean. Bluegreens (Cyanobacteria) in the Negev desert are considered to be keystone species by some ecologists. They comprise crusts that stabilize sand dunes. Hurlbert (1997) concludes that the keystone species notion has never been defined operationally and that no keystone species have been demonstrated. Keystone species such as *Pisaster ochraceus* produce marked changes in some populations of the mussel *Mytilus californianus* and deserve special heuristic status, if nothing else. See Power et al. (1996) for a review of the keystone species concept.

3.17   Homogeneity-Heterogeneity Indices

Homogeneity-heterogeneity indices have been used to quantify differences in communities or species groups over space and time. A community is homogeneous if it changes little over space and time; it is heterogeneous if it changes much over space and time, either relative to other communities or to specific criteria established by an ecologist.

(1) Margalef's Heterogeneity Index

This is a measure of the heterogeneity between two points (A and B), separated by the distance L (used in phytoplankton analysis - see Margalef, 1958).

Margalef uses the Gleason index of species diversity, as follows:

$$d = \frac{S-1}{Log_e N}$$

where

  S = total number of species

  N = total number of individuals

  d = species diversity

The value of d is then inserted in the following equation:

$$H = d_{AB} - \frac{\left[(d_A + d_B)/2\right]}{L}$$

where

  d_A = diversity index (d) of sample A

  d_B = diversity index of sample B

  d_{AB} = diversity index of samples A and B together

  L = number of samples

Example:

Phytoplankton station A: $d = 1.89$ (from the Gleason diversity equation above)

Phytoplankton station B: $d = 2.34$

Phytoplankton station A and B combined: $d = 2.51$

$$H = 2.51 - \left(\frac{1.89 + 2.34}{2}\right) \quad H = 0.40 \left(\text{low heterogeneity of species diversity}\right)$$

H ranges from 0 to ~5 (maximum heterogeneity)

To account for distance apart, express the heterogeneity value as a ratio: 0.40:2.3 km (from Margalef, 1958), The higher the heterogeneity over shorter distances, the higher the ratio. For example, $0.40/2.3 = 0.174$. If H is very high and the distance 1 km apart for the sampling stations, then $5.0/1.0 = 5.0$.

(2) Peterson's Homogeneity Index (modified)

This index is a measure of homogeneity of diversity within a community (a Monte Carlo simulation). In the brief example given below, the proportions for the three species are selected from a table of random numbers.

Basic Similarity Index:

$$HI = 1.0 - 0.5 \sum_{i=1}^{s} |a_i - b_i|$$

where

a_i = proportion of individuals in sample A that belong to species i.

b_i = proportion of individuals in sample B that belong to species i.

Example:

	Samples		
Species	A	B	C
1	0.55	0.30	0.15
2	0.35	0.50	0.25
3	0.10	0.20	0.60
Totals	1.00	1.00	1.00

$$HI : A - B = 1 - 0.5[0.25 + 0.15 + 0.10]$$
$$= 0.75$$

$$HI : A - C = 1 - 0.5[0.40 + 0.10 + 0.50]$$
$$= 0.50$$

$$HI : B - C = 1 - 0.5[0.15 + 0.25 + 0.40]$$
$$= 0.80$$

and so on in a pair by pair comparison when there are multiple samples

The HI value is influenced by:

(a) sample size

(b) differences in diversity between samples

(c) sampling error

The "Adjusted Similarity index" attempts to give a homogeneity value with these biases removed.

Steps

(a) Calculate crude HI for all the pairs and obtain the mean HI for the community. For the data above: Mean HI = 2.05/3 = 0.68.

(b) Make a simulated community census by pooling data and taking random samples (select species 1–3 data from a table of random numbers). The samples are of size n (same number of individuals as in each original sample) and K samples are taken (same number taken in the original census of the community). Recalculate HI for the simulated community.

For the data above: $1 = 0.75$

$$2 = 0.50$$

$$3 = 0.80$$

(c) Repeat the simulated community census a total of five times. For the data above, using a table of random numbers, you may select the following:

0.80	0.50	0.80	0.75	0.80
0.75	0.80	0.75	0.80	0.75
0.80	0.80	0.50	0.50	0.80
0.78	0.70	0.68	0.68	0.78

Median $= 0.70$

(d) Ratio $= \dfrac{\text{Actual Mean Similarity of Community}}{\text{Median of five simulation mean similarity indices.}}$

$$\frac{0.68(\text{mean})}{0.70(\text{median})} = 0.97$$

Index of Homogeneity $= 1 -$ Ratio value

Index of Homogeneity ranges from 0 (low) to 1 (high)

Index of Homogeneity for data above $= 1 - 0.97$

Index of Homogeneity for data above $= 0.03$ (very low homogeneity)

3.18   Niche Breadth

Niche breadth shows the variety of resources and how much of each resource is utilized by an animal. An example of this is the feeding niche shown in Fig. 3-13. **Measurements of niche breadth in animals should consider resource type use as well as resource availability.** See the BIODIV software program manual (Baev and Penev, 1993) and Krebs (2000a). Krebs (1999) recommends the Smith index as one of the best, as follows:

$$FT = \sum_{i-1}^{R} \left(\sqrt{p_j a_j} \right)$$

where

   $FT =$ Smith's measure of niche breadth

   $p_j =$ proportion of individuals found in or using the resource j

   $a_j =$ proportion of resource j to the total resources at the site

   $R =$ number of resources

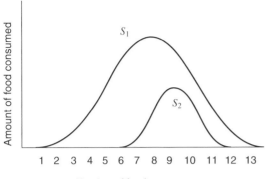

Number of food resources
S₁ = generalist   S2 = specialist

Figure 3-13.   Feeding niche of two species.

Values range from 0 (high degree of specialization) to 1 (extreme generalist). An approximate 95% confidence interval can be calculated (see Krebs, 1999). Niche breadth is the inverse of resource specialization (i.e., $1 - FT$). Krebs (2000a) presents a computer program for this. See also BIODIV (Baev and Penev, 1993).

Example: A large fish is feeding on marine organisms.

	Percentage in Gut	Abundance in the Field
Crustaceans	80%	50%
Polychaetes	15%	40%
Small Fish	5%	10%

$$FT = \sqrt{(0.8)(0.5)} + \sqrt{(0.15)(0.4)} + \sqrt{(0.05)(0.1)}$$
$$FT = 0.63 + 0.25 + 0.07$$
$$FT = 0.95 \, (\text{indicating a generalist feeder})$$

3.19   Niche Overlap

Niche overlap is an index that shows to what extent the same resources are utilized by two species, typically animals. An example of niche overlap are the foods eaten by two predators (Fig. 3-14).

Krebs (1999) considers Morisita's measure the best, as follows:

$$C = \frac{2\sum_{i}^{n} P_{ij}P_{ik}}{\sum_{i}^{n} P_{ij}\left[(n_{ij}-1)/(N_{j}-1)\right] + \sum_{i}^{n} P_{ik}\left[(n_{ik}-1)/(N_{k}-1)\right]}$$

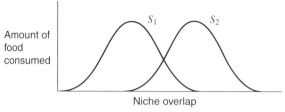

Amount of
food
consumed

Niche overlap
Number of food resources
S1 = species 1 S2 = species 2

Figure 3-14. Feeding niche overlap between two species.

where

C = Morisita's index of niche overlap between species j and k

P_{ij} = Proportion resource i is of the total resources used by species j

P_{ik} = Proportion resource i is of the total resources used by species k

n_{ij} = Number of individuals of species j that use resource category i

n_{ik} = Number of individuals of species k that use resource category i

N_j, N_k = Total number of individuals of each species in the sample

Example

Two species of fishes feeding on invertebrates. Modified from Krebs (1999). For example, Wrasse species 1 consumed seven items of Crustacean species 1. That represents 0.019 of the total items consumed.

Invertebrate species	Wrasse species 1	Wrasse species 2
Crustacean sp. 1	7 (0.019)	0 (0)
Crustacean sp. 2	1 (0.003)	0 (0)
Crustacean sp. 3	286 (0.784)	38 (0.160)
Crustacean sp. 4	71 (0.194)	24 (0.101)
Polychaete sp. 1	0 (0)	30 (0.127)
Polychaete sp. 2	0 (0)	140 (0.591)
Polychaete sp. 3	0 (0)	5 (0.021)
Total:	365 food items	237 food items

$$\sum P_{ij}P_{ik} = 0.145$$

$$\sum P_{ij}\left[(n_{ij}-1)/(N_j-1)\right] = 0.019\frac{7-1}{365-1} + 0.003\frac{1-1}{365-1}$$

$$+ 0.784\frac{(286-1)}{(365-1)} + \cdots = 0.652$$

$$\sum P_{ik}\left[(n_{ik}-1)(N_k-1)\right] = 0.160\frac{38-1}{237-1} + 0.101\frac{24-1}{237-1}$$

$$+ 0.127\frac{30-1}{237-1} + \cdots = 0.40$$

$$C = 0.28 \ (= \text{slight overlap in feeding on the same foods})$$

Niche overlap ranges from 0 (no invertebrates eaten by both species of fish) to 1 (all species of invertebrates consumed by both species of fish).

Krebs (2000a) presents a computer program for this. If data are in biomass units or proportions, use the Simplified Morisita Index (see Krebs, 1999). Niche overlap is strongly influenced by sample size (Southwood and Henderson, 2000).

3.20 Concordance

Concordance can be defined as the degree of agreement between sets of data. It is often expressed as a rank correlation in animal ecology.

(1) Concordance of Diversity

Concordance can be a measure of the degree of agreement among numbers of species of plants or animals sampled in a region, that is, a measure of species rich-ness homogeneity. **Samples are ranked for numbers of species** and nonparametric Kendall's statistic W is used to calculate the **homogeneity of diversity**.

Example: modified from Zar (1999)

$$W = \frac{\sum R_i^2 - \left(\sum R_i\right)^2 \big/ n}{M^2 \left(n^3 - n\right) \big/ n}$$

where

$\sum R_i$ = sum of ranks

M = No. of variables (e.g., site 1, 2 and 3 = three variables)

n = No. of data per variable (5 in this case)

Data = No. of species

Quadrats	Site 1 Data	Site 1 Rank	Site 2 Data	Site 2 Rank	Site 3 Data	Site 3 Rank	Sum of ranks
1	16	3	27	4	18	1	8
2	1	1	8	2	26	4	7
3	24	4	87	5	19	2	11
4	6	2	2	1	24	3	6
5	90	5	14	3	56	5	13

$$W = \frac{8^2 + 7^2 + 11^2 + 6^2 + 13^2 - \left[(8+7+11+6+13)^2 \big/ 5\right]}{3^2 \left(5^3 - 5\right) \big/ 5}$$

$$W = 0.16$$

W ranges from 0 (no association) to 1 (complete agreement among ranks).

To test the significance of W:

χ^2 of $r = M(n-1)\,W$ $\chi^2 = 3(4)\,0.16$

Calculated $\chi^2 = 1.9$

Table $\chi^2 = 9.5$ with 4 d.f. at $P = 0.05$

Therefore, the ranks of species richness are significantly different between the 3 sites with 95% confidence. Another example of using concordance is found above on p. 142.

See Chapter 2, Segal and Castellan (1988), and Zar (1999). Problems with using Kendall's W include that the test is sensitive to both sample size and the underlying species abundance distribution. Moreover, the presence of many rare species in samples results in creating concordance (Gotelli and Graves, 1996).

(2) Concordance of Dominance

Animals or plants are **ranked by abundance** in all samples and agreement among rankings tested by Kendall's statistic W which indicates the **homogeneity of abundance**. See Chapter 2, Segal and Castellan (1988), and Zar (1996).

> **Example:**
> Calculations are the same as above but abundances are ranked instead of species richness.

E. ADVANCED TOPICS

3.21 Game Theory

The theory of games began in economics in 1953 (Smith, 1982). Later game theory became more applicable to biology than economics. However, two economists won the Nobel prize in 2005 for using game theory to explain social, political, and business interactions indicating that this is still an active field of research in economics as well as biology.

Game theory is a method of modeling evolution and expressing this in the form of probabilities. It has been used in biology to try to understand how different behaviors (called strategies) or phenotypes (e.g., growth forms of a plant) have evolved. Game theory makes the following statements about the behavior of the model: (1) there is an evolutionary stable strategy for animal behavior, (2) there is an evolutionary stable phenotype for the growth form of a plant, (3) no mutant strategy can invade the population, and (4) the population genetic composition (i.e., gene frequencies) is restored by selection after a small to moderate disturbance.

The approach assumes that the different strategies or phenotypes occur as contrasting types, have a genetic basis, and that we can assign values of the various costs

Table 3-5. The Hawk-Dove Game. A payoff matrix for all possible pairwise interactions between hawks and doves. See table and text for further information.

	H	D
	$1/2\,(V-C)$	V
H	Each species has a 50% chance of injuring its opponent and obtaining the resource	The hawk obtains the resource
	0	$V/2$
D	The dove retreats – there is no change in fitness	The resource is shared between two doves

where

H = hawk

D = dove

V = gain in fitness (e.g., No. of offspring)

C = cost (e.g., reduction in fitness by an injury)

and benefits for each strategy and all the possible interactions between strategies. The net benefit (i.e., benefit less cost) represents the relative fitness of the competing strategies or phenotypes when they interact. The interaction can be analyzed mathematically to determine which strategy will prevail.

The simplest game theory example is the Hawk-Dove Game (Table 3-5).

Example:

It consists of two species contesting (i.e., competing for) a resource (e.g., food, mates), resulting in an increase in fitness (i.e., number of offspring). The matrix of possible interactions is called a payoff matrix (echoes from economics). The following is an explanation by Dick Neal (written communication). The symbols represent a change in fitness from the encounter. If V > C then the hawk will be a pure evolutionary stable strategy because it has the greatest payoff. It will resist invasion by a mutant dove strategy because invading doves have a payoff of zero, and any pure population of doves can be invaded and taken over by an invading mutant hawk strategy. The net result is that the hawk strategy will prevail, fighting and costs will escalate until the costs (C) begin to exceed the benefits (V). At this time there will no longer be a pure evolutionary strategy for hawks because the payoff for hawks will be negative and so an invading dove with a payoff of zero does better. A pure dove population can still be invaded by hawks because V > V/2. The frequencies of hawks and doves will stabilize at V/C hawks to $1 - (V/C)$ doves. This will result in a mixed evolutionary stable strategy. See Neal (2004) for further details. See Smith (1982), Weibull (1996), Dugatkin (2000), Gore and Paranjpe (2001), and Camerer (2003) for further information.

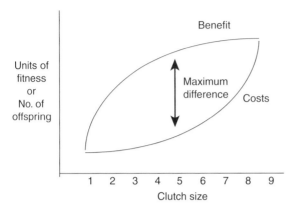

Figure 3-15. Cost-benefit model of the evolution of clutch size in birds (source: modified from Krebs, 2001).

3.22 Optimality or Optimization Models

Optimality models are cost-benefit models typically applied to populations and evolution. Figure 3-15 gives an example of a cost-benefit model for the evolution of clutch size in birds. If the bird lays more eggs (assuming a constant mortality rate), it will produce more descendents in the next generation and this will increase the biological fitness of the species. However, it takes more energy to lay more eggs. It also costs more energy to feed the young and this activity may reduce adult survival (i.e., from predation, accidents, getting lost, etc.). If there is insufficient feeding of the young, a potential exists for a reduction in the quality of the offspring. Finally, feeding more young may increase predation because the adult may be observed more often by predators. The absence of the adult on the nest may allow certain predators to enter the nest and consume the young.

Figure 3-15 shows the precarious balance between clutch size and units of fitness (i.e., number of offspring) in light of the costs and benefits just described. The optimal point (i.e., for clutch size) is reached when there is a maximum difference between benefits and costs. Optimality models are relatively common in ecology as they help to interpret evolutionary or population phenomena in an economic sense. See Cody (1974) and Gore and Paranjpe (2001) for further information.

3.23 Transition Matrices

Transition matrices deal with the analysis of sequences. Many sequences form Markov Chains where the state of the sequence is dependent on the immediately preceding state. A first order Markov chain shows sequence dependence one step ahead and a second order Markov chain two steps ahead. Markov chain analysis is especially important in pattern recognition. Markov chain analysis is available in Release 2 or later of Idrisi (see Appendix). Sequences are analyzed with a transition frequency matrix. This is a matrix of one state being succeeded by another (e.g., substratum on a beach changing from rock to tidepool to sand to algal cover). The total number of times rock is followed by sand, algae, tidepool, and so forth are entered into the matrix. The frequency matrix is converted to decimals or percentages to standardize the data. The decimal matrix is converted to a transitional proportional matrix (i.e., each row number is divided by the total for the row) or a transitional probability matrix. The transition probability matrix is then compared with the expected transition probabilities using a χ^2-test. This can indicate whether the sequences are random or follow some pattern. If there is a pattern, then the final step is to investigate the reasons for the sequence pattern. Davis (1986) explains transition frequency analysis in detail.

Bakus and Nishiyama (1999) recorded marine organisms and substrata along transect lines on a coral reef in Cebu, Philippines. The null hypothesis was that the sequence of occurrences of organisms and substrata were random. The results were tested for a Markov chain by the nonparametric statistic χ^2. The data suggested that the sequences were close to random thus supporting the null hypothesis. Among the major causes proposed for this random small scale distribution are high species richness, predation on invertebrate larvae and plant spores by plankton-feeding fishes (e.g., damselfishes such as *Chromis*), and the grazing and browsing activities of fishes on the benthos.

An example of how these types of data are handled is as follows:

Imagine that a satellite has photographed a tropical coastal area. You examine the photograph and record the sequence of community types by drawing two parallel transect lines across the photo. This is the sequence you get where:

A = seagrass, B = sand, C = coral, D = mangroves

CCCAAAAAACCCCCAACCDDCCBBBBBCCCB
BCCCCCAAAAACCAAACCDCCDDCAAAAAAC

1. Construct a transition frequency matrix (e.g., A is followed by A 18 times, etc.).

		A	B	C	D	Row Totals
	A	18	0	5	0	23
	B	0	5	2	0	7
From	C	5	2	18	3	28
	D	0	0	3	2	5
Column Totals		23	7	28	5	63 Grand Total

To (header over A B C D)

2. Convert the numbers to decimals.

		To				Row Totals
		A	B	C	D	
From	A	0.78	0	0.22	0	1.00
	B	0	0.71	0.29	0	1.00
	C	0.18	0.07	0.64	0.11	1.00
	D	0	0	0.60	0.40	1.00

3. Determine the relative proportions of the four community types by dividing the row totals by the total number of transitions.

$$
\begin{array}{c}
A \\
B \\
C \\
D
\end{array}
\begin{bmatrix}
0.37 \\
0.11 \\
0.44 \\
0.08
\end{bmatrix}
$$

4. Calculate an expected transition probability matrix.

		To			
		A	B	C	D
From	A	0.37	0.11	0.44	0.08
	B	0.37	0.11	0.44	0.08
	C	0.37	0.11	0.44	0.08
	D	0.37	0.11	0.44	0.08

5. Multiply the expected transition probabilites by the row totals (see No. 1) to obtain the expected frequencies.

Expected Transition Probabilities				Totals		Expected Frequencies			
0.37	0.11	0.44	0.08	× 23 =		8.5	2.5	10.1	1.8
0.37	0.11	0.44	0.08	× 7 =		2.6	0.8	3.1	0.6
0.37	0.11	0.44	0.08	× 28 =		10.4	3.1	12.3	2.2
0.37	0.11	0.44	0.08	× 5 =		1.9	0.6	2.2	0.4

6. Compare the observed frequencies with the expected frequencies using the χ^2 equation (or the G-test), as follows:

$$\chi^2 = \sum \frac{(O-E)^2}{E}$$

$$\chi^2 = \frac{(18-8.5)^2}{8.5} + \frac{(5-10.4)^2}{10.4} + \cdots$$

$$\chi^2 = 20.99 \ \text{(calculated)}$$

$$\chi^2 = 16.92 \ \text{(from table, } P = 0.05, 9 \text{ d.f.)}$$

Conclusion: Because the calculated chi-square is greater than the table chi-square, the sequence of community types is not random. Certain of the community types listed (i.e., seagrass, sand, coral, or mangroves) are repeatedly followed by certain other of these community types with 95% confidence. See p. 84 in Chapter 2 for a discussion of chi-square.

Transition frequency analysis does not seem to work well on the species level. Larger scale studies (e.g., probability of transitions between community vegetation types, associations or assemblages) are better suited for this kind of analysis. Isagi and Nakagashi (1990) used transitional frequency analysis for predicting fire frequency and plant community successional change. Licuanan and Bakus (1994) and Bakus and Nishiyama (1999) studied transitional frequencies on coral reefs. Transitional frequency analysis could be used to predict community changes caused by typhoons, Crown-of-Thorns (*Acanthaster*), and pollution (see Bradbury and Young, 1983).

Nishiyama et al. (2004) present a method of analyzing contact encounters on the species level, based on interactions between three species of tropical Philippine toxic sponges and five hard coral species. Although there are other association methods to determine the presence and strength of interactions between species, these methods were often more complicated and time consuming than the method presented here. The utility of this method is still being tested.

Gregory Nishiyama Encounter Index:

$$\text{Encounter Index} = \frac{\text{Observed Number of Encounters}}{\text{Expected Number of Encounters}}$$

The expected number of contact encounters (between plants and/or animals) if no association existed is obtained by determining the probability that each individual would encounter another if no association existed. The number of random contact encounters is contingent upon the sizes and densities of both organisms.

$$\text{Expected Number of Encounters} = 2(\pi)(r_A + r_B) \times (D_B) \times \text{Dens}_A \times \text{Dens}_B$$

where
species B is the smaller of the two species

π = pi (or 3.14159)

r_A = radius of species A

r_B = radius of species B

D_B = diameter of species B

$Dens_A$ = density of species A

$Dens_B$ = density of species B

Encounter Index of $0 - 0.49$ = negative associations

Encounter Index of $0.5 - 2.49$ = random associations

Encounter Index of ≥ 2.5 = positive associations

Example

The following is an example of the use of the Encounter Index to determine the type of interaction between two organisms. With this index the user is not required to know the interactions of every organism within an area, but can determine the association with only the size and density data of two species. The following example uses interaction data between foraminiferans and algae which settled on settlement plates set out on a coral reef off Mactan Island, Philippines. The plates were exposed to allelochemicals found and exuded from the sponge *Plakortis lita*. When using the association equation, organism B is always the organism with the smaller diameter. This particular Index deals with contact interactions between two organisms. There is another equation that deals with not only contact data but also interactions between organisms at a distance. However, this equation is more complicated and not presented here.

	Organism B (Smaller diameter) Algae	Organism A (Larger diameter) Foraminiferan
Density	$1062.5/m^2$	$562.5/m^2$
Diameter	0.00121 m	0.00245 m
Radius	0.000605 m	0.001225 m

Expected Number of Encounters

$$= 2(3.1415)(0.001225\,m + .000605\,m)(0.00121\,m) \times (562.5/m^2)(1062.5/m^2)$$

$$= 8.31 \text{ encounters per } m^2$$

Observed number of Encounters = 22.95 encounters per m^2

$$\text{Encounter Index} = \frac{\text{Observed number of encounters}}{\text{Expected number of encounters}}$$

$$\text{Encounter Index} = \frac{22.95}{8.31}$$

$$= 2.76$$

An Index value of 2.76 indicates a positive association between both organisms (Nishiyama et al., 2004).

Correction factor: space in the study area that is unavailable for the settlement of the species.

$$\frac{1}{1-q}$$

where

q = percentage of unavailable space in the study area (e.g., sand)

The value of q can be obtained by line transect data, and so forth.

Example

If 33% of a study site is unavailable to the organisms due to the presence of sand, then the correction factor is:

$$\frac{1}{1-0.33} = \frac{1}{0.66} = 1.52$$

The equation becomes:

Expected Number of Encounters $= 2(\pi)(r_A + r_B) \times (D_B) \times \text{Dens}_A \times \text{Dens}_B (1/1-q)$

If q consists of discrete organisms or substrates (e.g., boulders) where other organisms cannot settle, the equation $1/1-q$ can be replaced by:

$$\frac{1}{1 - \sum(\text{Dens}_i \times A_i)}$$

where

Dens_i = density of different species, rocks, and so forth in the study site which are unavailable to the organism

A_i = the areas of each different species, rocks, and so forth in the study site which are unavailable to the organism.

The equation then becomes:

$$2(\pi)(r_A + r_B) \times (D_B) \times \text{Dens}_A \times \text{Dens}_B \frac{1}{1 - \sum(\text{Dens}_i \times A_i)}$$

3.24   Fractals

Fractals are geometrical objects with irregular (non-Euclidian) boundaries (e.g., coastlines, clouds, alveoli, foam, ferns, snowflakes). Fractals represent the discipline of irregular geometry. This has given rise to interesting and colorful applications in the art world (Fig. 3-16). Fractals began with von Koch (1904) and were later fully developed by Mandlebrot (1982), now considered to be the father of fractal geometry. Many fractals are repeated patterns of individual components that are identical except for *scale*. This characteristic is called **self-similarity** (Fig. 3-17 and see Costanza et al., 1993.).

The complexity of a boundary can be described by an index called D, as follows:

$$D = \frac{2 \ln P}{\ln A}$$

where

   D = fractal dimension

   ln = log normal

   P = perimeter

   A = area

$D = 1$ for a line, $D = 1.26$ for a Koch snowflake, $D = 2$ for a random boundary (the most complex type of boundary), and $D = 3$ for a circle. Landscape ecologists use the index D in the range of $D = 1$ to 2 in order to describe physical complexity in the environment. A graph of temperature up a mountain slope may have a fractal dimension of approximately 1.1 whereas a graph of soil pH in an agricultural field may be close to 1.9, indicating considerably greater irregularity or heterogeneity in soils (Palmer, 1988).

Figure 3-16.   A fractal image.

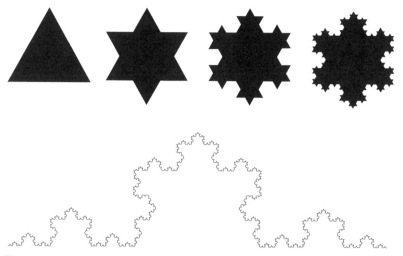

Figure 3-17. An example of self similarity and scale. A triangle is replicated to arrive at a Koch snowflake (source: Gleick, 1987; and Benoit Mandelbrot).

Fractal geometry has been applied to biology in several ways: (1) To study the hierarchical scaling in nature. For example, with coral reefs the complexity of corals increases with scale, from branch to colony, coral patch, group of coral patches, buttresses, and the entire reef or atoll. See Bradbury et al. (1984) and Brown and West (2000). (2) An index of plant succession. The longer succession proceeds, the more complex is the structure of the plant community, thus D increases over time. (3) To measure surface areas to save considerable time over conventional (i.e., Euclidian) methods (Fig. 3-18). (4) To measure how far apart organisms should be sampled. Fractal theory may help to integrate systems across levels of organization at many scales.

Two specific examples of the use of fractals in ecology are presented below. One method is to express decreasing species richness of plants or animals with depth in fractal dimensions (D), the other is to ascertain how far apart sampling sites should be spaced. Spatial variation in geology has traditionally been characterized by using the variogram in place of the covariance function (Thompson, 2002). The variogram is defined as the variance of the difference of y-values at separate sites.

Examples

$$D = (4 - m)/2$$

where

D = fractal dimension

m = slope of the double log plot of a semivariogram

Figure 3-18.   Fractal image of a fern. Produced by a model consisting of four simple transformations each having only six parameters. By contrast, a Euclidian description of this complex shape might be a polynomial involving thousands of fitted parameters (source: Sugihara and May, 1990).

Semivariance: measurement of the variance among scales or the sum of the squared differences between all possible pairs of points separated by a chosen distance.

$$\text{semivariance (yh)} = \frac{1}{2N(h)} \sum \left(z(i) - z(i+h)^2 \right)$$

where

$N(h)$ = No. of pairs of points separated by distance h

$z(i)$ = dependent variable (e.g., species richness) at point i

$z(i+h)$ = species richness at other points

As an example, suppose we plot species richness (No. species/m^2) versus depth:

where

SV = semivariance

$$SV(1) = \frac{\left[(100-80)^2 + (80-60)^2 + (60-40)^2 + (40-20)^2\right]}{2 \times 4} = 200$$

$$SV(2) = \frac{\left[(100-60)^2 + (80-40)^2 + (60-20)^2\right]}{2 \times 3} = 800$$

$$SV(3) = \frac{\left[(100-40)^2 + (80-20)^2\right]}{2 \times 2} = 1800$$

$$SV(4) = \frac{\left[(100-20)^2\right]}{2 \times 1} = 3200$$

The semivariogram would look like Fig. 3-19.

To calculate the slope: m = [log(y(2h)) − log (y(h)/log 2)]

where

2*h* means twice the given log

(*y*(*h*) = semivariance)

m = [log 800] − [log 200]/log 2 = 2.0 therefore, *D* = 1.00
(log of SV2 versus log of SV1)

m = [log 3200] − [log 800]/log 2 = 1.34 therefore, *D* = 1.33
(log of SV4 versus log of SV2), and so forth.

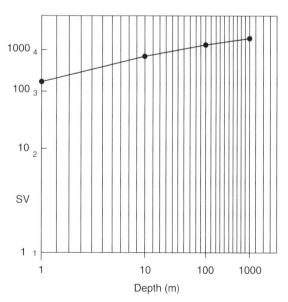

Figure 3-19. Semivariogram of species richness with depth. See text for further information.

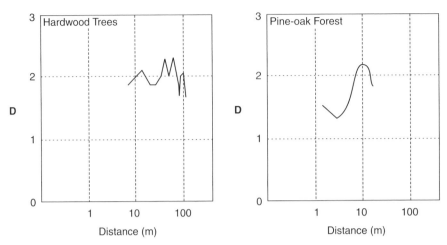

Figure 3-20. Fractogram plotting fractal dimension (D) vs. distance (m). (source: Palmer, 1988).

To calculate semivariance and fractal dimensions, develop an ASCII data file then use it in the program "semivar.exe" in the AUTOMATA folder on the accompanying CD-ROM. These programs were developed by Wilfredo Licuanan.

In Summary:

(1) The fractal dimension (D) varies as a function of scale.

(2) The fractal dimension (D) varies between 1 and 2 for terrestrial vegetation.

(3) The fractal dimension (D) varies with the type of variable used (e.g., temperature, pH, species richness, percentage cover, etc.).

(4) Because statistics assumes that replicate samples are independent, samples can be spaced by distances at which the fractal dimension is determined from a fractogram.

Figure 3-20 shows a fractogram (plotting D against distance) with many samples. It shows that quadrats should be sampled 10–100 m apart (where $D = 2$) for hardwood forests and about 10 m apart (where $D = 2$) for pine-oak forests. Stuart Hurlbert (pers. comm.) states that it is not clear that fractal theory is appropriately used to determine spacing of samples and that spacing is irrelevant to statistical independence.

Further information on fractals is found in Palmer (1988), Sugihara and May (1990), Milne (1991), Hastings and Sugihara (1993), and Sornette (2004). A discussion of variograms is found in Legendre and Legendre (1998) and Dale (1999). Also see Ver Hoef and Cressie (2001).

3.25 Deterministic Chaos

Chaos has been described as random, noise, stochastic, and unpredictable (Delisi et al., 1983; Kolata, 1986). A different type of chaos was first described by Hadamard (1898). This was later called deterministic chaos. Deterministic chaos developed in the physical

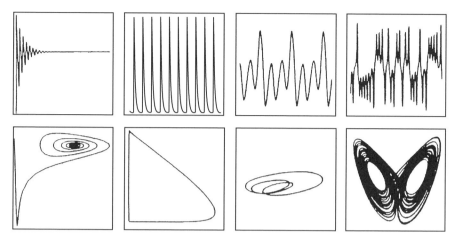

Figure 3-21. Two-dimensional (2D) plots above and phase space (third dimension) plots below. See text for explanation (source: Gleich, 1987 and Irving Epstein).

sciences from classical mechanics. Later it was described in the chemical sciences and then in biology by May (1974). Deterministic chaos can be defined as: A trajectory is chaotic if it is bounded in magnitude, is neither periodic nor approaches a periodic state, and is sensitive to initial conditions (Cushing et al., 2003). Patterns of deterministic chaos are not random but generated by a simple deterministic equation (Krohne, 1998).

Let us first examine portraits of time series (Fig. 3-21 and Chapter 6). Shown are four patterns of time series culminating in chaos. Below them are the same trajectories in phase space, as if you were looking at the upper 2-D series from the left side, showing a third dimension. Note that the chaotic pattern above (extreme right) looks like a butterfly below. Figure 3-22 shows three trajectories in space that converge to a *strange attractor (X)*, a single point or steady state (Gleich, 1987). Chaotic attractors are fractals, according to mathematicians. Chaotic attractors are one of four possible

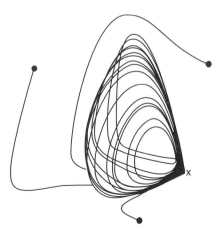

Figure 3-22. Three trajectories converging to a strange attractor (x).

types used to classify system dynamics (Costanza et al., 1993). The other three in-clude: (1) point attractors (indicating stable, nontime varying behavior), (2) periodic attractors (indicating periodic time behavior), and (3) noisy attractors (indicating stochastic time behavior).

Now we discuss different, real examples of deterministic chaos. When a faucet is gradually turned on, there are drops, then linear flow, followed by spattering or chaos. When water is heated there are convection currents, small bubbles, then boiling or chaos. Because the same "chaotic" patterns can be recreated by identical initial conditions, the phenomenon is called deterministic chaos. In a similar manner, animal populations (or plants such as phytoplankton) can show deterministic chaos. May (1974) studied the 17-year cicada on the east coast of the United States, using field data for simulation on a computer. This cicada appears in enormous numbers within a few weeks, followed by mating, the females laying eggs in the ground, then dying. The population disappears lo-cally for 17 years then repeats the cycle. The population bursts are not synchronized over the east coast. The latest population outburst occurred in 2004. If the biotic potential (r) on the x-axis is plotted against population size (N) on the y-axis (Fig. 3-23), the popula-tion growth can be described by the equation (Dick Neal, written communication):

$$N_{r+1} = N_t \left[1 + r \left(1 - \frac{N_t}{K} \right) \right]$$

where

N = number of individuals

r = biotic potential or intrinsic rate of natural population increase

t = time

K = carrying capacity of the environment

If the rate of population growth (r) is too small, the population becomes extinct (because mortality exceeds natality). If the population growth rate is adequate, the population reaches a steady state or equilibrium. **If the rate of population growth is somewhat high, Fig. 3-23 shows a bifurcation (i.e., division) where the popula-tion can alternate between two states or population levels**. For the 17-year cicada, this occurs when r = 2.692. If the rate of population growth increases even more, the bifurcations occur faster followed by chaos, a point at which different population levels are possible but **the specific population level is unpredictable**. The point at which the population curve first bifurcates is predictable, afterward it is unpredict-able, thus the name deterministic chaos.

Deterministic chaos was thought to have been demonstrated in large, long-term data bases such as the Canada Lynx population (data over 200 years) and measles epidemics in New York. To date, only measles epidemics are thought to demonstrate deterministic chaos (Cushing et al., 2003), other than 17-year cicadas. Although other populations are reported to be chaotic (e.g., rodents, insects – see Rickleffs and Miller, 2000), it is very difficult to find deterministic chaos in natural popula-tions. Insufficient time series data, noise in ecological data, and the similarity of

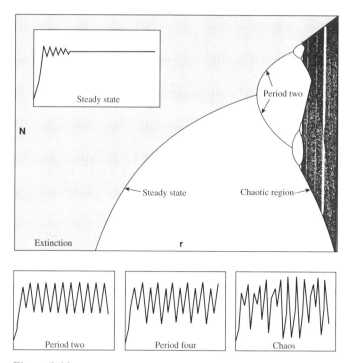

Figure 3-23. Deterministic chaos in the 17-year cicada (source: modified from Gleich, 1987).

deterministic and stochastic events preclude the presence of chaos (Cushing et al., 2003). However, chaos has been convincingly demonstrated in laboratory experiments with the flour beetle *Tribolium castaneum* (Cushing et al., 2003). The importance of deterministic chaos is in realizing that the populations of some species can be predicted initially but become chaotic later and therefore unpredictable. Consequently, monies for research on certain commercial fishes, and so forth may be wasted beyond a certain point in time. Also, because small differences in initial conditions expand exponentially over time, precise long-term predictions of population levels over time are impossible. This is known as the "butterfly effect" in meteorology where minute air currents created by the butterfly expand exponentially and result in a storm! See Gleich (1987) and Roughgarden (1998) for further information. Sugihara and May (1990) suggest how true noise in data (i.e., stochastic events) can be distinguished from deterministic chaos. Also see Schaffer and Kot (1986), Schaffer and Truty (1988), Crutchfield et al. (1986), Williams (1997), May (2001), Turchin (2003), and Sornette (2004).

3.26   Artificial Neural Networks

Artificial neural networks (ANN) evolved from neurobiology (Lawrence, 1992) as an attempt to solve problems by simulating in a simple manner how the brain functions.

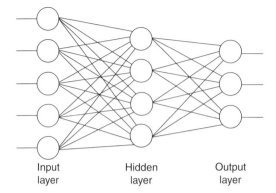

Figure 3-24. A simple artificial neural network (source: Lawrence, 1992).

ANN can handle large, difficult data sets and even intractable data sets attempted with conventional statistics. In many cases ANN gives equal or better results than do standard statistical procedures with complex data (e.g., multiple linear regression), provided that large data sets are used.

A simple neural network consists of an input layer, a hidden layer, and an output layer (Fig. 3-24). There can be more than one hidden layer. The hidden layer is essential for the calculations that occur behind the scene (e.g., matrix analysis hidden from view). The hidden variables are functions of the weighted sum of the input variables (Chester, 1992). The input layer receives the input data (e.g., a retina in the eye or numbers from a data file), the hidden layer processes the information, and the output layer produces an output (i.e., visual image or a number on the computer screen). The output variables are functions of the weighted sum of the hidden variables. Artificial neural network logic values are combined in a process similar to that of fuzzy logic (Chester, 1992).

Figure 3-25 shows how the information is processed. Data (a-f) are input to the input layer of neurons. Weights (in this case ranging from −1 to +3) are assigned randomly to the input layer which controls the strength of the signal. The weights are multiplied by the input values and summed. When the sum strength of the signal (usually squared) reaches a threshold level, the hidden layer neurons activate the

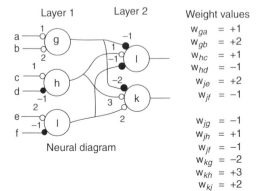

Figure 3-25. Data processing in an artificial neural network. See text for explanation (source: Lawrence, 1992).

output layer neurons, which then produce an output value (e.g., a number). If the output number differs from the correct number (i.e., compared with the original data base), back propagation occurs. Back propagation is a feedback training technique whereby the weights are altered and a recalculation occurs to produce a new output number, which is now closer to the correct number (i.e., value from the original data base). Back propagation is commonly known as feedback in biology and is similar to recursion in computer algorithms. Back propagation is a training process whereby the neural network continues learning by recursion until the output number equals or is close to the correct value (i.e., approximating the average of the values in the original data base).

Artificial neural networks can be used to predict the outcome of horse races, football games, and so forth. They are also used by some stock market companies to predict future stock values and in pattern recognition to recognize letters and words. However, 50–200 sets of data are required for adequate training and setting up the program is rather tricky (see below).

The major steps involved in using artificial networks are as follows: (1) input the data and create a data file, (2) train the neural network, and (3) make a prediction based on new input data. As an example, data were collected on oceanographic parameters at Pt. Arguello, California by Richard Dugdale. Pt. Arguello is a region of complex water masses and upwelling. Table 3-6 indicates the field data collected at Station 7 on 8 April 1983. Note that 19 parameters or variables are listed. My first step took data from six stations at six depths and analyzed them with the artificial neural networks program called Brainmaker. The results showed no pattern. A graduate student (Wilfredo Licuanan) suggested selecting parameters at a single depth (i.e., 10 m). This made good sense as organisms, light intensity and wave length, and so forth can change rapidly with depth, causing confounding effects when using all the depth data. Selected parameters (12) were retested for a depth of 10 m. The results indicated that temperature, oxygen and ammonia (dependent variables) were critical in predicting the chlorophyll concentration (Chl a or phytoplankton density, the independent variable; Fig. 3-26). Temperature was the most important dependent variable.

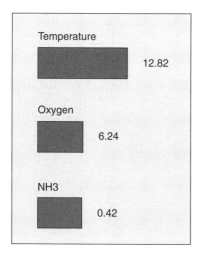

Figure 3-26. Visual and numerical displays of the importance of each variable in estimating the chlorophyll a concentration of phytoplankton abundance. Temperature is the most important and ammonia the least.

Table 3-6. Oceanographic data from Point Arguello, California in 1983 (source: Richard Dugdale).

STATION	DATE	TIME	LATITUDE	LONGITUDE	STATION NAME	DRIFTER NUMBER
7	8 APR 83	06:45:00	34 24.8 N	120 48.1 W		5209

DEPTH	TEMPERATURE	SALINITY	OXYGEN	NITRATE	NITRITE	AMONIA	PHOSPHATE	SILICATE	PARTICLE VOLUME
0	12.105	33.578	6.061	5.888	0.146	0.456	0.892	7.836	1.240
(10)	(12.065)	33.566	(6.022)	6.238	0.140	(0.448)	0.892	7.814	1.483
25	12.102	33.563	5.972	6.786	0.140	0.468	0.892	8.433	1.428
50	10.744	33.635	4.227	17.365	0.140	0.438	1.739	20.372	1.056
75	10.417	33.706	3.889	19.411	0.112	0.481	1.829	22.053	0.829
100	9.906	33.757	3.396	21.053	0.062	0.456	2.007	23.318	0.309
125	9.492	33.874	3.064	23.852	0.056	0.456	2.185	27.563	0.319
150	9.298	33.957	2.690	25.948	0.084	0.462	2.342	30.536	0.242
175	8.947	34.004	2.490	27.245	0.070	0.447	2.431	33.397	0.213
200	8.675	34.032	2.379	28.543	0.056	0.433	2.565	36.045	0.234

DEPTH	CH A	PHAEOPIGMENT	BACT. CELLS PRODUCTION	BACT. CELLS 0.2–0.6 µm	BACT. CELLS 0.6–1.0 µm	BACT. CELLS 1.0–1.4 µm	BACT. CELLS 1.4–2.0 µm	BACT. CELLS 2.0–2.8 µm	TOTAL BACT CONC.
0	4.189	0.253	8.947	—	—	—	—	—	—
(10)	(7.015)	0.000	5.525	—	—	—	—	—	—
25	5.407	0.924	18.020	—	—	—	—	—	—
50	22.506	4.411	2.962	—	—	—	—	—	—
75	61.380	31.620	0.285	—	—	—	—	—	—
100	5.237	9.638	2.192	—	—	—	—	—	—
125	2.728	2.089	—	—	—	—	—	—	—
150	1.754	1.277	—	—	—	—	—	—	—
175	7.989	1.249	—	—	—	—	—	—	—
200	—	—	—	—	—	—	—	—	—

Training by recursion took only one minute (i.e., because of a small data base coupled with a fast computer) and over 10,000 iterations or recalculations were made. New data on temperature, oxygen and ammonia were input and the computer predicted a Chl a value of 1.4033. This gave an error of 6% when compared with the actual value for Chl a that was 1.3155. This error was principally a function of using a very small data base (six sets), a minimum of 50 data sets being required for a proper training and prediction.

Recently, artificial neural networks and advanced image analysis techniques were combined to produce DiCANN, an advanced pattern recognition dedicated system. Some 23 species of marine dinoflagellates were characterized. Training with an early prototype included 100 specimens per species. DiCANN processing is invariant to specimen rotation and allows up to 10% variation in specimen size. The most accurate artificial neural network (Radial Basis Function or RBF) identified the 23 species with a best performance of 83% whereas an expert panel of taxonomists returned a best performance of 86% (Culverhouse et al., 2002).

For examples of using artificial neural networks in modeling, see the special issue on marine spatial modelling (Høisaeter, 2001). See the internet for neural network and neuro-fuzzy techniques.

3.27 Expert Systems

Expert systems began in 1943 as a part of artificial intelligence (AI). It consists of computer programs developed by so-called AI engineers to simulate the knowledge of an expert, based on interviewing one or more experts in the field. Expert systems have been used to prescribe antibiotics, classify chemical compounds, diagnose equipment failures, classify patterns in images, and for taxonomic identifications in biology. For example, the commercial program PROSPECTOR identifies mineral types, DENDRAL identifies organic compounds, and MYCIN classifies infections and provides diagnoses and theraputics. MYCIN includes probabilities from -1 (certainly false) to $+1$ (certainly true).

Most expert systems are rule-based systems. They are based on "If/Then" rules. An example would be: If the tropical nemertean is red and black, then it belongs in the genus X. Expert systems usually consist of 4 to more than 1000 rules of this type. An example of the use of expert systems in biology (i.e., biodiversity) is a program developed by Dustin Huntington for identifying shorebirds (sandpipers) in the United States. Shorebirds, especially sandpipers, are difficult to identify, particularly in winter plumage. The program trains the bird watcher to be aware of key characters in the field (e.g., bill length). It includes uncertainty (i.e., probabilities with scales ranging from 0 to 10 or 0 to 100) and tries to present results with the most probable or likely species and its probability value. The computer program **editxs** (not included in the CD-ROM with this book) is used to construct the expert system (i.e., the series of rules) and the computer program **exsys** is used to run the completed expert system file. Use **exsys** to call up the program *birddemo.txt* when asked. More recent expert system programs, such as Exsys CORVID (see Appendix), allow one to include visual information (e.g., photos, line drawings, etc.), which greatly enhance taxonomic expert systems. Carden Wallace has developed a CD to identify the hard coral *Acropora* and Charles Veron is developing a CD for general hard coral identification (Wilfredo Licuanan, pers. comm.).

Expert systems are reported to be about 75–85% as good as an expert in the discipline. Expert systems are not applicable to situations that evolve or change rapidly (as new rules would be required to deal with them). Consulting with more experts in the development of the expert systems program and the inclusion of considerable visual information with the rules would increase the accuracy of the expert system programs.

3.28 Digitization, Image Processing, Image Measurement, and Image Analysis or Pattern Recognition

The taking, analyzing, and processing of images have become important tools for the biologist as they cope with aerial photographs, satellite images, and GIS information. There are three major methods of pattern recognition: (1) artificial neural networks, (2) syntactic pattern recognition, and (3) statistical or mathematical pattern recognition. Artificial neural networks have been discussed (see above). This technique has become increasingly popular over the past decade. With syntactic pattern recognition (also called descriptive scene analysis) primitives (i.e., simple subpattern units) are identified, a syntax analysis is conducted, and a grammar developed to describe the object.

Example:

A heartbeat is described as an ECG or EKG (Fig. 3-27). The primitives are the upslope (P), large negative (QRS), and the trailing edge (T). The pattern is P → QRS → T, repeated each time the heart beats. For speech, the primitives are equal to the letters of the alphabet, the pattern is equal to sentences, the grammar is equal to rules (i.e., syntax, syntactic), and the semantics are equal to context. Syntactic pattern recognition is the least common of the three types of pattern recognition.

For many years statistical pattern recognition was very popular but artificial neural networks is now gaining rapidly. Image analysis or pattern recognition (statistical) consists of first enhancing the image (see below) then segmenting (i.e., separating) the image into its component parts and finally classifying the parts. Segmentation involves at least two major operations: (1) thresholding and (2) edge detection. Thresholding is simply constructing a histogram of the grey level pixel intensities (ranging from 0 or white to 255 or black) along a line crossing the image then selecting the range of pixels desired by setting a threshold level (Fig. 3-28). This process eliminates the pixels you do not desire, leaving the pixels you want.

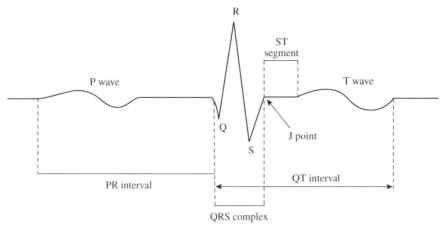

Figure 3-27.   Electrocardiogram (ECG or EKG) recording of a heartbeat (source: http://lanoswww .epfl.ch/personal/schimmin/uni/ecglex/waves2.gif).

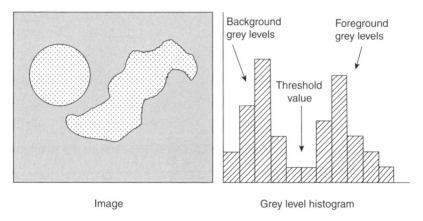

Figure 3-28.   Thresholding an image with a histogram.

There are many algorithms that detect edges. Perhaps the best known edge detector is the Sobel operator (Table 3-7). Once the objects in the scene (e.g., objects in a photo, etc.) are separated or defined by segmentation, the segments are then classified (i.e., analyzed by PCA, etc. [see Chapter 5] then given names or symbols). Once the objects in the scene are classified then the same objects in similar scenes can be identified (e.g., lakes in satellite images).

Many biologists are interested in digitizing objects followed by image enhancement then image measurement. Images have been digitized with a frame grabber, an expensive board placed in a computer that allows one to capture an image with a television camera (RC-170 – see Fig. 3-29) or a CCD. Nowadays one can use a 6+ megapixel resolution SLR camera for the same purpose. Once the image is captured

Table 3-7.   The Sobel operator.

1. Apply the following changes to a 3 × 3 matrix of pixels:

−1	0	+2
−2	0	+2	then
−1	0	+1

2. Apply the following changes to the same 3 × 3 matrix of pixels:

+1	+2	+1
0	0	0
−1	−2	−1

3. Repeat this process for the remaining pixels in the image.

4. Apply an algorithm that computes the strength and direction of each edge.

Explanation: The Sobel operator changes the grey levels in a black and white image as described above. For example, pixel intensities range from 0 (white) to 255 (black). If the pixel intensity in a block of pixels is the following:

137	140	170
136	141	169
134	139	171

Applying the second change above (No. 2), we get:

138	142	171
136	141	169	etc.
133	137	170

This process enhances and detects edges if they should occur.

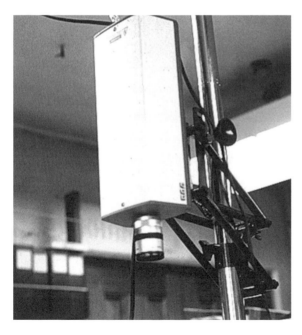

Figure 3-29.   RC-170 television camera used in image processing.

(Fig. 3-30), a small part of the total image on the monitor is selected to save time in processing the image and others like it. This is sometimes called the Area of Interest (AOI) (Fig. 3-31). The image can be enhanced by many algorithms (Fig. 3-32). However, several critical steps must be taken before image enhancement (Table 3-8) or the resulting image may be of poor quality. Finally, objects in the image can be measured manually or automatically (Fig. 3-33).

Figure 3-30. Captured image of fouling tubeworms on a wood panel.

Figure 3-31. Framegrabber image showing white area of interest (AOI).

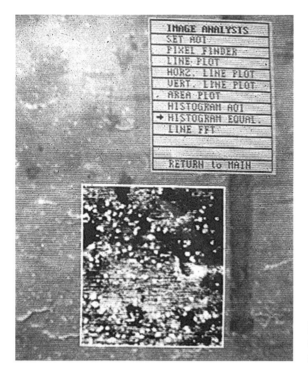

Figure 3-32. Image enhancement by histogram equalization.

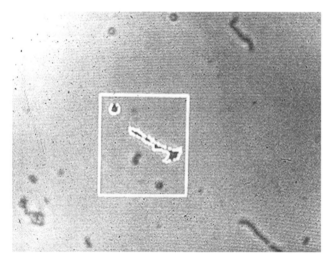

Figure 3-33. Automatic count and measurement of a marine bacterium (surrounded by white) within an area of interest (white box).

Manual measurements consist of setting up the correct scale then clicking on the ends of the object (e.g., photo with eggs or pollen grains, or eggs or pollen grains viewed directly in a microscope with a television camera or CCD attached to it) to obtain the diameter or length. Automatic measurements may be more complex

Table 3-8. Image processing techniques. See table for further information.

1. Enhancing Contrast

   (a) *Avoiding Glare*: Avoiding reflections or glare during the frame-grabbing process is the single most important step for a good image. Use filters.

   (b) *Using Filters*: Use red, blue, green, and Polaroid (one on TV camera, one rotated over specimens) filters alone or in combination.

   (c) *Averaging Images*: Take multiple images (e.g., 6) of the test specimen (e.g., animal) and blank (background) then average them to reduce effects of electronic noise. See subtracting images below.

   (d) *Uniformity*: Use a background that is uniform in color and texture (e.g., do not use a grainy wood).

   (e) *Binary Images*: Keep the contrast maximum between the background and the object. If possible, create a binary image (black and white image) (e.g., cells white and background black). To do this – do a histogram of the scene then set a threshold pixel value equal to the value of the "histogram valley". Assign all gray values less than the threshold value to zero and all gray values greater than the threshold value to one.

   (f) *Subtraction of Images*: Sometimes you can improve the image by subtraction. For example, subtract the original background (e.g., wet wood panel before adding chemicals) from the image taken later (e.g., wet wood panel after immersion in the sea for a month). Subtract an image of your object (e.g., monitor scene) illuminated from one side from the image of your object illuminated from the other side.

   Subtraction can be used to analyze changes over time. For example, subtracting the image of a growing cell at time t from the same cell at time $t + 1$ can give you the growth rate of the cell in percentage cover.

2. Choice of Threshold

   In order to save considerable time in certain image – enhancing operations – use a constant threshold (not an adaptive threshold) in the object-spotting routine.

3. Separation of Objects

   Separation of objects with similar pixel intensities but different shapes (e.g., barnacles and worm tubes) can be accomplished simply by measuring roundness. Sophisticated separation techniques include PCA (Principal Components Analysis – ratio of eigenvalues along two axes) and neural networks.

4. Counts of Animals versus Percentage Cover

   Use counts for separated objects and percentage cover or areal measurements for overlapping objects (e.g., worm tubes). Special techniques exist to convert overlapping objects such as worm tubes (which cannot be distinguished by the computer) into separate countable tubes.

An example of how fouling tubeworms on wood panels were counted automatically, using many of the techniques mentioned above, is found in Wright et al. (1991).

depending on the software program. Edges of the desired object are detected by the computer then the object is counted and the area measured (Fig. 3-33). The computer program SIGMASCAN does both manual and automatic measurements. A simplified process of using SIGMASCAN was developed by my former student Gregory Nishiyama and is presented in the Appendix. Other much more expensive

Table 3-9.   Example of a script. The image is a satellite image of the coast. The objective is to enhance this image then use the script to enhance other similar satellite images overnight.

Initialize the screen then call up the image number (automatic)
Set AOI: 0 0 511 511 – this sets the desired initial screen image size
Sharpen image – this uses the Sobel operator algorithm to sharpen the image
Select/Modify LUT:2.0 – this brings up changes from a look-up table
Erode: 137 1 1 1 1 1 1 1 1 1 – this erodes the image
Dilate: 52 1 1 1 1 1 1 1 1 1 – this dilates the image
Histogram AOI:- this defines an area of interest and draws a histogram of pixel intensities in a cross-section of this area
Line Plot: 341 197 236 268 – this plots a line at specific pixel sites
Scale AOI: 2 2 – this sets a scale for the area of interest
Zoom AOI: 2 2 177 130 378 321 0 0 511 511 – this zooms certain places in the area of interest
Area Plot: 323 153 340 163 – this plots a specific area of the image on the screen
Roberts Edge – this applies a Roberts Edge algorithm to the area plot
Color Edges – this colors the edges of the coastal water masses
Show AOI – this shows the final modified image

computer programs (i.e., dedicated programs) can detect, separate, and count overlapping organisms (e.g., serpulid polychaete worm tubes). Image-Pro, a popular image processing program is distributed by MediaCybernetics (see www.mediacy. com on the internet). A public domain image analysis program called ImageJ (see http://rsb.info.nih.gov/ij/ on the internet) runs on a variety of platforms, including Windows, MacIntosh, and Linux. The program was developed by the National Institutes of Health (NIH). A script (i.e., menu of operations) can be created with more expensive software programs. This allows one to enhance dozens to hundreds of images overnight. The script directs the computer to automatically carry out image enhancement procedures such as sharpening, coloring, scaling, and edge enhancement, followed by even counting and measuring (Table 3-9). For further information on image processing see Jähne (1997), Russ (1998), Smith and Rumohr (2005), and the monthly journal Advanced Imaging.

3.29 Multimedia Development

Multimedia development primarily is not only for entertainment (90%, principally for movies, TV and games) but also for education (10%). PowerPoint presentations for classes and presentations during conferences often involve at least limited multimedia development. Multimedia may take an inordinate amount of time because one is required to learn perhaps a dozen sometimes complex software programs and the process of development is tedious and time-consuming. My first multimedia program took two years to complete and contained a minimal amount of animation!

Today, with information at hand, I can complete a draft of a large multimedia program (300–500 MB) with a minimum of animation in about six months working 20 h/week. Reviews and corrections extend the time to one to two years. Extensive animation would easily double or triple the working time for an individual multimedia developer.

The minimum requirements include a computer with a fast Intel Core 2 Duo or AMD chip, CD-ROM drive, pair of speakers, a large and fast hard disk (100+ GB, 7200+ rpm), videocapture card, sound card (with wave synthesis), microphone, and a multimedia software development computer program. Firewire or USB connections to digital videocams have become standard. It is also useful to possess either an external hard drive (e.g., Maxtor 300 GB) or a CD or DVD burner because videos take considerable hard disk space (e.g., 1 min of video requires about 10 MB of disk storage space).

The program components are as follows:

Title Page – includes the title of the product with the name of the multimedia author and institutional or business affiliation.

Requirements – a list of the minimal and preferred computer component requirements such as type of microprocessor, memory, and so forth (Fig. 3-34).

Introductory Page – this gives directions for using the mouse, interactive buttons, and so forth (Fig. 3-35).

Menu – a main menu is needed to move rapidly to other sections of the CD-ROM. A large, complex program may have a series of submenus. The menus have interactive buttons that take you quickly to other parts of the program (Fig. 3-36).

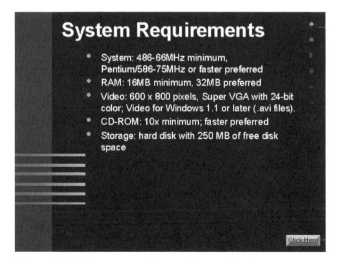

Figure 3-34.   Computer requirements for a multimedia program.

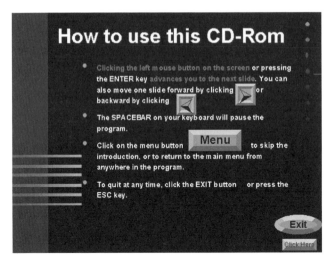

Figure 3-35. Instructions on operation of the multimedia program.

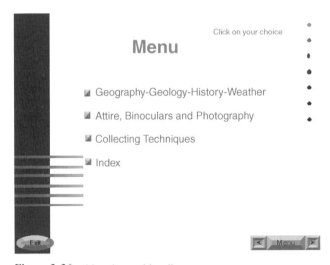

Figure 3-36. Menu in a multimedia program.

Text Page – a template page can be developed for multiple text or photo (etc.) pages. It might include a repeated background scene or pattern; menu, forward, backward and quit buttons; and perhaps music or animations (Fig. 3-37). **It is preferable that you use an embeddable font (e.g., Arial) as it can be used for display (as text) on any computer**.

Scanning – pages of text or figures are scanned with a flatbed scanner (Fig. 3-38) and **preferably saved as JPEG or TIFF (best quality retention) files because they are compatible with both PCs and Macs**. Use a slide scanner

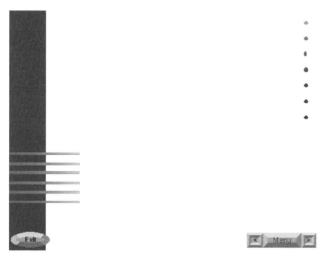

Figure 3-37. A template page for a multimedia program.

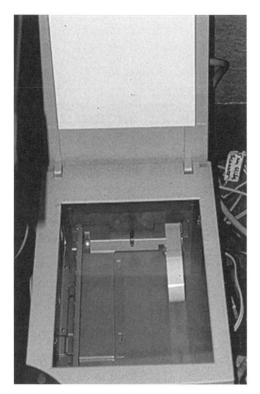

Figure 3-38. Epson flatbed scanner.

to scan 35 mm slides (Fig. 3-39). Slides can also be taken to a local drugstore to be scanned by Kodak and placed on a CD-ROM (Fig. 3-40). The photos are then imported into the multimedia program. Alternatively, use a digital camera and download the photos directly to the hard disk, for later incorporation into the multimedia program.

Figure 3-39. Minolta 35 mm slide scanner.

Figure 3-40. Kodak Photo CD disk.

OCR – optical character recognition software is used to translate scanned text into English (or other) language. Accuracy is now around 95% but the procedure takes considerable time. It is most efficiently used when scanning and translation of a large number of pages is required. These programs often strongly corrupt equations and visually poor text and considerable time is spent in correcting them.

Warping and Morphing – warping takes an image and creates strange shapes (e.g., a twisted image). Morphing takes an image and transforms it into another image (e.g., changing a human head into a cat's head).

Animation – is used to create action or humorous scenes (e.g., bouncing balls, running dogs, etc.). It is also a valuable teaching tool used to demonstrate difficult concepts. Animation is difficult and time-consuming and requires special skills or training, and expensive software programs for the creation of captivating scenes.

Video Capture – In the recent past, a special video capture board was needed to capture analog scenes from a videocamera. Scenes from digital videocameras can now be acquired directly by software programs (e.g., Pinnacle Studio). The video is then edited with the software program and eventually transferred to the production CD-ROM.

3D – Most of the images on a computer today are 2-D images. Soon 3-D images will become the standard (running in real time) but may be costly. Meanwhile, a 2-D object (e.g., beach ball) can be converted with computer software to a 3-D object but this takes time, thus the need for the fastest computer in the marketplace. The process is called rendering (i.e., converting a 2-D image to a 3-D image).

Music and Sounds – Prerecorded music and sounds can be purchased on CD-ROM's. The desired files are imported into your multimedia program. Some software programs allow you to create your own music (e.g., Sonic Foundry).

CD-R, CD-RW, and DVD – CD-R disks record multimedia (etc.) permanently and are inexpensive. CD-RW files can be erased and rewritten. A CD-R or CD-RW burner is required for this. CD-R disks are pretty much universal but not all CD players will accept CD-RW disks. A full disk (700 MB) can now be recorded in less than 5 min with the fastest (most expensive) CD burners. DVD burners are now available but are more expensive (Fig. 3-41). DVD disks can hold several gigabytes of information. Some people prefer to place

Figure 3-41. CD-RW and DVD burner.

the information on the internet where it can be accessed by others, typically in PDF format. However, downloading PDF pages is exceedingly slow without a DSL or fast cable or fiber optics delivery system.

It is useful to maintain a careful record of the resources used in producing a CD-ROM in order to locate the resource rapidly, should the original become corrupted (Table 3-10). Also, a record of resources is needed to pay the producers of music and art their required royalties. Multimedia sources (music, photos, etc.) are often free if produced in small numbers for educational purposes and not sold commercially. Some of the available multimedia software development programs are listed in Table 3-11.

Microsoft PowerPoint is popular with students and for use in conferences (Fig. 3-42). It is relatively simple to learn and helpful for scientific presentations. The steps for using PowerPoint are summarized in Table 3-12. Macromedia Director (Fig. 3-43) and Macromedia Authorware Professional (Fig. 3-44) are used by professionals worldwide. They are difficult to learn and slow to use but can do many things in detail. One major advantage in using Director is its bi-platform nature, that is, it works both on a PC and a Mac (although separate

Table 3-10.   Multimedia record keeping

Assigned Name or Number	Scanned Image & Source	Photo- CD Number & Source	Sound Name & Source	Video Name & Source	Animation Name & Type	Text, Morph, Distort, and so forth	Misc. – Desc. (file = mmtable)

Table 3-11.   List of selected multimedia development programs.

1. Introductory to Intermediate:
     PowerPoint
     Ulead DVD Workshop
2. Advanced:
   A. Animation & General:
     Director
     Quest
   B. Education & Training:
     Authorware Professional
     IconAuthor

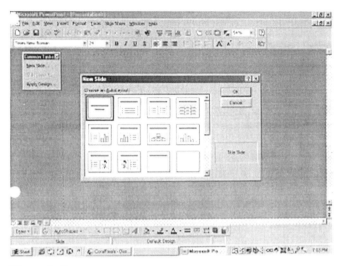

Figure 3-42. Microsoft PowerPoint desktop view.

Table 3-12. Directions for using Microsoft PowerPoint

POWERPOINT

Open Microsoft PowerPoint – then

1. Click on Blank Presentation (or a Template if you desire)
2. Choose on Autolayout design for the slide (e.g., Bulleted List). For plain text only – choose Blank then add text to it.
3. Type in what you want (Insert – Text Box) then select another New Slide (Insert – New Slide).
4. You can add text, clipart (Insert – Picture – Clipart), scanned photos (Insert – Object), video and sound (Insert – Movies and Sounds) and charts (Insert – Chart).
5. You can apply slide transitions (Slide Show – Slide Transition), action buttons to take you to another slide (Slide Show – Action Buttons), and simple animations (Slide Show – Preset Animation).
6. If you want to see all you slides click on View – Slide Sorter. Double-clicking on a sorter slide takes you to that slide.
7. Examine the icons (top and bottom taskbars on the screen) by placing the pointer on each icon, thus activating the icon to change to explanatory text.
8. When you are done with your program you can review the entire presentation (Slide Show – View Show) by clicking on the screen or clicking on the right mouse button (PC's only) then Next, followed by clicking no the screen, moving from one slide to another slide.

Don't forget to save your work!

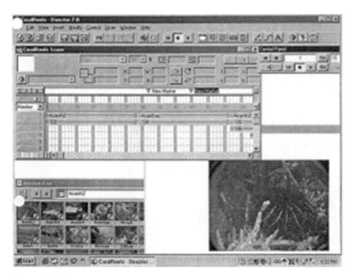

Figure 3-43. Macromedia Director desktop view.

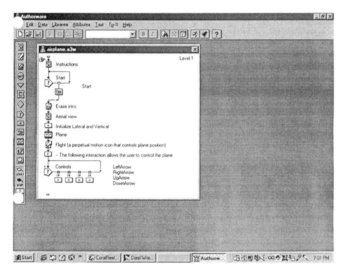

Figure 3-44. Macromedia Authorware Professional desktop view.

software development programs are required, one for creating the executable program [i.e., .exe file for the PC], the second for making it playable on the other
platform [i.e., creating an executable program for the Mac]). Most of these development programs now have the capability of converting the file for use on the
internet. See Simonson (1986).

3.30   Landscape Ecology

Landscape ecology is a relatively new discipline that has evolved rapidly during the past decade. **A landscape can be defined as a spatially heterogeneous area, an environmental mosaic, or a nested hierarchy of scales**. The structure, function and change of landscapes are scale dependent (Turner and Gardner, 1991). Landscape ecology uses numerous mathematical techniques and models. The important technical tools of landscape ecology include (1) remote sensing, aerially and by satellite, (2) a geographical positioning system or GPS that indicates location with between 1 and 30 m accuracy, depending on the cost, and (3) a geographical information system or GIS that ties together all the information. For further information see Turner et al. (2001).

Remote sensing is carried out by aircraft or satellite fitted with special instruments that work like cameras to map land or water using specified light spectra (e.g., to measure sea surface temperature or chlorophyll a). The images are enhanced by image processing and measurements are made (e.g., the sizes of surface water masses characterized by specific temperatures).

The **Geographical Positioning System (GPS)** (Fig. 1-29) is a small, hand-held instrument used in landscape ecology for ground-referencing (e.g., identifying on a detailed map the exact geographical position of the edge of a forest). It relies on contacting 3–12 satellites to determine its position, with the best accuracy within a meter of the true position. The GPS indicates the latitude and longitude, contains maps at different scales (showing highways and cities), velocity and direction of travel in a car or an airplane, the elevation, routes traversed when traveling cross-country, and so forth.

The **Geographical Information System** (GIS) is an information management or computerized mapping system for the capture, storage, management, analysis and display of spatial and descriptive data (Coulson et al., 1991). These systems comprise complex computer programs that are artificial intelligence or AI-based knowledge systems (i.e., rule-based expert systems). They are often used in land management to predict the consequence of a proposed action (i.e., typically a development such as a new harbor). The discipline of GIS evolved from cartography. The basic elements of a GIS are shown in Fig. 3-45.

Digitized maps occur in two formats: (1) DLG (data line graph) of the U.S.G.S. and (2) TIGER of the U.S. Census Bureau. The data are either raster-based (i.e., a grid of pixels) or vector-based (i.e., points, lines, polygons, coordinates). Digital maps exist for many parts of the world, the major U.S. sources being the U.S.G.S., N.O.A.A., and the U.S. Census Bureau (check the internet for these agencies and their products). Maps are digitized by three methods: (1) tracing the map with a cursor using a digitizing table, (2) scanning a map with a flatbed scanner, and (3) using digital video to copy a map (resolution may be too low, producing poor quality maps). The variety of maps that are used are listed in Table 3-13. Overlays (i.e., layers of maps) are commonly used in GIS.

There are many types of GIS. The most widely used computer software program in North American universities is **IDRISI**, developed in the Department of Geography, Clark Labs, Clark University, Worcester, Maine. It is a complex program yet

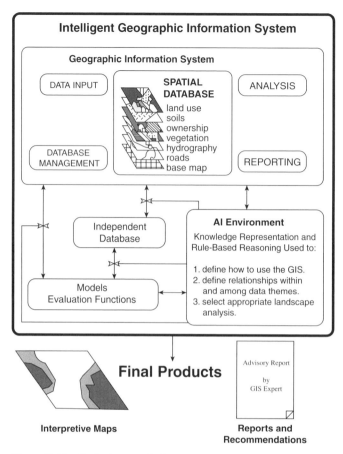

Intelligent Geographic Information System

Geographic Information System

DATA INPUT

SPATIAL DATABASE
land use
soils
ownership
vegetation
hydrography
roads
base map

ANALYSIS

DATABASE MANAGEMENT

REPORTING

Independent Database

AI Environment

Knowledge Representation and Rule-Based Reasoning Used to:

1. define how to use the GIS.
2. define relationships within and among data themes.
3. select appropriate landscape analysis.

Models Evaluation Functions

Final Products

Advisory Report by GIS Expert

Interpretive Maps

Reports and Recommendations

Figure 3-45.   Basic elements of a Geographical Information System (GIS) (source: Coulson et al., 1991).

Table 3-13.   Examples of types of maps used (source: Clarke, 1999).

Reference map
Topographic map
Dot map
Picture symbol map
Graduated symbol map
Network map
Flow map
Chloropleth map
Area qualitative map (land use)
Stepped statistical surface map
Contour (isoline) map
Hypsometric map
Gridded perspective (fishnet) map
Realistic perspective view map
Hill-shaded relief map
Image map

relatively easy to use as it was designed for students. The most popular American commercial program is the **ArcGIS** family of software by ESRI (Environmental Systems Research Institute), located in Redlands, California (check the internet under ESRI for information). Australians at AIMS have used the GIS software program microBRIAN for many years for studies on the Great Barrier Reef.

An example of the use of GIS is that of Leal Mertes of the University of California at Santa Barbara (Mertes et al, 1998). A study was made of the southern California Channel Islands, a group of eight islands. Information was compiled included the biota (kelp, seagrass, shellfish, seabird colonies, and sea lions), location of sensitive archeological sites, location and dimensions of sea caves, shipping lanes, oil platforms, bathymetry, geology, vegetation cover, soils, and topography. The study concluded that the greatest terrestrial influence on coastal waters of Santa Barbara were river-derived freshwater plumes following winter storms with precipitation greater than 3 cm. The second most important effect was the location of the coastal river outlet. More recently, she measured flood output from 110 coastal watersheds in California (Mertes and Warrick, 2001). For further information on landscape ecology and GIS see Clarke (1999), Turner et al. (2001), Hunsaker et al. (2001) Bissonette and Storch (2002), Gutzwiller and Forman (2002), Jongman and Pungetti (2003), Reiners and Driese (2004) and the internet. Dale Kiefer (kiefer@usc.edu), Department of Biological Sciences, University of Southern California, has developed a GIS system (called Easy) for the marine environment. It is written in C++ and compiles and manipulates environmental and marine data in 3-D. See Wadsworth and Treweek (1999) for an introduction to geographical information systems for ecology and Costanza and Voinov (2003) for landscape simulation modeling (also see Unwin, 2005) and GIS information in the Appendix.

3.31   Aquatic Ecotoxicology

For information on quantitative methods in aquatic ecotoxicology see Newman and Adriano (1995), Moriarty (1999), Walker (2001), and Newman (2002).

3.32   Coastal Zone Management

For information on methods used in coastal zone management see Beatley et al. (2002). EcoSpace (http://ecospace.newport.ac.uk) is a training program in environmental management and ecology.

3.33   Conservation and Environment

Many quantitative field methods used in conservation and environment are similar to those discussed in this book. Additional recent information resources include the following. Ferson and Burgman (1998) discuss quantitative methods for conservation biology and Anderson et al. (2002) for environmental issues. Feinsinger (2001) discusses the design of field studies for biodiversity conservation. Stein and Ettema (2003) discuss sampling design for comparing ecosystems and for land use studies. Natural resource modeling is presented by Shenk and Franklin (2001). Modeling as a guide to policy and management decisions is discussed by van den Belt (2004). Monitoring ecosystems is discussed by Busch and Trexler (2002), Artiola et al. (2004), and Spellerberg (2005). A guide to marine reserves is authored by Sobel and Dahlgren (2004) and marine protected areas for whales, dolphins and porpoises by Hoyt (2005). Selection and management of coral reef preserves are discussed by Bakus (1983) and White et al. (1994). Management of severely damaged coral reefs is discussed by Westwacott et al. (2000). A compendium of information on the seas is presented by Sheppard (2000). Among the newest publications are a book on marine conservation biology (Norse and Crowder, 2005) and an ocean coastal conservation guide (Helvarg, 2005). Also see Mann (2000), Williams et al. (2002), Hempel and Sherman (2003), and Hennessey and Sutinen (2005).

3.34   Environmental Impact Assessments

Environmental impact assessments (EIAs) have become an important part of applied ecology in the United States and many other countries. Conducting EIAs entails obtaining information from environmental, social, and economic sources. It is a detailed process that is described on a CD-ROM developed by the author (see the EIA folder in the Appendix).

The quantitative techniques that are used in EIAs are summarized here:

Data must be obtained on attributes (e.g., ozone, pH, biotic diversity, fatalities, employment, health costs, etc.) as a fundamental part of the EIA process. Information is typically derived from the literature but data collection may also be required from the field or from laboratory studies.

Risk assessment requires an estimation of environmental risks by the following methods: (1) data collection, (2) experiments and testing, (3) modeling, and (4) probability analysis. Probability analysis can be simple, such as the probability of an oil spill and sewage overflow (from a sewage treatment plant that breaks down) occurring simultaneously at a site along the coast. It can be estimated by using accident trees. Finally, probabilities can be updated then estimated using Bayesian statistics (e.g., conditional probabilities widely used in epidemiology). See p. 119 in Chapter 2, EIA folder in the Appendix, Suter (1993), and Brebbia (2004).

Perhaps the most interesting phase of conducting an EIA is the final step, selecting the best site for development or the best policy for management (i.e., costing the least while creating the fewest negative impacts). This is a decision making process that requires a mathematical treatment known as decision analysis. There are many types of decision analysis. Some involve complex mathematics or probabilities such as game theory or Pareto optimality but these are not especially popular. The most widely used group of decision analysis techniques in North America are called Multi-attribute Utility Measurements (MAUM). One of the most popular simple MAUM techniques is SMART (Simple Multiattribute Rating Technique – see Bakus et al., 1982). SMART uses ratings and rankings with alternatives (usually sites) and attributes (criteria).

Decision analysis originated in calculus, based on the pioneering works of Edwards, Howard, Keeney, and Raiffa in the 1960s. By the 1980s there was considerable controversy concerning quantitative methods in decision analysis. At that time, this author surmised that the behavioral interaction part of the decision making process would come to dominate over decision analysis. This has come to pass some 20 years later. Yet problems persist in the quantitative sector of decision making.

The ranking process in decision analysis (e.g., SMART method) is seldom used by decision analysts today (Detlof von Winterfeldt, pers. comm.) because in-depth appraisal of values are not done. Judgments of rank by the public, concerned groups (e.g., environmental organizations), and legislators are then used by decision makers in inappropriate ways (Keeney, 1992). So which methods are decision analysts using today?

Direct ratio methods, swing weighting, even swaps, and other analytical methods (e.g., analytic hierarchy process) may be of use to the specialist (i.e., decision analyst) but are often too complex for the layperson. The swing weighting method is better than direct ratio judgments (von Winterfeldt, 1999). Both are not sufficiently range sensitive. Swing weights and ratio weights show a strong splitting bias.

Utility measurements today commonly use only ratings and ranges or natural scales. These are often normalized or rescaled to a 0 (worst) to 100 (best) scale. An example of this would be the pH of freshwater which ranges from about 4 to 8. It is important to use ranges since ranges should strongly control the weights (von Winterfeldt and Edwards, 1986). High values (e.g., pH 8) and low values (e.g., pH 4) are presented for an attribute (i.e., pH) then the value of this attribute (e.g., pH 6) for a specific site (or alternative) is calculated. An example of this is as follows:

Assume that the range of pH of freshwaters is pH 4–8. Assume that a pH measurement taken at a site is pH 6. Therefore,

$$\frac{4 \ (\text{range})}{100 \ (\text{scale})} : \frac{2 \ (\text{the pH value above pH 4})}{X}$$
$$X \ (\text{the new rescaled value}) = 50$$

The value of 50 (on a scale of 0–100) is now entered for this attribute-alternative combination. It should be emphasized that in some cases (e.g., pH) a low value is harmful for most organisms, in other cases a low value is beneficial for organisms (e.g., low levels of lead). This must be addressed when rescaling.

The fact that simple additive utility models perform just as well as more complex models (von Winterfeldt, 1999) strongly suggest that ranges and simple addition of ratings (i.e., parsimony) would be preferable for environmental impact assessments that include laypersons and nontechnical participants (e.g., some historians, environmentalists, politicians, artists, etc.). Table 3-14 shows a common current MAUM technique using natural ranges and rescaling.

Table 3-14.   Example of a Multi-attribute Utility Measurement (MAUM) technique using natural ranges, rescaled values, probabilities and experience values.

	Attributes		
pH (environmental)	Aesthetics (social)	Employment (economics)	Utility Value
Alternatives			
Alt 1 (site 1)			
Range = 4–8	Range = 80–150	Range = 100–1000	
Measurement = 6	Measurement = 130	Measurement = 396	
Rescaled value = **50**	Rescaled value = **71**	Rescaled value = **44**	
	EV = 10% of 71		
	EV = 7		
	71 + 7 = **78**		
		50 + 78 + 44	= 172
Alt 2 (site 2)			
Range = 4–8	Range = 80–150	Range = 100–1000	
Measurement = 4.1	Measurement = 100	Measurement = 700	
Rescaled Value = **2.5**	Rescaled value = **29**	Rescaled value = **67**	
	EV = 10% of 29	probability (p) = 0.8	
	EV = 3	(0.8) 67 = **54**	
	29 − 3 = **26**		
		2.5 + 26 + 54	= 83

A weighting percentage of 0–20% of the range can be added to or subtracted from the scores for experience. The percentage value added (if better or > 50) or subtracted (if worse or < 50) is decided by the decision making group. In this example, an experience value (EV) of 10% of the rescaled value was added (by the decision analyst) to the social attribute score of the sociologist for Alternative 1 and 10% of the rescaled value was subtracted from the rescaled value for the social attribute score of the sociologist for Alternative 2. A probability value (p) of 0.8 or 20% uncertainty altered the economic value for employment in alternative 2. The higher utility value of 172 indicates that Alternative 1 (site 1) is preferable for the development.

A computer program (called EIA) with SMART on CD-ROM, written in Java, and a newer approach to decision analysis, using natural ranges and rescaling of values, is presented on this CD-ROM. The technique is programmed in C++ and is called *Utility 2003*. In addition to the basics described above, uncertainty (i.e., probabilities) and experience value (e.g., a weighting of ±0 to 20%) can be included in the quantitative decision making process.

Schmitt and Osenberg (1996) discuss ecological impacts in coastal habitats. See also Bartell et al. (1992), Jensen and Bourgeron (2001), Downes et al. (2002) and Brebbia (2004). Recent information on risk assessment and decision making can be found in the journals Medical Decision Making, Operations Research and Risk Analysis.

3.35   Analysis of DNA/RNA Sequences

Recent developments in population genetics, ecology, and biogeography include the analysis of genetic sequences (i.e., base pairs) as an important tool. For example, populations of fish or of plants in different geographical localities can be studied to determine if the subpopulations are reproductively isolated or if gene exchange is occurring between them (i.e., metapopulations). A table compiled by my former graduate student Domingo Ochavillo summarizes information on sequence alignment, frequency analysis, distance analysis, analysis of molecular variance, phylogenetic analysis, and correlation analysis (Table 3-15). Recommended general

Table 3-15.   Analysis of DNA sequences (source: Domingo Ochavillo).

I. Alignment of Sequences

   A. CLUSTAL W (e.g. www.clustalw.genome.ad.jppbil.ibcp.fr)
     1. Online server
     2. Progressive alignment approach, using a global alignment algorithm
     3. Provides relatively large power with low confidence
     4. Performs best with sequences of similar length

   B. PILE-UP (e.g. www.gcg.com)
     1. Online server
     2. Progressive alignment approach, using a global alignment algorithm

   C. DIALIGN (e.g. www.bibiserv.techfak.uni-bielefeld.de)
     1. Online server
     2. Local alignment approach, to compare sequences of high similarity
     3. Useful for finding blocks of highly conserved regions in sequences

   D. MAP (e.g., www.dot.imgen.bcm.tcm.edu:9331/multialign/options/map.html)
     1. Online server
     2. Progressive alignment approach, using a global alignment algorithm
     3. Provides relatively large power with low confidence

   E. PIMA (www.dot.imgen.bcm.tcm.edu:9331/multialign/options/map.html)
     1. Online server
     2. Progressive alignment approach, using a global alignment algorithm
     3. Provides relatively large power with low confidence

   F. MSA (e.g., www.alfredo.wustl.edu/msa/html)
     1. Online server
     2. Progressive alignment approach, using a global alignment algorithm
     3. Provides relatively large power with low confidence

   G. BLOCK MAKER (e.g., www.blocks.fhere.org/blockmkr/make_blocks.html)
     1. Online server
     2. Local alignment approach
     3. Power and confidence highly unpredictable, and unrelated to the rate of sequence identity

(Continued)

Table 3-15.   (*Continued*)

    H. MEME (e.g., www.sdsc.edu/MEME/memel.4/meme.nofeedback.html)
        1. E-mail server
        2. Local alignment approach
        3. Power and confidence highly unpredictable, and unrelated to the rate of sequence identity
    I. MATCH-BOX (e.g., www.fundp.ac.be/sciences/biologie/bms/matchbox submit.html)
        1. E-mail server
        2. Local alignment approach
        3. Very reliable when aligning structurally conserved/similar blocks with groups of protein sequences of known structure
        4. Unique feature: provides reliability score for each position of alignment
        5. Provides large confidence with variable power

II. ANALYSIS OF POPULATION GENETIC STRUCTURE
    A. Haplotype frequency analysis (http://bioweb.wku.edu)
        1. Analyzes departure from homogeneity using chi-square analysis
        2. Tests significance using Monte Carlo permutations

    B. Analysis of molecular variance (AMOVA Excofffier et al. 1992)
        1. Analyzes distance among DNA sequences using different approaches (MEGA, PAUP, NJbot)
        2. Calculates variance components in different hierarchical levels
        3. Tests significance using Monte Carlo permutations

    C. Phylogenetic Analysis
        1. Analyzes evolutionary relationships using shared-derived characters (PAUP)
        2. Can be used together with analysis of molecular variance

    D. Correlation Analysis
        1. Analyzes correlation between geographic and sequence distance matrices
        2. Tests significance using Mantel test program

references on this topic include molecular evolution (Avise, 1994; Page and Holmes, 1998; DeSalle et al., 2002) and molecular genetics (Hillis et al., 1996). Statistical methods in molecular biology are discussed by Looney (2002).

Examples of Research Projects Using DNA Sequencing:

The intention of summarizing the work here is to indicate the importance of DNA sequencing in various marine ecological studies. Ochavillo and Geiger received their doctorate degrees in biology and Vogel is near completion.

Augustus Vogel (Dissertation Research)   The kelp bass, *Paralabrax clathratus*, was studied (Vogel, 2006 - see Addendum) because of its importance to the southern California sportfishing community (Love, 1996). Seven, nuclear, micro-satellite markers and the control region of the mitochondrial genome were analyzed for evidence of spatial and temporal genetic variance. Spatial structure was low, yet significant (P=0.0005); unpredictable patterns and differences between adult and recruit samples were indicative of genetic variance between cohorts. Although temporal analysis, which utilized the annual banding in otoliths (i.e., ear bones) to group samples into year classes, did not find significant evidence of variance, other patterns of linkage disequilibrium, low allelic richness in a recruit sample, and calculation of a small effective population size indicated that temporal variance was indeed important.

These results presented a number of conclusions. First, there was no evidence of isolated populations that should be specially managed. Second, temporal variance and a small effective population size are indicative of a species that requires a large reproductive population found in a diversity of locations.

Domingo Ochavillo (Dissertation Research)   Rabbitfishes (Family Siganidae) are widely distributed in tropical and subtropical areas of the Indo-Pacific region. The rabbitfish *Siganus fuscescens* is the most abundant fish and the main fishery on a fringing reef on N.W. Luzon Island, Philippines (Ochavillo, 2002). It is a schooling species that feeds on benthic algae in a seagrass community. It is also found on two offshore fringing reefs, each about 200 km away. The question that arose was to what extent were the fish populations of these three sites interbreeding? Mitochondrial DNA from fish muscle tissue was sequenced from the three populations. Each population had a distinct sequencing pattern, suggesting little genetic mixing among them. Late-stage fish larvae settling on patch reefs shared the same genetic pattern as the local adults. Some 500 fish larvae that were released 2 km offshore swam towards where they were born (Normile, 2000). This suggests that tropical fish larval dispersal may not be as widespread as is usually assumed (see p. 87).

Daniel Geiger (Dissertation Research)   Specimens of abalone (Family Haliotidae) were examined in major museums in Europe and the United States and from private collections. Statistical analysis was performed using shell morphometric data. Sequence alignment of 16S mitochondrial DNA among species was compared with the objective of distinguishing homologies. More than 200 described species were evaluated and 56 species were retained, distribution maps of which were presented. Both area cladograms and taxon cladograms were constructed. Allozyme frequency data were recoded for cladistic analysis. Morphological characters, DNA sequence data, and allozyme frequency data were combined into a total evidence cladistic analysis. The genus *Haliotis* was suggested for all abalones (Geiger, 1999).

Genetic Analyses – A Discussion (Augustine Vogel)   There exists a diversity of tools for many types of genetic analyses. When using direct sequencing, genes such as the 12S RNA gene of the nuclear genome have been found useful in phylogenetic studies of divergent phyla and kingdoms. Different areas like the internal transcribed spacer (ITS) region and the COI or the mitochondrial control region, because of their typically larger number of polymorphic sites, have been more useful at the species or population level (Stepien and Kocher, 1997). Still other sequences, such as those found in satellite DNA, have been used to distinguish even as far as individual parentage (Ward, 2000).

Even without the use of direct sequence data, researchers have been able to ascertain valuable information through assays that infer underlying genetic information. Allozymes (Edmands et al., 1996), restriction fragment length polymorphisms (RFLPs) (Gold et al., 1994) and polymerase chain reaction single-strand conformational polymorphisms (PCR-SSCP) (Li and Hedgecock, 1998) are just a few of the dozens of options available to the investigator.

A researcher about to start a potentially interesting project must make sure to use genetic markers that will most directly (and cheaply) answer the question at hand. Reasons for studying a particular organism must be clear; the use of genetic tools is not the final goal of the project, but rather the method for asking ecological questions concerning migration, larval dispersal or perhaps parentage. The constraints of the potential genetic markers must also be understood, as well as anthropogenic and environmental considerations, so as to create a meaningful and informative project.

The first decision to be made in the project is to determine the organism and relevant question to be studied. This process may start at either the question or the organism, depending on what interests the investigator more. Someone may be primarily interested in a species of butterfly, for example, and then develop analyses relevant to the organism's biology. Perhaps they are interested in relationships within the species family, or maybe the number of parents represented in a clutch of eggs.

On the contrary, an investigator could be interested in ocean currents or the effects of certain land masses on local populations, and then choose an organism that they think would be useful in the detection of possible patterns. When studying the evolutionary effects of the Panama isthmus for example, comparison of fishes such as the Pacific Porgy (*Calamus brachysomus*) with Caribbean species (such as the Jolthead *Calamus bajonado* and the Silver *Diplodus argenteus*) could prove useful, whereas a more circumglobal species such as the Blue Marlin (*Makaira nigricans*) might not.

The researcher should also remember that many times money will be most available for species of economic interest, perhaps a fishery species or an organism endangered by development. The source of funding can drastically cull species that do not have pragmatic or political interest, or that may be too expensive to properly sample.

Once an organism and a broad question have been established, a winnowing process must occur to create a project with a manageable number of attainable results. Through literature research, an investigator can begin to speculate which peculiar aspects of a species' biology will affect the outcome of the genetics project. Potential hypotheses will slowly begin to crystallize, as various relevant processes and interactions are taken into account.

In aquatic genetics, for example, the ability of an organism to translocate either through adult migration or larval dispersion appears to be one of the most important parameters affecting the genetics of a species (Bowen and Avise,1990). Marine fish have been found to have a median Fst value (a measure of variance that assigns a value to the amount of genetic variation between populations, as opposed to between individuals (Wright, 1951)) of only 0.020 (Waples, 1998). More restricted freshwater species have a much higher mean Fst of 0.222 (Ward et al., 1994). Intertidal species and organisms that have some barrier to gene flow will also display more structure between populations (higher Fst values). Organisms without barriers to gene flow (geographic or behavioral) will require an analysis different than a more restricted species. Larger sample sizes (Baverstock and Moritz, 1996), more polymorphic markers (Banks et al., 2000), and analysis of temporal variance (Johnson and Black, 1982; Lenfant and Planes, 2002) can all influence and clarify interpretation of extremely subtle structure.

The researcher should also keep in mind the pragmatic limitations of various genetic techniques while studying the available literature resources (Baverstock and Moritz, 1996). A student might be more drawn to RFLPs or allozyme analysis because of the low overhead cost; for allozymes though, all tissues must be fresh or have been frozen in the field, not a simple task if sample collection must occur in remote sites. For RFLPs, loss or gain of a cut site is not always indicative of a single nucleotide polymorphism, nor are all polymorphisms recorded with a screening of cutting enzymes. Direct sequencing will have its own problems, including cost, fidelity of Polymerase Chain Reaction (PCR) primers, and ambiguities in the interpretation of mutation sites.

After comparison of genetic techniques and potential results, it is appropriate to commence a pilot project. To avoid spending considerable money on a molecular technique that may not work, a variety of possible options should be tried so as to affirm the existence of the right amount of variation (with relevance to the question) in the molecular markers. If it is found that the chosen markers work well, then the main project can be commenced.

These main projects are potentially very important, because most fishery management plans are not constructed on genetic analysis. Certain genetically unique populations, or populations that rely heavily on self-recruitment may happen to exist in a location convenient to fishery fleets. The broadbill swordfish (*Xiphias gladius*) for example, exhibits population structure (separation) within and between ocean basins (Reeb et al., 2000) but the most serious effects of fishing pressure are apparent mainly on Atlantic and Mediterranean populations (Berkeley, 1989).

A heavily used fishery can also potentially have unique structure that more represents artifacts of historic gene flow than current restrictions on patterns. For example, the sardine *Sardinops ocellatus* off the west coast of Africa has had populations with different migratory habits. Both have at times crashed, leaving only scattered individuals that eventually join other, still viable populations. Although these losses of population size may not be completely attributed to fishing pressure (Bakun, 1996), a researcher studying the genetics of the remaining population must recognize the potentially unique history of each individual sample collected.

Considerable work should be put into the decision of how to use a genetic marker. Simply choosing a convenient assay and an accessible organism will in many cases not support the creation of useful information. Poor data are especially insidious when connected to species in need of management. Incorrect project design can potentially support programs that harm more than help. Species need to be properly identified. Voucher specimens need to be deposited in permanent collections.

3.36   Fuzzy Logic

Fuzzy logic was first introduced as a discipline within mathematics in 1965 by Lofti Zadeh, an electrical engineer at UC Berkeley. Fuzzy logic is based on a fuzzy set. Fuzzy sets differ from standard set theory in that objects belong to a degree rather than belonging to a set or not belonging to a set. Statements are true to various degrees, ranging from completely true to completely false. Reasoning can occur in words rather than in numbers; ambiguities, contradictions, and uncertainties can be easily accommodated. Fuzzy sets have overlapping boundaries, simulating how humans react to parameters such as temperature.

An example of fuzzy logic would be a fuzzy thermometer that contains three categories, cool (~15 °C), just right (~20 °C) and warm (~25 °C). Another example is that of speed: slow–medium–fast. A fuzzy number 2 might look like this (Fig. 3-46). Note that the numbers range from 100% confidence (completely true) to 0% confidence (completely false) with the number 2 being completely true but 1.5 and 2.5 being completely false. Numbers >1.5 and <2.5 have greater degrees of truth. Fuzzy logic allows simpler and faster calculations than solutions with differential equations, with effective results. Expert knowledge is used in place of differential equations.

Fuzzy controllers (algorithms that control things – see below) are the most important applications of fuzzy theory, such as the control of small devices (toasters, cameras), large systems (subways, nuclear power plants), computer programs that learn, and pattern recognition (speech, handwriting). Fuzzy logic has applications in engineering, industry, computing, pest management, business administration, finance, economics, psychology, criminology, and so forth. Fuzzy logic research is most active in Japan and China.

Until about the last decade, fuzzy logic was not applied to biological problems. Some early studies were done on fuzzy ordination with plants (see fuzzy ordination on the internet). Scientific Fishery Systems, Inc. developed SciFish 2000, a

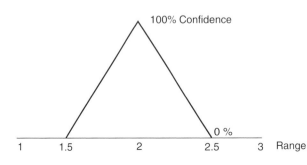

Figure 3-46.   Example of a fuzzy number using fuzzy logic.

shipboard fish detection system that identifies swimming fish species and their sizes in real time (Fireman, 2000). This involves the use of broadband sonar transmission, spectral processing using spectral decomposition, and a **fuzzy neural network classifier**. Signatures of fish species (fuzzy neural network coefficients) are stored in a database and later recalled for classification (i.e., to identify fish detected below the sea surface by sonar). SciFish 2000 uses Windows NT and a DAP board. It has been used in the U.S. Great Lakes region (freshwater fishes) and Prince William Sound, Alaska (marine fishes) and gives >80% correct species identifications with more than 20 species of fishes. For further information contact Jerry Fireman on email at: editor@scimag.com. See the internet for other fuzzy and neuro-fuzzy techniques.

3.37   Meta-Analysis

Meta-analysis is a statistical method used to compare and synthesize the results of many independent studies. It is the quantification of a literature review for a review paper. In this way it differs from the classical narrative approach of literature reviews. Meta-analysis originated and was developed in the social sciences. In biology it is most useful in ecology or evolution where there is a moderate to large number of studies on a topic.

The following steps are involved in meta-analysis. Studies are gathered from the literature that have similar designs and address a common question or hypothesis, such as: Does competition effect biomass? The literature must be carefully selected and must have quantitative information or data of the proper type. Historically, especially in the social sciences, the difference between the means of experimental and control groups have been divided by the pooled standard deviation (from experimental and control groups) in order to obtain a standardized measure of effect size. However, Hurlbert (1994) states that this approach does not yield meaningful results for most types of biological data. He suggests that the standardized effect size be measured as either mean 2/mean1 or (mean 2 – mean 1)/mean 1. Cote et al. (2001) use the ratio of the means. The "effect size" documented in the various studies are all converted into the same units. To assess the evidence of the studies considered collectively, a weighted average effect size is computed based on the sample size. The idea behind this is that studies with more data should receive more weight or importance than studies with fewer data. Cote et al. (2001) weigh each abundance and richness estimate by the natural logarithm of the total area covered by the census. Effect sizes are combined into a common estimate of the magnitude of the effect. The significance level of the overall size effect is computed using randomization techniques. The homogeneity of effect sizes is calculated (e.g., for terrestrial, freshwater and marine environments).

Example:

Six scientific journals were selected that had published the most recent field experiments on competition (Gurevitch et al., 1992). Means, standard deviations, and sample sizes were obtained for experimental and control groups from papers judged suitable in the six journals. Data were analyzed on Microsoft Excel spreadsheets.

Larger studies (i.e., those with more experiments) were weighted more heavily than smaller studies. The effect size in terms of biomass responses to competition were calculated. The authors concluded that competition had a large effect on biomass with a grand mean effect size of $d_t = 0.80$ (= large effect size). See Gurevitch et al. (1992).

Most meta-analyses have been based on fixed effects models, that a class of studies share a common effect size. In the mixed model there is random variation among studies in a class in addition to sampling variation (Gurevitch and Hedges, 2001). Mixed models in meta-analysis are analogous to mixed models in ANOVA (p. 91 in Chapter 2). Random effects models are not of general interest for ecological research. Statistical tests of significance can be carried out using parametric or randomization approaches. Bootstrapping can be used to calculate confidence intervals around mean effect size.

Problems with meta-analysis include (1) Nonrepresentative studies may be included, constituting a bias, (2) and a normal distribution is assumed. If there are sufficient data and the distribution is not normal then bootstrapping techniques must be used. (3) An independence of elements is assumed when some dependent relationships occur; (4) there may be a lack of uniformity among studies; (5) there may be a mixing of good and bad studies.

See Gurevitch at al. (1992) for an example how met-analysis has been used in biology and Arnqvist and Wooster (1995) for a general review. Computer software called Metawin (Version 2.0 or greater) is available from Sinauer Associates (use the internet). See Cooper and Hedges (1994) and Gurevich and Hedges (2001) for further information.

Chapter 4

Community Analyses: Similarity–Dissimilarity Indices, Cluster Analysis, Dendrograms, Analysis of Similarities, Indicator Species

4.1 INTRODUCTION

Communities (Assemblages, Associations, Vegetation Types, Vegetation Series) are groups of species that have been defined by environmental factors (biotope) or by organisms and their interactions (biocoenose), by dominance, by constancy and fidelity, by trophic functions, by recurrent species groups (Fager, 1957), by delineating a portion of a continuum, or by a combination of these (see Whittaker, 1956; Stephenson, 1973; Underwood, 1986; and Ornduff et al., 2003 for a discussion of community concepts). Community analyses today deal primarily with species abundances and sites. These analyses represent a type of multivariate analysis. For a summary of data analysis in community and landscape ecology, see Jongman et al. (1995).

    The major steps in the analysis of communities are as follows: (1) construct a two-way table of species (X-axis) versus sites (Y-axis), (2) use a similarity (S) or dissimilarity ($1 - S$) index on the collected data, (3) use a cluster analysis technique on those data to construct a dendrogram (i.e., visual tree representation of the data), and (4) select the boundaries to delineate species-site groups. This can be done visually in clear-cut cases but may require the semi-quantitative choice of pseudospecies (i.e., abundance classes) cutoffs (see p. 227).

Quantitative Analysis of Marine Biology Communities: Field Biology and Environment
by Gerald J. Bakus
Copyright © 2007 John Wiley & Sons, Inc.

4.2   METHODS OF HANDLING DATA

Data Reduction

Excessive data add time and costs to the processing of calculations. The following methods are some ways of deleting data without affecting too seriously the results.

(1)   Reduction of Species

We can eliminate any species that occur only once in a quadrat sample or eliminate species with low variance contribution, every tenth sample, and so forth. Some workers use arbitrary levels of elimination to decrease computation and to clarify the form of dendrograms, and so forth. Field et al. (1982) suggest the removal of species representing $X\%$ of the total abundance, biomass or percentage cover. Clarke and Warwick (2001) removed those making up less than 3% of soft-sediment benthic data. Other community analysts prefer to use the entire data base, as computer power is no longer a limiting factor. However, time is lost entering data that will be automatically eliminated by some computer programs (so try to understand how your computer program operates). In addition, too many rare species may result in a messy dendrogram or ordination, or problems may occur when computing eigenvalues (see Chapter 5, p. 244, and Clarke and Warwick, 2001). **However, other important problems (e.g., major ecological information loss) may occur if a low density keystone species or top predator is eliminated from the list of species**. For example, the predaceous snail *Polinices* may occur in low densities yet may control community structure on soft bottoms in certain localities (e.g., Morro Bay, California, ~30 years ago, before they were over-collected, Bakus, pers. observ.).

(2)   Pooling of Data

The average value of a group of pooled data may be used in place of the raw data. We can also lump data into higher taxonomic categories, for example, to genera, family, and so forth. Genera are more appropriate than species for some environmental studies. **Genera appear to reflect more large-scale environmental patterns while species reflect micro-environmental patchiness** (Green, 1979). For routine monitoring, pooling by families (e.g., polychaete families) may be acceptable (Clarke and Warwick, 2001).

Cleaning Data

1) Elimination of Data Collected by Inappropriate Sampling Gear   Grab sample A is incomplete because a shell caught in its jaws, allowing part of the sample to be washed out. We, therefore, do not use that sample in our computations.

2) Elimination of Unidentifiable Material   Sixty species are collected, five of which cannot be identified nor are they numerically important. This would be typical of collections in the tropics. We disregard them in our calculations. Once again, this procedure may result in an ecologically important species being eliminated.

For information on Data Standardization and Data Transformation see sections 2.3 and 2.4, respectively.

4.3 MEASURES OF SIMILARITY AND DIFFERENCE (SIMILARITY AND DISSIMILARITY)

Over 60 measures of similarity or dissimilarity have been described (Legendre and Legendre, 1998). Only a few have been examined for their performance in different situations. Dissimilarity or distance measures can be categorized as metric, semi-metric, or nonmetric (McCune et al., 2002). A metric distance is based on the following rules: (1) the minimum value is zero when two items are identical, (2) when two items differ, the distance is positive (negative distances are not permitted), (3) symmetry: the distance from object A to object B is the same as the distance from B to A, and (4) triangle inequality axiom: with three objects, the distance between two of these objects cannot be larger than the sum of the two other distances. Semimetrics (e.g., Bray–Curtis) violate the triangle inequality axiom. Semimetrics are extremely useful in community ecology. Nonmetrics violate two or more of the metric rules and are seldom used in ecology. The following section summarizes some of the preferred similarity and difference techniques.

Similarity (Dissimilarity) Coefficients

1) Binary or Presence-Absence Data

There are numerous coefficients of association based on binary data (i.e., presence or absence data; Legendre and Legendre, 1998). We will use two methods that are considered to be preferable because they are simple and give intuitively good results. The Jaccard and Czekanowski coefficients range from 0 (no similarity) to 1.0 (highest similarity).

(a) Jaccard (1908) similarity coefficient

$$J = \frac{c}{a+b+c}$$

where

J = Jaccard coefficient

a = number of occurrences of species a alone

b = number of occurrences of species b alone

c = number of co-occurrences of the two species (a and b).

Example:

$a = 16, \ b = 8, \ c = 26$

$$J = \frac{26}{50}$$
$$J = 0.52$$

Note that simultaneous nonoccurrences of species b and c are not included in the equation (certain coefficients give results that are weakened because they include non-occurrences). Ecostat (a computer program – see Appendix) has a routine which calculates the Jaccard coefficient for comparing communities. Krebs (2000a) also presents a computer program for the same coefficient.

   (b) Czekanowski (1913) similarity coefficient (also known as the Sorensen (1948) Coefficient)

$$C = \frac{2c}{2a+b+c} \qquad \text{where: (same as in Jaccard)}$$

Example:

$a = 16, \ b = 8, \ c = 26$

$$C = \frac{52}{66}$$
$$C = 0.79$$

Note that the number 2 in the equation enhances (doubles) co-occurrences. Simultaneous non-occurrences are not included in the equation. Janson and Vegelius (1981) recommend the use of the Jaccard, Ochiai and Dice coefficients. See MacKenzie et al. (2005) for methods in analyzing presence-absence data surveys.

2) *Meristic or Metric Data*

Dissimilarity coefficients based on meristic or metric data are numerous and have the advantage over presence-absence data of providing more information on the association. Most of these coefficients range between 0 and +1. There are two coefficients that are widely used today in marine ecology.

   a) Bray–Curtis (1957) dissimilarity coefficient. This is an extension of the Czekanowski coefficient and was initially developed by Motyka et al. (1950). It is a standardized Manhatten metric or city-block metric (Krebs, 1999). Because this coefficient is considerably influenced by dominance, the raw data can be reduced by $\sqrt[n]{\text{raw}}$ datum (e.g., $n = 2$ for square root, $n = 3$ for cube root, etc.) or even further reduced. Some investigators use $\log_{10}(n + 1)$ in order to decrease the importance of species having large numbers. Krebs (2000a), PC-ORD, and PRIMER (Clarke and Gorley, 2001) have computer programs which do this (see the Appendix). Bray–Curtis is recommended by Clarke and Warwick (2001) and others as the most appropriate dissimilarity coefficient for community studies.

Bray–Curtis:

$$BC = \frac{\sum_{j=1}^{n} |X_{1j} - X_{2j}|}{\sum_{j=1}^{n} (X_{1j} + X_{2j})}$$

where

Σ = sum (from 1 to n)

X_{1j} = # organisms of species j (attribute) collected at site 1 (entity).

X_{2j} = # organisms of species j collected at site 2.

BC = Bray–Curtis coefficient of distance

$\|$ = absolute value

j = 1 to n

n = number of species

Figure 4-1 gives the **"normal" or Q mode**, that is, distances between sites. The calculation of distances between species or the **"inverse pattern" or R mode** requires vertical data pair analysis, as follows:

$$\frac{[10-3]+[5-6]}{(10+3)+(5+6)}$$ for distance between species 1 and 2

b) Canberra metric coefficient. This coefficient of Lance and Williams (1966–1971) can be handled efficiently with a computer program developed by them, by Krebs (2000a), and so forth. To reduce the importance of zeros, some investigators replace raw data zeros (0) by 1/5 of the smallest number in the data matrix (often 1/5 of 1 individual or 0.2). $\sqrt[2]{\text{raw}}$ data and $\log_{10}$ transformations are also used (see p. 66 in Chapter 2). Once again, conjoint absences are avoided. Because both Bray–Curtis and Canberra Metric coefficients are strongly affected by sample size (i.e., the

	Sites	
	1	2
1	10	5
2	3	6
3	0	8
4	0	0

Species

$$\frac{|10 - 5| + |3 - 6| + |0 - 8|}{(10 + 5) + (3 + 6) + (0 + 8)} = 0.5$$

← – – – – – – – conjoint absences are not used.

Figure 4-1.   Data matrix for Bray-Curtis showing the distance between sites 1 and 2.

coefficients change with the number of samples taken – see Krebs, 1999), Morisita's similarity index has been recommended (see below).

$$CM = \frac{1}{N} \sum \frac{\left|X_{1j} - X_{2j}\right|}{\left(X_{1j} + X_{2j}\right)}$$

where

N = number of samples with organisms

CM = Canberra metric coefficient (see p. 213 for other symbols)

c) Morisita's Index

Krebs (1999) considers Morisita's index as the best similarity index, as follows:

$$C_\lambda = \frac{2 \sum X_{ij} X_{ik}}{(\lambda_1 + \lambda_2) N_j N_k}$$

$$\lambda_1 = \frac{\sum \left[X_{ij} \left(X_{ij} - 1 \right) \right]}{N_j \left(N_j - 1 \right)}$$

$$\lambda_2 = \frac{\sum \left[X_{ik} \left(X_{ik} - 1 \right) \right]}{N_k \left(N_k - 1 \right)}$$

where

Σ = sum

C_λ = Morisita's Index of Similarity

X_{ij} = No. individuals of species I in sample j

X_{ik} = No. individuals of species I in sample k

N_j = Total No. individuals in sample j

N_k = Total No. individuals in sample k

Example

How similar are the abundances of species on North-facing and South-facing slopes in a submarine or terrestrial canyon?

Species	No. Individuals N-facing Slope	No. Individuals S-Facing Slope
1	35	22
2	32	16
3	16	14
4	8	18
5	6	2
Total	97	72

$$\lambda_1 = (35)(34) + (32)(31) + (16)(15) + (8)(7) + (6)(5)$$

$$\lambda_1 = 1190 + 992 + 240 + 56 + 30 = 2508$$

$$\lambda_1 = \frac{2508}{(97)(96)} = \frac{2508}{9312} = 0.27$$

$$\lambda_2 = (22)(21) + (16)(15) + (14)(13) + (18)(17) + (2)(1)$$

$$\lambda_2 = 462 + 240 + 182 + 306 + 2 = 1192$$

$$\lambda_2 = \frac{1192}{(72)(71)} = \frac{1192}{5112} = 0.23$$

$$C_\lambda = 2(35)(22) + (32)(16) + (16)(14) + (8)(18) + (6)(2)$$

$$C_\lambda = \frac{1540 + 512 + 224 + 144 + 12}{(0.27 + 0.23)(97)(72)} = \frac{2432}{(0.5)(6984)} = \frac{2432}{3492}$$

$$C_\lambda = \frac{2432}{3492} = 0.70 \quad \text{(moderately high similarity in abundance between N-facing and S-facing slopes)}$$

The Morisita Index varies from 0 (no similarity) to ~1.0 (complete similarity). It is almost independent of sample size except for very small sample sizes. For data on biomass, productivity, or percentage cover see Krebs (1999) for a modified equation. To obtain dissimilarity or distances: 1 – Morisita's similarity index value.

Computer programs for similarity or dissimilarity indices are provided by Ecostat (range 0–1 where 1 is the greatest similarity), Krebs (2000a), PC-ORD, and PRIMER. The pros and cons of a large variety of similarity and dissimilarity measures are summarized by Legendre and Legendre (1998). Also see Ellis (1968).

Coefficients of Association   These coefficients are applicable to binary and continuous data. They are of two types, ranging from -1 to $+1$ (e.g., Pearson Correlation Coefficient) and from 0 to X (e.g., χ^2). Although many coefficients of association exist, the commonest ones include χ^2 (chi square – Chapter 2, p. 84) and the Pearson Correlation Coefficient (Chapter 2, p. 105).

The Pearson product moment correlation coefficient (also called the correlation coefficient, Pearson coefficient, Pearson correlation coefficient, Phi coefficient, and product moment coefficient) uses conjoint absences, the use of which is inappropriate for comparing sites (Clarke and Warwick, 2001), although it is appropriate for comparing species. It has been applied to both species/species and site/site analyses. It gives peculiar results if there are more than 50% blank data entries. Cassie has used this coefficient with ordination in plankton studies since 1961. See Cassie and Michael (1968).

Normal Pattern

Sites

		1	2	3	4
	1	2000	1000	1500	500
Species	2	20	10	15	5
	3	—	5	—	—

Distance between sites 1 and 2 is:

$$E\,D_{1,2} = \sqrt{(2000 - 1000)^2 + (20 - 10)^2 + (0 - 5)^2}$$

Inverse Pattern

Distance between species 1 and 2 is:

$$E\,D_{1,2} = \sqrt{(2000 - 20)^2 + (1000 - 10)^2 + (1500 - 15)^2 + (500 - 5)^2}$$

Figure 4-2.   Data matrix for Euclidean Distance

Euclidian Distance   Euclidean distance is another measure of distance that can be applied to a site by species matrix (Figure 4-2). It has been widely used in the past because it is compatible with virtually all cluster techniques.

$$\text{E.D.} = \sqrt{\sum_{i=1}^{n} \left(X_{ij} - X_{ik} \right)^2}$$

where

E.D. = Euclidean distance between samples j and k

X_{ij} = No. individuals of species i in sample j

X_{ik} = No. individuals of species i in sample k

n = Total No. species

The examples given include Euclidean Distance (E.D.) between sites and between species. **The main criticism of Euclidean Distance is that it overemphasizes outliers and dominance thus giving poor ecological results**. It assumes that variables are uncorrelated within clusters (McGarigal et al., 2000), a situation seldom encountered in wildlife research. **Double zeros lead to a reduction in distances with Euclidian Distance and with Manhattan Metric** (another popular distance measurement). Despite this, many statistical programs include Euclidian distance because it is used in numerous non-ecological studies. It is also often appropriately used to calculate distances on the basis of environmental variables (Clarke, 1993). When using Euclidean distance, it is often useful to standardize the raw data by conversion to percentages, geometric mean, σ, variance, or double standardized by the geometric mean to put variables on the same scale before going on to do clustering or ordination (see p. 65 in Chapter 2). See Noy-Meir (1971) as an example of an exemplary study using Euclidean distance.

Information Theory Measures   Information analysis measures were developed by Williams et al. (1966). For details see Poole (1974), Clifford and Stephenson (1975), Van der Maarel (1980), and Chapter 3, p. 145.

4.4   CLUSTER ANALYSIS

The main purpose of cluster analysis in ecology is to organize entities (e.g., sites or species) into classes or groups such that within-group similarity is maximized and among-group or between-group similarity minimized (McGarigal et al., 2000). Cluster analysis is used to find species groups or site groups in ecology. The analysis is followed by a visual presentation of the data in a tree diagram or dendrogram. A number of decisions are required before choosing a clustering strategy. These choices are represented in Fig. 4-3.

All clustering strategies can be compared with one another according to their "space-conserving", "space-dilating", or "space-contracting" characteristics. Space-contracting cluster formation (e.g., nearest neighbor) may result in chaining, space-dilating (e.g., furthest neighbor) may result in intense clustering, and space-conserving (e.g., group average) is generally somewhere in between these two extremes. There are many types of clustering techniques. Most of these are sensitive to the presence of outliers (McGarigal et al., 2000). The value for percentage chaining can be calculated using PC-ORD. Extreme chaining (>25%) should be avoided (McCune et al., 2002).

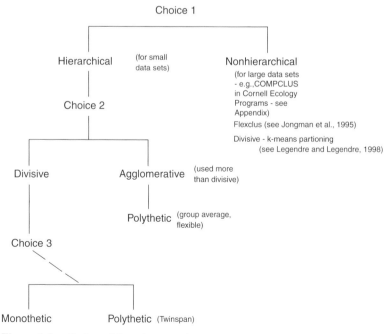

Figure 4-3.   Choices of clustering strategies

Hierarchical clustering is one in which members of a lower class are also members of a higher class, allowing similarity relationships to be shown. It is used in ecology and taxonomy (Gauch and Whittaker, 1981). Nonhierarchical methods do not impose a hierarchical structure on the data and are used for data reduction (Jongman et al., 1995). Agglomerative clustering is a classification procedure by progressive fusion beginning with individual members of a population and ending with the complete population. All agglomerative strategies are polythetic (see Fig. 4-3). Divisive methods classify by progressive division, beginning with the complete population and ending with individual members (Legendre and Legendre, 1998). Polythetic clustering is a measure of similarity applied over all attributes considered (e.g., species) so that an individual (i.e., at a site) is grouped with individuals that, on the average, it most resembles. Monothetic methods are those in which divisions are made on the basis of one attribute (e.g., species or character) (Jongman et al., 1995). Monothetic techniques can only be divisive (McGarigal et al., 2000).

Hierarchical clustering arranges clusters into a hierarchy and is suitable for small or large data sets. Nonhierarchical clustering assigns each entity to a cluster and is generally used for large data sets or for classes of data (e.g., a vegetation map of a park with 5 plant associations). Agglomerative methods tend to be used more frequently than divisive methods.

Nonhierarchical polythetic divisive clustering (e.g., K-means clustering) is seldom used in ecology. It assumes equal covariance matrices among clusters and is strongly biased toward finding elliptical and spherical clusters (McGarigal et al, 2000). K-means is the most widely used method of nonhierarchical clustering (McCune et al., 2002). It is useful for large data sets.

There are at least nine major types of hierarchical, agglomerative, polythetic clustering strategies, as follows:

group average

nearest neighbor

furthest neighbor

flexible

centroid

median

incremental sum of squares

minimal variance

Ward's method

Two popular agglomerative polythetic techniques are Group Average and Flexible. McCune et al. (2002) recommend Ward's method in addition. PC-ORD has computer programs for these. For further information see Anderberg (1973), Everitt (1980), and Späth (1985) for algorithms. Gauch (1982) preferred to use divisive polythetic techniques such as TWINSPAN. It is one of many such programs found in PC-ORD. Also see Bloom et al. (1977), the discussion below, and Bortone et al. (1991).

Group Average Method

Originated by Sokal and Mitchener (1985) as the "unweighted group mean method," it is also known as the "unweighted pair group clustering method". It is essentially the average (dis)similarity between all possible pairs of members such as species, groups of species, or sites (Jongman et al., 1995). Fusion of a member with a group occurs where the mean distance between them is least.

The example to be presented below assumes that animals have been collected from quadrats in the field. The data (species presence or absence) are used in calculating Jaccard (or some other preferred similarity or dissimilarity) coefficients. Computer programs for these are found in Ecostat, Krebs (2000a), PC-ORD (Version 4 or later) and PRIMER (see Appendix). The group average method (sorting strategy) is then employed to form a dendrogram (based on frequency of co-occurrences). By convention, dendrograms are constructed from the bottom up (agglomerative) in community ecology and from the top down (divisive) in taxonomy.

After calculating similarity (or dissimilarity) coefficients (e.g., Jaccard) between species pairs (or site pairs) (Fig. 4-4), one can illustrate the results by fitting the data into a dendrogram. We then arbitrarily choose a cutoff point on the coefficient axis (e.g., 0.5 or 50% co-occurrence for temperate latitudes – e.g., see Y-axis in Fig. 4-5). in order to obtain groups or clusters from the dendrogram output.

We can construct the group average dendrogram in the following manner:

(**I**) Arrange species by name or number (easier) on the X- and Y-axes and include coefficients for each species pair (already calculated – see Figure 4-4).

(**II**) Change all similarity coefficients to dissimilarity coefficients by subtracting each value from 1. For example, if using the Jaccard Coefficient:

$$1 - \frac{a}{a+b+c}$$

	Species				
	1	2	3	4	5
1		0.2 (0.8)	0.3 (0.7)	0.4 (0.6)	0.14 (0.86)
2			0.7 (0.3)	0.24 (0.76)	0.6 (0.4)
3				0.34 (0.66)	0.42 (0.58)
4					0.9 (0.1)
5					

Species (label on left, rows 1–5)

Figure 4-4. Jaccard similarity coefficients and corresponding dissimilarity coefficients ( ). This is an example of a trellis diagram.

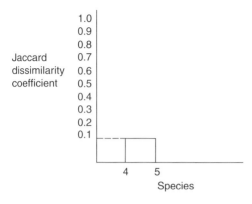

Figure 4-5. First step in the construction of a dendrogram using dissimilarity coefficients from Fig. 4-4.

This alteration is made in order to construct a dendrogram where dissimilarity = distance. Place the dissimilarity coefficients in parentheses ( ) directly below the similarity coefficients (see Fig. 4-4). All further work will proceed only with dissimilarity coefficients.

(III) Choose the species that show the smallest dissimilarity (4 and 5 have 0.1) and begin constructing the dendrogram. See Fig. 4-5.

(IV) Determine which of the remaining species or pairs (1, 2, 3, 1-2, 1-3, 2-3,) show the smallest average dissimilarity with species pair 4-5 or with each other.

$$1\begin{cases}1 \text{ and } 4 = 0.6 \\ 1 \text{ and } 5 = 0.86 \quad \bar{x} = 0.73\end{cases}$$

$$.2\begin{cases}2 \text{ and } 4 = 0.76 \\ 2 \text{ and } 5 = 0.40 \quad \bar{x} = 0.58\end{cases}$$

$$3\begin{cases}3 \text{ and } 4 = 0.66 \\ 3 \text{ and } 5 = 0.58 \quad \bar{x} = 0.62\end{cases}$$

$$1-2\{1 \text{ and } 2 = 0.8 \quad \bar{x} = 0.8$$

$$1-3\{1 \text{ and } 3 = 0.7 \quad \bar{x} = 0.7$$

$$2-3\{2 \text{ and } 3 = 0.3 \quad \bar{x} = 0.3$$

(V) Connect species 2-3 at the 0.3 dissimilarity level (Figure 4-6).

(VI) Determine where species 1 connects either with species group 2-3 or with species group 4-5 and where species group 2-3 connects with species group 4-5

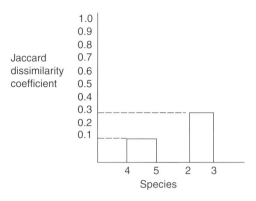

Figure 4-6. Second step in the construction of the dendrogram based on data in Fig. 4-4.

$$
\begin{aligned}
1 \text{ and } 2 &= 0.8 \\
1 \text{ and } 3 &= 0.7
\end{aligned} \Bigg\} \quad \bar{X} = 0.75
$$

$$
\begin{aligned}
1 \text{ and } 4 &= 0.6 \\
1 \text{ and } 5 &= 0.86
\end{aligned} \Bigg\} \quad \bar{X} = 0.73
$$

$$
\begin{aligned}
2 \text{ and } 4 &= 0.76 \\
2 \text{ and } 5 &= 0.40 \\
3 \text{ and } 4 &= 0.66 \\
3 \text{ and } 5 &= 0.58
\end{aligned} \Bigg\} \quad \bar{X} = 0.60
$$

(VII) Connect species 2-3 with 4-5 at the 0.6 dissimilarity level (Figure 4-7).

(VIII) Determine at what dissimilarity the remaining species (i.e., species 1) combines with species group [2,3,4,5]. Connect species 1 with the species group [2,3,4,5] at dissimilarity level 0.73 (Fig. 4-8).

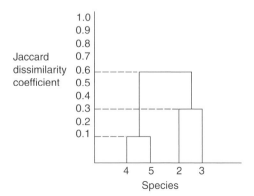

Figure 4-7. Third step in the construction of the dendrogram based on data in Fig. 4-4.

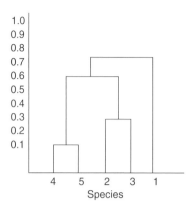

Figure 4-8. Completion of the dendrogram based on data in Fig. 4-4.

The completed dendrogram shows that two distinct species groups occur (4-5 and 2-3) and that the species in group 4-5 are only a small distance apart (i.e., have a high degree of association between one another or a very low level of dissimilarity). Species 1 is alone (**an outlier**) and some distance from the other two species groups.

To determine species groups with even more precision one would use the raw data from the frequency of co-occurrence (e.g., Bray–Curtis dissimilarity index) in place of the Jaccard (presence-absence) equation. The number of resulting species groups may be based on an arbitrary cut-off point. At a dissimilarity level of 0.2, there would be one species group (4-5) and three individuals species (1-2-3); at a level of 0.7 there are two species groups (4-5 and 2-3) and one individual species. Ecologists in temperate latitudes (e.g., see Fager, 1957) often use 0.5 as a cut-off (i.e., 50% co-occurrence) whereas in the more diverse tropics they have used 0.25 (i.e., 25% co-occurrence - Gilbert Voss, pers. comm.), giving them numerous groups consisting of a few species each (Interoceanic Canal Studies, 1970).

Another method of determining species and site groups is to examine data patterns in a two-way table, where the order of species and sites have been arranged according to their sequences in a dendrogram (Fig. 4-9). The associated dendrogram is presented in Fig. 4-10. As mentioned previously, cutoffs can be created visually (Fig. 4-9) or by arbitrary dissimilarity coefficients (i.e., distances on the Y-axis). See p. 227 below for further details. There are no completely satisfactory methods for determining the number of clusters. One useful approach in ecology is to examine the dendrogram (see Fig. 4-13 below) for large changes between fusions and to evaluate the changes in interpretability of cluster compositions in relation to varying numbers of clusters (McGarigal et al., 2000). Another possibility is to use bootstrapping (see McGarigal et al., 2000).

Remember that analyses with more than a few species (typical of most biological studies) require that all possible combinations be arranged, a task requiring ample computer storage capability. Remember also that one can use group average techniques to portray the relationships (dissimilarities) either among species or among sites (or both). See Breiman et al. (1984).

Sites (normal pattern)

Species (inverse pattern)	3	8	6	4	7	10	1	2	9	5	
7								3	3	2	
11								4	3	2	= raw data (i.e., no. of
4								5	4	3	individuals of species 11 at site 5)
1								2	3	4	
3	4	3	3								
6	3	3	4								
8				4	5	3	2				
10				4	4	3	5				
2				3	1	2	4				
5	3	4	1	3	2	5	1	3	2	1	
12	3	2	4	4	3	3	2	3	2	2	
9	3	4	2	3	4	1	5	1	2	3	

Figure 4-9. A two-way table of species versus sites. Species and sites are rearranged in the table according to dendrogram sequences (see Fig. 4-10). Site numbers 2,9, and 5 are characterized by species 7,11,4 and 1, thus forming one distinct group of sites with similar recurrent species (i.e., species that frequently co-occur at specific sites). Other sites and species form further groups.

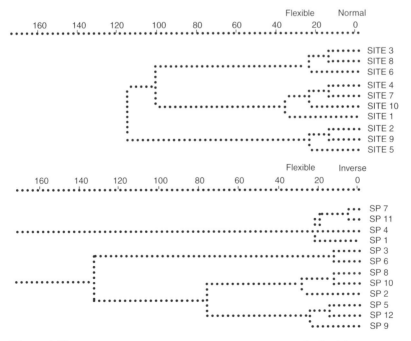

Figure 4-10. Dendrogram of species and sites resulting from data in Fig. 4-9.

223

Flexible Method (Flexible Group Method)

Lance and Williams (1966–1971) have made major contributions to quantitative community ecology. Lance and Williams (1966b) described a unifying method in which, by using a single equation and adjusting four parameters, some of the more commonly used clustering strategies could be approximated. The equation is a simple linear one.

Lance and Williams (1966) give the following parameter values to approximate five of the following well-known clustering strategies (see Fig. 4-11):

$$\text{Nearest-neighbor: } \alpha_i = \alpha_j = +1/2; \ \beta = 0; \ \gamma = -1/2$$

$$\text{Furthest-neighbor: } \alpha_i = \alpha_j = +1/2; \ \beta = 0; \ \gamma = +1/2$$

$$\text{Group average: } \alpha_i = n_i/n_k; \ \alpha_j = n_j/n_k; \ \beta = \gamma = 0$$

where

$n_i = $ # of elements in the ith group

$n_j = $ # of elements in the jth group

$n_k = $ # of elements in the kth group

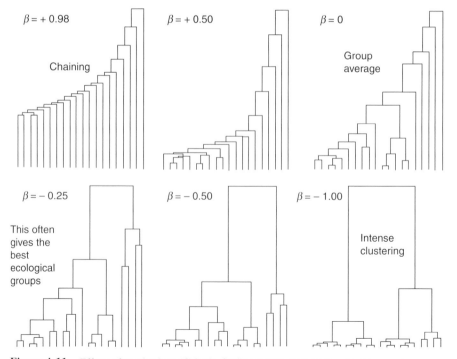

$\beta = +0.98$

Chaining

$\beta = +0.50$

$\beta = 0$

Group average

$\beta = -0.25$

This often gives the best ecological groups

$\beta = -0.50$

$\beta = -1.00$

Intense clustering

Figure 4-11.   Effects of varying beta (β) in the flexible sorting strategy (source: Lance and Williams, 1967).

Median: $\alpha_i = \alpha_j = +1/2; \beta = -1/4; \gamma = 0$

Centroid: $\alpha = n_i/n_k; \alpha_j = n_j/n_k; \beta = -\alpha_i\alpha_j; \gamma = 0$

Further variables to be used for the following strategies are given by Clifford and Stephenson (1975).

Incremental sum of squares:
$$\alpha_i = \left(n_h + n_j\right)/\left(n_h + n_k\right)$$
$$\alpha_j = \left(n_h + n_j\right)/\left(n_h + n_k\right)$$
$$\beta = -n_h\left(n_h + n_k\right)$$
$$\gamma = 0$$

How to use flexible group average

Equation: $d_{hk} = \alpha_i d_{hi} + \alpha_j d_{hi} + \beta d_{ij} + \gamma\left|d_{hi} - d_{hj}\right|$

Constants: $\alpha_i = n_i/n_k ; \alpha_j = n_j/n_k ; \beta = \gamma = 0$

Using a site matrix (Fig. 4-12), proceed as follows:

(1) **Fuse sites with the minimal distance**. Sites 3 and 5 are fused at distance 304. **Place this fusion on the dendrogram** (Fig. 4-13).

(2) Determine the **minimal distance** between all other sites (2, 4, 5) and 3–5 and between all other sites.

$$1-3,5 : 1/2\ 1069 + 1/2\ 969 + 0 + 0 = 1019$$
$$2-3,5 : 1/2\ 741 + 1/2\ 621 = 681$$
$$4-3,5 : 1/2\ 1105 + 1/2\ 946 = 1025$$
$$1-2 : 803$$
$$1-4 : 945$$
$$2-4 : 647$$

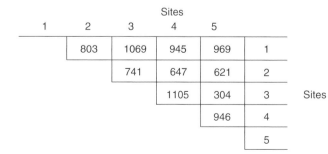

		Sites				
1	2	3	4	5		
	803	1069	945	969	1	
		741	647	621	2	
			1105	304	3	Sites
				946	4	
					5	

Figure 4-12.   Site matrix generated by the use of Euclidean Distance.

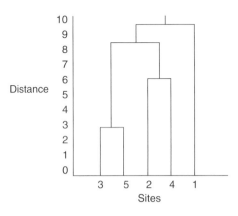

Figure 4-13.   Dendrogram for sites using flexible-group average clustering data. Data from Fig. 4-12.

The **minimal distance** here is between sites 2 and 4 at a distance of 647. This fusion is added to the dendrogram.

(3) Determine the **minimal distance** between 3,5 and 2,4 (2-3,5; 4-3,5 as previously computed and 2-2,4.

$$2,4-3,5: 1/2\ 681 + 1/2\ 1025 = 853$$
$$1-3,5: 1019$$
$$1-2,4: 1/2\ 803 + 1/2\ 945 = 874$$

The **minimal distance** here is the fusion between 2,4 and 3,5 at the distance 853. This fusion is added to the developing dendrogram.

(4) Determine the fusion distance between 1 and 3,5 (1-3,5; 1-2,4 as computed above).

$$1-3,5\ 1/2\ 1019 + 1/2\ 874 = 946$$

This fusion is at 946, which is added to the dendrogram to complete it (see Fig. 4-13).

A computer program for Flexible group average clustering is found in PC-ORD.

Many computer programs that construct dendrograms have poor quality graphics. The computer program Clustan Graphics was designed to overcome these deficiencies. It is compatible with several of the popular larger statistical programs (e.g., Statistica, SAS, etc.). See the Appendix.

4.5   SPECIES-SITE GROUPS

TWINSPAN (Two-way Indicator Species Analysis) takes species-site abundance data and then often rescales the abundance to a 1–5 scale, 5 being the most abundant class. Other abundance scales may range from 1 up to 9. TWINSPAN ordinates the species-site abundance scale data by correspondence analysis (see p. 249 in Chapter 5) and produces a re-ordered two-way table of species (Y-axis) and sites (X-axis). Details of this process are presented in Jongman et al. (1995). Table 4-1 shows 58 species of plants (names abbreviated) in a two-way table. The numbers on the top are the sample numbers (plots, quadrats, sites). The first plot from left to right is 27 then comes 34, and so forth. The numbers on the left (Y-axis) are the species code numbers followed by their abbreviated names. The numbers within the matrix (2 to 7) represent the abundance classes for each species at each plot. The numbers on the right (Y-axis – species groups cut-off) and at the bottom (X-axis – site groups cut-off) are the cut-off numbers (the cut-off levels representing standard deviations). Note that there are different columns and rows of these numbers. At the first cut-off level for species (Y-axis), the species group would include No. 21 through No. 45 and another species group for No. 22 through No. 58. The second cut-off level produces four species groups, that is, No. 21 through No. 40, No. 1 through No. 45, No. 22 through 57, and No. 15 through No. 58. For sites, the first cut-off level is No. 27 through No. 28 and No. 30 through No. 23 (a total of two site groups). The second cut-off level produces four site groups, and so forth. The selection of the cut-off level is arbitrary. Tropical latitude communities are often cut-off at higher levels than for temperate communities, showing a greater diversity of tropical species-site groups. Thus, a dendrogram produces either species groups (one figure) or site groups (another

Table 4-1.   TWINSPAN 2-way table of species (y-axis) and sites (x-axis)

TWO-WAY ORDERED TABLE

```
                        2341234   1     11 12333112231123322233   12
                        7405251347216834504128185969639018072973
21    Epiang            333- - - - - - - - - - - - - - - - - - - - - - - - - - - - - - - - - - -2      000
 9    Aranud            35423233323342323333- - - - - - - - - - - - - - -2- - -      00100
28    Linbor            33333- 223- 2- 22- - - - 2- - 2- - 2- - - 23- - - - - - 2- - - -      00100
 5    Adebic            - - - 2222- - - - - - - 2- 2- - - - - - - - - - - - - - - - -      001010
13    Ceasan            - - - - - - - 22223- - - - - - - - - - - - - - -2- - - - - - - - -      001010
38    Shecan            - - - 4477- - - - - - - - - - - - - - - - - - - - - - - - - - - - -      001010
47    Vaccap            - - - - - - - - 2- 222- - 2- - - 2- - - - - - - - - 2- - - - - - - -      001010
10    Arcuva            - - - - 264- - - - - - - - - - - - 22- - - - - - - - - - - -2- - - - -      001011
16    Cliuni            23432- 3- - - 3- - 232- 222- 22- - - - - - - - - - - - - - -      001011
17    Corcan            45423- - - 23- - - - 23- - - - - 2- - - - - - - - - - 22- - - - - -      001011
30    Lonuta            - - - - - - - - - 2- - - - - 2- 2- - - - - - - - - - - - - - - - -      001011
37    Senpse            334- - 34- - - - - - - - - - - - - - - - - - - - - -2- - - - -      001011
 8    Apoand            - - - - - - - - 2- 2- - - - - - - - - 2- 2- 2- - - - 2- - - - - - - - -      001100
19    Dishoo            - - - 2- 23- - - - - - - - - 2- 33- - 2- - - - - - - - - - - - - -      001100
39    Smiste            - - 3- - - 22- - 2- - 222- 22322- - 222- - 2- - - - - - - - -      001100
11    Astlae            2- - - - - - - - - - - - - - - - - 2- - - - - - - - - 2- - - - - - - - -      001101
34    Rosa              - - - 233332323- 2- 45334232232322333423- - - - -2      001101
49    Viola             - - - - - - 22- 222- - - - - - - - 2- - - - - - - - 2- 332- 3- - - - - -      001101
56    Oryasp            656- 3- 3- 2- - 22- - - 2- 33- - - - - 33223332- - - - - - - -      001101
```

(Continued)

Table 4-1.   (*Continued*)

40	Spibet	33444333633222263332- 3222233333454332- - - -	00111
1	Piceng	- 2222- - - - - - - - - -	010000
3	Pinpon	- 33- - - - - - - - -	010000
27	Genama	- 2- - - - - - - 2- - - - -	010000
53	Carcon	- - - - - - - - - - - - - - - - - - 3- 2- - - 22- - 234- - - - - - - - -	010000
24	Filarv	- - - - - - - - - - - - - - - - - - - 2- - - - - - - - - 2- - - - - - - - -	010001
20	Dissp	- 22- - - - - - - - - - - - - - - 22- - - - - - - 23- - - - - - - - -	01001
36	Rumace	- - - - - - - - - - - - - - - - - - - 33- - 2- - - - - - - - - - - - - -	01001
43	Thalic	- 23- - - - - - - - - - - - - - - -	01001
50	Xerten	- - - - 23342- - - - - - - 3366777774- - - 3- - - - - - - - - - -	01001
6	Amealn	- - - - - - - - - - - 2- 2- - 2- - - 332232- 2- 3333- - - - 2- - -	0101
52	Calrub	- - - - 233- - - 2- - - - 2- - 4- 33- 23- - 234- - 324- - - - - 3	0101
12	Berrep	43334343534343333576333336646656424434222-	0110
41	Symalb	444- 334- - - 2- - - - - 2- 4466344333336532333- - - 4	0110
54	Cargey	- - - 2- - 3- - - - - - 222- - - - - - - - 3453- - 33- 2322- - - -	0110
35	Rubpar	- - - - - - - - - - - - - - - 63- 333633- - - - - 334- 223- - - 2	0111
45	Triagr	- 2- - - - - - - - - - - - - - - 33- 2- - - - 2- - - - - - - 32- - - - -	0111
22	Epipan	3- - - 222- - - - - - - - - - - 42- 2- - 322- - 434- - 722- - - 6	100
26	Fravir	233- - - - - 2223- 2- - - - - - 333233322222- - 344- 222	100
48	Vertha	222- - 33- - - - - - 2- - - 22322222332- 23222332- 234	100
55	Fes/De	2- - - - - 2- - - - - - - - - - - - - - 32- - - 2- - - 2- - - - - 3	100
31	Pacmyr	- 22222- - 22222-	101
44	Tradub	- 2- - - 22- - - - 2- - - -	101
46	UnkCom	- - - - - - - - - - - - - - - - - - 2- - - - - - 2- - - - - - - - 22- - - -	101
57	Phleum	- 44- - - - 2- 2- 24- - - -	101
15	Cirvul	33- 423- 2222426632542- - 2223263333445323775	11000
14	Cirarv	333- - 33- - - - - - - - - - - 2- - - - - 32- - - 23- - - 64- - 23	11001
23	Epiwat	222- - - - - - - - - - - - 2- - - - - - - - - - - - - 222- 2- 222	11001
29	Loncil	- - - - - - - 33- 24- 2- - 2- - - - - - - - 33223- - - - - 33323-	11001
42	Taroff	2- 2- 2- - - 2- - - 22- - - -	11001
7	Antluz	- - - - - - 2- - - 2- - - - - - 22- - - - - - 2- - - - - 2- - 22- - 2-	1101
25	Fraves	- - - 2223- 223- - - -	1101
2	Pincon	- 2- - - 3- - - -	11100
18	Cyn	- 233- - - -	11100
32	Pruvul	- 3- - - 33- - - -	11100
4	Acergl	- 223323	11101
33	Rhamnu	- - - - - - - - - - - - - - - - - - - 2- - - - - - - - - - - 322222	11101
51	Agr.sp	- 32- - - - - - - - - - - - - - - 2- - - - - - - - - - - 2- - - 2-	1111
58	Poapra	334- 2- - - 55- - - 2	1111

```
        00000000000000000000000000000000000000111111
        0000000000000000011111111111111111111001111
        00011111111111110000000001111111111
          000011111111100001111110000000111
            000000111      001110000011
            000011         00111
        *********** TWINSPAN Completed *********
```

figure) whereas a two-way table produces both groups in one table. The main criticisms of TWINSPAN are: (1) the cut-off levels for species or sites are arbitrary, (2) sites close in species composition may be separated based on CA, DCA, and so forth. (see p. 249 in Chapter 5), and (3) TWINSPAN assumes that a strong gradient dominates the structure (Legendre and Legendre, 1998). Twinspan performs poorly with more than one important environmental gradient (McCune et al., 2002). Typical community data sets are often better represented by cluster analysis than with TWINSPAN.

4.6 MANTEL TEST

The Mantel Test (Mantel, 1967) **compares two dissimilarity (distance) matrices** using Pearson correlation (e.g., comparison of the morphological characteristics of populations of crabs from two regions using the modified Jaccard index $[1 - J]$). The matrices must be of the same size (i.e., same number of rows). **It would appear that the Mantel Test could be used to evaluate the internal structure in two sets of samples (e.g., pre-storm vs. post-storm beach communities) by comparing the two (dis)similarity matrices (and, thus, indirectly compare the bases for two dendrograms)**. This has not been done yet, so far as is known. The computer program Statistica runs a Sorenson (= Bray–Curtis) distance measure on the matrices *before* doing the Mantel test. Other distance measures can also be used. McCune and Mefford (1999) have a computer program for performing the Mantel test. After the Mantel statistic has been calculated, the statistical significance of the relationship is tested by a permutation test or by using an asymptotic *t*-approximation test. The minimum number of permutations (randomizations) recommended by Manly (1997) is 1000. The Mantel and ANOSIM (see p. 230 below) procedures (where distances are transformed into ranks) produce similar probabilities (Legendre and Legendre, 1998). Some of the problems associated with the Mantel test include (1) weakness in detecting spatial autocorrelation where the spatial pattern is complex and not easily modeled with distance matrices, (2) a larger number of data points may be needed for field experiments than is usually obtained, and (3) multivariate data are summarized into a single distance or dissimilarity and it is not possible to identify which variable(s) contributed the most (Fortin and Gurevitch, 2001). This lattermost criticism is characteristic of the majority of clustering and ordination techniques.

The use of Partial Mantel tests can distinguish the relative contributions of the factors of a third matrix considered as covariables. Mantel and partial Mantel tests can produce complementary information that ANOVA cannot provide and may do a better job than ANOVA in detecting block effects (Fortin and Gurevitch, 2001). Thus, it is possible to distinguish the effects of spatial pattern from those of experimentally imposed treatment effects with these Mantel tests.

Example:

The following example is taken from McCune and Mefford (1999). The data consist of two matrices, one with 50 species of plants, the second with six environmental factors (e.g., temperature, soil pH, etc.). The data are entered as listings and either Mantel's asymptotic approximation method or a randomization method is used to formally test for a significant relationship between the species data (based on Bray–Curtis dissimilarities) and the standardized environmental data (based on Euclidean distance) (Table 4-2).

Table 4-2. Mantel's asymptotic approximation method (source: McCune and Mefford, 1999).

```
DATA MATRICES
------------------------------------------------------------------------------------
 Main matrix:
   19 STANDS (rows)
   50 SPECIES (columns)
      Distance matrix calculated from main matrix.
      Distance measure = Sørensen
 Second matrix:
   19 STANDS (rows)
    6 ENVIRON (columns)
      Distance matrix calculated from second matrix.
      Distance measure = Sørensen
  LISTING OF FIRST DISTANCE MATRIX
    1 STAND1
  0.000E + 00 4.286E − 01 2.308E − 01 8.182E − 01 6.296E − 01 4.286E − 01 5.333E − 01
  8.667E − 01 6.000E − 01 6.774E − 01 5.625E − 01 7.931E − 01 6.000E − 01 6.129E − 01
  6.774E − 01 6.667E − 01 7.500E − 01 5.556E − 01 7.222E − 01
    2 STAND2
  4.286E − 01 0.000E + 00 5.652E − 01 1.000E + 00 9.167E − 01 6.800E − 01 7.037E − 01
  8.519E − 01 6.296E − 01 8.571E − 01 7.241E − 01 8.462E − 01 6.296E − 01 7.143E − 01
  7.857E − 01 8.519E − 01 8.095E − 01 7.576E − 01 7.576E − 01
  .
  .
  etc.
  .

  LISTING OF SECOND DISTANCE MATRIX
    1 STAND1
  0.000E + 00 4.636E − 03 3.652E − 03 8.246E − 02 8.383E − 02 8.366E − 02 7.859E − 02
  2.197E − 01 2.143E − 01 2.235E − 01 2.151E − 01 2.777E − 01 2.615E − 01 2.631E − 01
  2.653E − 01 2.154E − 01 2.095E − 01 2.069E − 01 2.027E − 01
    2 STAND2
  4.636E − 03 0.000E + 00 9.846E − 04 7.784E − 02 8.775E − 02 8.757E − 02 8.254E − 02
  2.153E − 01 2.172E − 01 2.191E − 01 2.174E − 01 2.734E − 01 2.582E − 01 2.638E − 01
  2.632E − 01 2.110E − 01 2.124E − 01 2.025E − 01 2.056E − 01
  .
  .
  etc.
  .

  TEST STATISTIC: t-distribution with infinite degrees of freedom
            using asymptotic approximation of Mantel (1967).
            If t< 0. then negative association is indicated.
            If t> 0. then positive association is indicated.
```

Table 4-2.   (*Continued*)

```
STANDARDIZED MANTEL STATISTIC: .481371 = r
            OBSERVED Z = .2838E + 02
            EXPECTED Z = .2645E + 02
        VARIANCE OF Z = .1222E + 00
  STANDARD ERROR OF Z = .3496E + 00
                    t = 5.4969
                    p = .00000005

MANTEL TEST RESULTS: Randomization (Monte Carlo test) method
---------------------------------------------------------------------
        -.054504 = r = Standardized Mantel statistic
     .246553E + 02 = Observed Z (sum of cross products)
     .256640E + 02 = Average Z from randomized runs
     .264561E + 01 = Variance of Z from randomized runs
     .198977E + 02 = Minimum Z from randomized runs
     .299859E + 02 = Maximum Z from randomized runs
             1000 = Number of randomized runs
              723 = Number of runs with Z > observed Z
                0 = Number of runs with Z = observed Z
              277 = Number of runs with Z < observed Z
          .278000 = p (type I error)
---------------------------------------------------------------------
p = proportion of randomized runs with Z more extreme or equal
    to the observed Z: for positive association.
p = (1 + number of runs >= observed)/(1 + number of randomized
    runs)
Positive association between matrices is indicated by observed
    Z greater than average Z from randomized runs.
```

Note that in this case, the association is negative: the observed Z is less than the average Z from randomized runs and the sign of the standardized Mantel statistic is negative. Because it is negative, the p value is calculated using the number of randomized runs with Z less than or equal to the observed value.

4.7 ANALYSIS OF SIMILARITIES

The Analysis of Similarities (ANOSIM) is a multivariate test that compares multi-species abundance or biomass data. ANOSIM is a simple nonparametric permutation (randomization) procedure applied to a rank similarity matrix for comparing groups of sites (Clarke, 1993; Clarke and Warwick, 2001). The ANOSIM procedure is described as follows:

(1) From site by species data, calculate similarities between every pair of sites (e.g., using Bray–Curtis or other measures).

(2) Rank the similarities, with 1 being the highest similarity between two sites or replicates.

(3) Using the rank similarity matrix (Table 4-3) compute R.

$$R = \frac{\left(\bar{r}_B - \bar{r}_w\right)}{1/2\,M} \qquad M = n(n-1)/2$$

where

R = degree of separation among groups

$\bar{r}_B$ = average of rank similarities among replicates between different groups

$\bar{r}_W$ = average of rank similarities among replicates within groups

n = total number of replicates

R ranges between −1 and +1, with 0 indicating that similarities between and within groups are the same and $R = 1$ indicating replicates within groups are more similar to each other than those from different groups. Negative values for R can occur, although they are rare (see Chapman and Underwood, 1999).

(4) The replicates are randomly reshuffled across the groups, R recalculated, and the process repeated for all possible combinations. When the number of possible combinations is very large, a random subset (e.g., 5000) of all possible permutations is used.

(5) The significance level is calculated by referring the calculated value of R to its permutation distribution.

Table 4-3. Rank similarity matrix – Frierfjord macrofauna (source: Clarke and Warwick, 2001).

	B1	B2	B3	B4	C1	C2	C3	C4	D1	D2	D3	D4
B1	—											
B2	33	—										
B3	8	7	—									
B4	22	11	19	—								
C1	66	30	58	65	—							
C2	44	3	15	28	29	—						
C3	23	16	5	38	57	6	—					
C4	9	34	4	32	61	10	1	—				
D1	48	17	42	56	37	55	51	62	—			
D2	14	20	24	39	52	46	35	36	21	—		
D3	59	49	50	64	54	53	63	60	43	41	—	
D4	40	12	18	45	47	27	26	31	25	2	13	—

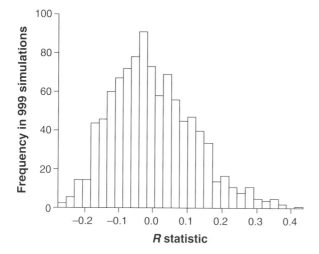

Figure 4-14. Frierfjord macrofauna. Simulated distribution of the test statistic R under the null hypothesis of no site differences. This contrasts with an observed value of 0.45 for R. Source: Clarke and Warwick (2001).

The resulting calculated value for R in the above example is 0.45. Because 0.45 is outside the range for permuted R values (Fig. 4-14), the null hypothesis (H_0) is rejected (i.e., the groups are dissimilar).

One can then compare specific pairs of sites if desired. PRIMER (see Appendix) has a computer program for ANOSIM. ANOSIM for more complex field and laboratory designs (e.g., two-way nested and two-way crossed cases) are also discussed by Clarke and Warwick (2001). Also see Clarke (1993).

4.8 INDICATOR SPECIES ANALYSIS

There are many types of indicator species ranging from individual species (e.g., some sponges) that are very sensitive to sewage effluents (i.e., they disappear) to species groups (e.g., the polychaete family Capitellidae) whose members often become abundant when bathed in sewage effluents. This analysis (Dufrêne and Legendre, 1997) identifies indicator species based on abundance and fidelity (relative frequency) of occurrence in a community. **It requires data from two or more sample units**. Indicator values (range = 0 for no indication to 100 for perfect indication) are presented for each species. The statistical significance of the maximum indicator value recorded for a given species is generated by a Monte Carlo test.

Example

Indicator species are sought from a large group of plants. The following example shows just nine of 35 species and their values, taken from McCune and Mefford (1999). The results indicate that of the species shown, species 1-3, 33 and 35 (MaxGrp) are the best indicator species (Table 4-4).

Input data: 35 stands by 35 Species

RELATIVE ABUNDANCE in group. Percentage of perfect indication (average abundance of a given Species in a given group of Stands over the average abundance of that Species in all Stands expressed as a Percentage)

				Group		
			Sequence:	1	2	3
			Identifier:	2	13	3
			Number of items:	15	5	13
Column	Ave	Max	MaxGrp			
1 Bry	33	90	13	10	90	0
2 Bryfus	33	41	13	25	41	35
3 Cndcon	33	54	3	0	46	54
4 Cetfen	33	100	2	100	0	0
5 Cetpin	33	100	3	0	0	100

etc.

.

.

32 Usnhir	33	69	2	69	0	31
33 Usnlap	33	43	13	30	43	27
34 Xanfal	33	40	2	40	40	20
35 Xanmon	33	36	3	30	35	36
Averages	30	56		32	26	31

RELATIVE FREQUENCY in group. Percentage of perfect indication (Percentage of Stands in given group where given Species is present)

				Group		
			Sequence:	1	2	3
			Identifier:	2	13	3
			Number of items:	15	5	13
Column	Ave	Max	MaxGrp			
1 Bry	9	20	13	7	20	0
2 Bryfus	32	40	13	27	40	31
3 Cndcon	12	20	13	0	20	15
4 Cetfen	2	7	2	7	0	0
5 Cetpin	5	15	3	0	0	15

etc.

.

.

32 Usnhir	23	47	2	47	0	23
33 Usnlap	51	60	13	47	60	46
34 Xanfal	34	40	2	40	40	23
35 Xanmon	75	80	13	67	80	77
Averages	32	32		22	25	21

Table 4-4. (*Continued*)

```
INDICATOR VALUES (Percentage of perfect indication.
based on combining the above values for relative
abundance and relative frequency)
```

				Group		
		Sequence:		1	2	3
		Identifier:		2	13	3
	Number of items:			15	5	13
Column	Ave	Max	MaxGrp			
1 Bry	6	18	13	1	18	0
2 Bryfus	11	16	13	7	16	11
3 Cndcon	6	9	13	0	9	8
4 Cetfen	2	7	2	7	0	0
5 Cetpin	5	15	3	0	0	15

```
           etc.
            .
            .
```

32 Usnhir	13	32	2	32	0	7
33 Usnlap	17	26	13	14	26	12
34 Xanfal	12	16	2	16	16	5
35 Xanmon	25	28	13	20	28	27
Averages	9	17		9	11	8

```
MONTE CARLO test of significance of observed maximum
indicator value for Species
 1000 permutations.
Random number seed:    1878
```

		IV from		
	Observed	randomized		
	Indicator	groups		
Column	Value (IV)	Mean	S.Dev	$p*$
1 Bry	18.0	11.2	6.23	0.086
2 Bryfus	16.4	22.1	8.41	0.792
3 Cndcon	9.3	13.3	6.59	0.848
4 Cetfen	6.7	9.2	4.77	1.000
5 Cetpin	15.4	11.4	6.62	0.316

```
           etc.
            .
            .
```

32 Usnhir	32.4	21.7	7.89	0.162
33 Usnlap	25.9	28.0	7.93	0.466
34 Xanfal	15.9	23.2	8.73	0.727
35 Xanmon	27.7	33.5	5.76	0.963

*proportion of randomized trials with indicator value
equal to or exceeding the observed indicator value.
$p = (1 + \text{number of runs} \geq \text{observed})/(1 + \text{number of}$
randomized runs)

SUMMARY

The similarity or dissimilarity measures, forming the backbone of most multivariate clustering and ordination techniques, which are favored by many ecologists are Jaccard and Bray–Curtis. The most popular clustering techniques in marine field studies are Group Average and Flexible group average clustering. For many of these techniques, it is impossible to choose a "best method" because of the heuristic nature of the methods (van Tongeran, in Jongman et al., 1995; Anderson, 2001). The choice of an index must be made based on the investigator's experience, the type of data collected, and the ecological question to be answered. When comparisons are made, the Jaccard index is among the least sensitive of the similarity (or dissimilarity) indices (van Tongeran, in Jongman et al. 1995). The Bray–Curtis coefficient satisfies at least a half-dozen criteria (see Clarke and Warwick, 2001). Similarly, hierarchical agglomerative techniques (e.g., particularly Group Average) have proven to be very useful to ecologists in the construction of dendrograms.

Chapter 5

Community Analysis: Ordination and Other Multivariate Techniques

5.1 INTRODUCTION

Quantitative field sampling and direct community gradient analysis began in 1930 (Gauch, 1982). Computer analysis of communities commenced in 1960 and modern quantitative community analysis in 1970 (Gauch, 1982). Multivariate analysis is used extensively in community studies. It helps ecologists discover structure and provides a relatively objective summary of the primary features of the data for easier comprehension. However, it is complicated in theoretical structure and in operational methodology (for a recent review see Jongman et al., 1995; Legendre and Legendre, 1998; McGarigal et al., 2000; and McCune et al., 2002). The following pages briefly summarize the discipline of multivariate techniques.

Analysis of community data is accomplished by (1) direct gradient analysis, (2) ordination, and (3) classification (Gauch, 1982). Direct gradient analysis uses environmental factors directly (e.g., wet to dry, clay to sand) and often assumes a normal or Gaussian distribution along one or more dimensional environmental axes (for example, distribution of plants along a wet to dry gradient – see Figs. 5-1 and 5-2). Direct gradient analysis is common in terrestrial ecology but is seldom used in marine ecology. **Ordination is a process of ordering species or samples (as points in space) along a dimensional line in multidimensional space (ter Braak, in Jongman et al., 1995). It is a data reduction method. Ordination typically produces a two-dimensional graph (scatter diagram – Axis 1 versus Axis 2, Axis 1 versus Axis 3, Axis 2 versus Axis 3, etc.) in which similar species or samples are near one another and dissimilar ones apart (Figs. 5-3 and 5-4). Ordination in ecology commonly uses species abundances as a data base**. Interpretation of relationships of axes with environmental variables are typically a second, separate step

Quantitative Analysis of Marine Biology Communities: Field Biology and Environment
by Gerald J. Bakus
Copyright © 2007 John Wiley & Sons, Inc.

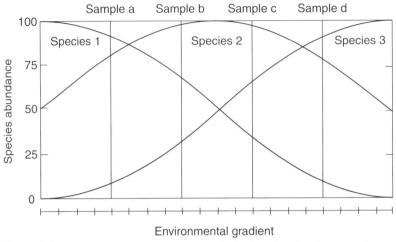

Figure 5-1.   Theoretical Gaussian curve for the distribution of species along an environmental gradient (source: Gauch, 1982).

in community analysis, redundancy analysis (RDA) and Canonical Correspondence Analysis (CCA) being the most notable exceptions. Classification groups similar entities (species or sites) into clusters. These clusters can then be displayed visually in the form of a dendrogram (see p. 222 in Chapter 4 and Fig. 5-5). Both ordination and classification use a two-way data matrix (i.e., a two-way table) as input data (i.e., plots, quadrats, or samples, and species). **The results from ordination often provide useful information with typically from one to three dimensions (axes).**

Reasons for using multivariate analysis are presented in Table 5-1. The most frequently encountered multivariate methods in ecology are listed in Table 5-2. Principal Component Analysis (PCA) was one of the first multivariate analyses to be used in biology. It is the basic eigenanalysis technique (McCune and Mefford, 1999). **PCA reduces many variables (e.g., 20 or 30 variables, as points in space) to a few principal components (e.g., 2 or 3).** Some species are normally distributed along an environmental gradient, others are not (Gauch, 1982). Figure 5-2 shows several species of plants that are not normally distributed (e.g., *Artemesia frigida, Utricularia vulgaris*, etc.). Important environmental gradients are often intercorrelated (Barbour et al., 1999). **Environmental gradients may not be explainable solely in terms of environmental variables**. The effects of competition or predation may produce species zones or more complex patterns on environmental gradients (Green, 1979). Although studies of biological interactions are strongly desirable, there may be neither time nor money to study the effects of biological controlling factors by conducting experiments in many pelagic and deep sea marine environments (e.g., community studies by ship or by using submersibles).

Eigenanalysis produces eigenvectors and eigenvalues. **Eigenvectors are values that specify the direction of points along an axis in species space. Eigenvalues**

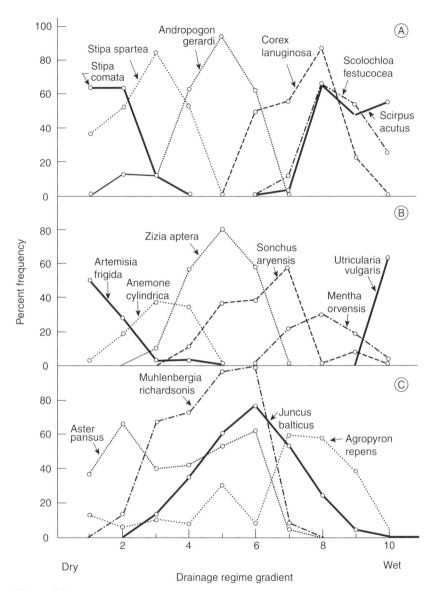

Figure 5-2. Distribution of 16 plant species along a drainage regime gradient in Nelson County, North Dakota. Fig. (A) shows graminoids (grasses) and sedges, (B) forbs (non-grass herbaceous plants), and (C) graminoids and forbs of ubiquitous distribution (source: Gauch, 1982).

describe the variance or scatter of points along an axis (Tables 5-3 and 5-4; Fig. 5-6; see Legendre and Legendre, 1998 for details). The components (PC scores) are produced as a linear combination of the original variables by reference to their eigenvector weights (Legendre and Legendre, 1998). PCA is an excellent

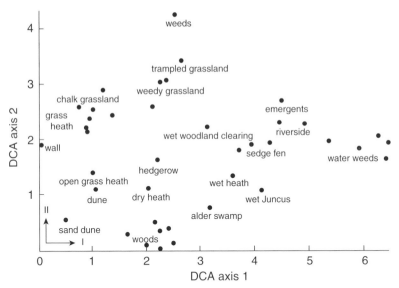

Figure 5-3. Detrended correspondence analysis ordination of a vegetation survey of southeast England. The first axis extends from dry to wet conditions, the second axis from woodland to weed communities. The axis scales are in units of average standard deviation of species turnover (source: Gauch, 1982).

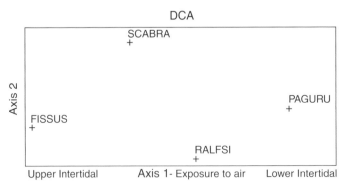

Figure 5-4. DCA ordination of four intertidal (littoral) species on a rocky beach, Royal Palms, Palos Verdes Peninsula, southern California.

FISSUS = *Chthamalus fissus* (small acorn barnacle)
SCABRA = *Lottia scabra* (limpet)
RALFSI = *Ralfsia* sp. (encrusting red alga – black color)
PAGURU = *Pagurus samuelis* (hermit crab)

multivariate analysis for typically normal distributions (e.g., morphological characters in living organisms and in fossils). However, PCA and several other multivariate techniques produce an "arch effect" or mathematical anomaly (Fig. 5-7). The arch effect occurs because of the assumption that species distributions change

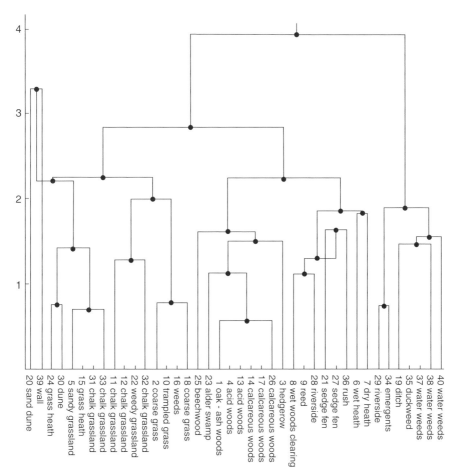

Figure 5-5. Two-way indicator species analysis (TWINSPAN) showing a dendrogram of 40 composite samples from a vegetation survey of southeast England. The sample sequence reflects moisture status from xeric (sand dune and wall) to aquatic (water weeds). The ordinate is the average Euclidean distance in the detrended correspondence analysis sample ordination. Similar samples are joined at a low level in the dendrogram, whereas dissimilar samples are not joined until high levels (source: Gauch, 1982).

linearly along an environmental gradient, often not met in nature. For this and other reasons, ecologists have sought different approaches. Gauch (1982) recommended Detrended Correspondence Analysis (DCA) because it goes one step further by diminishing the "arch effect". Smith and Bernstein (1985) and Wartenberg et al. (1987) discuss the merits of DCA. The ideal ordination is presented in Fig. 5-8a. A comparison of different ordination techniques, using the same data base, is shown in Figs. 5-8-bc.

For a review of multivariate analysis, principal component analysis, factor analysis, clustering techniques, dendrograms, morphometrics and FORTRAN programs

Table 5-1.   Reasons for conducting community studies using multivariate analysis (source: Gauch, 1982).

Description

    Description of a given community
    Delimiting and naming of communities
    Mapping of communities within a region
    Identification of recurring species groups
    Assignment of new community samples to previously defined community types

Understanding

    Structure of communities
    Regulation and maintenance of communities
    Distribution of species and communities along environmental gradients
    Competitive interactions of species
    Species niches and habitats

Prediction and management

Prediction of community from environment or environment from community

Prediction of course of succession or response to disturbance

Land use recommendations

Management of grazing, forest, and recreational areas

Relating community data to other data bases (fire, harvesting, weather)

Data reduction for inventory

---

for several of these see Orloci (1967, 1975), Noy-Meir (1970, 1971, 1973), Blackith and Reyment (1971), Cooley and Lohnes (1971), Allen and Skagen (1973), Beals (1973), Davis (1986), Mahloch (1974), Clifford and Stephenson (1975), Morrison (1976), Smith (1976), Gauch (1977), Kendall (1980), van der Maarel (1980), Janson and Veglius (1981), Smith (1981), Gauch (1982), Shin (1982), Cureton and D'Agostino (1983), Pielou (1984), Smith and Bernstein (1985), Manly (1986), Digby and Kempton (1987), Peet et al. (1988), Podani (1994), Jongman et al. (1995), Statsoft, Inc. (1995), Legendre and Legendre (1998) McCune and Mefford (1999), McGarigal et al., (2000), Clarke and Warwick (2001), Paukert and Wittig (2002), Leps and Smilauer (2003), and Miller (2005). Barbour et al. (1999) give an excellent brief summary and McCune et al. (2002) an excellent detailed discussion of ordination techniques. A brief description of the software program CANOCO by ter Braak is presented in the Appendix.

    Outliers (i.e., data points far removed from the majority of points) need to be removed before proceeding with an ordination because they distort the results. A good example of this is shown by McCune et al. (2002) where the correlation coefficient (r) is $+0.92$ with the outlier included and -0.96 with the outlier excluded. Outliers can strongly effect the results from PCA. The simplest method for removing outliers is to standardize the data by subtracting the mean and dividing by

Table 5-2.   Multivariate and hypothesis testing methods used in ecology (source: Jongman et al., 1995; Legendre and Legendre, 1998; Barbour et al., 1999; McCune and Mefford, 1999; McGarigal et al., 2000; Clarke and Warwick, 2001; among others).

1. Principal Coordinate Analysis
   Step-across Principal Coordinates
   Gower Coordinate Analysis

2. Multiple Regression and Correlation (MR and MC)

3. Canonical Correspondence Analysis (CCA)

4. Nonmetric Multidimensional Scaling (MDS)

5. Factor Analysis
   Principal Component Analysis
   Correspondence Analysis (Reciprocal Averaging)
   Correspondence Analysis (Detrended) (DCA)
   Redundancy Analysis

6. Direct Ordination or Weighted Averages

7. Polar Ordination

8. Gaussian Ordination

9. Canonical Correlation Analysis

10. Discriminant Analysis (Canonical Variate Analysis) (DA)

11. Multivariate Analysis of Variance

12. Non-Parametric Multiple Analysis of Variance

13. Multiple Analysis of Covariance

14. Analysis of Similarities

15. Multiresponse Permutation Procedures

Table 5-3.   Principal component axes, the species and their eigenvectors.

		Principal component axes		
		I	II	III
Eigenvectors	1	−0.22	−0.03	−0.04
Species	2	0.11	−0.23	−0.01
	3	0.20	−0.07	−0.06

the standard deviation for each variable. This is followed by an inspection of the data for entities with any value more than 2.5 standard deviations, for example, from the mean on any variable (McGarigal et al., 2000). See McCune and Mefford (1999) for further details and for a PC-ORD computer program that removes outliers.

Table 5-4. Example of eigenvalues*. The eigenvalue of Principal Component # 1 = 50% of the variance. This is the most important axis. Principal Component # 2 = 20% of the variance, the second most important axis. Principal Component # 3 is almost as important (18%) as # 2. Principal Component # 4 is not used because it represents such a small proportion (8%) of the total variance.

Principal Component # 1
Eigenvalue = 20.1
Percent of Total Variance = 50
Cumulative Percent of Total Variance = 50
Principal Component # 2
Eigenvalue = 8.2
Percent of Total Variance = 20
Cumulative Percent of Total Variance = 70

Principal Component # 3
Eigenvalue = 7.1
Percent of Total Variance = 18
Cumulative Percent of Total variance = 88

Principal Component # 4**
Eigenvalue = 3.3
Percent of Total Variance = 8
Cumulative Percent of Total Variance = 96
Residue Percent Variance = 4

---

*Sequence of axes of diminishing importance. Typically, the first three eigenvalues are 40–90% of the total variance.

**Discarded because it represents only 8% of the total variance.

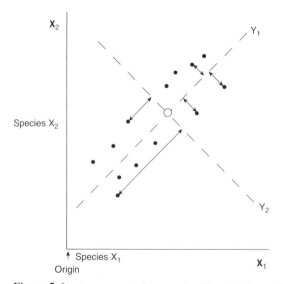

Figure 5-6. Arrangement of two species (X_1 and X_2) in multidimensional space. One of the earlier stages in the ordination process.

o = mathematical center of all data
←→= distance between site and mean for Y_1 axis
←- - - - - - - - - →= distance between site and mean for Y_2 axis

244

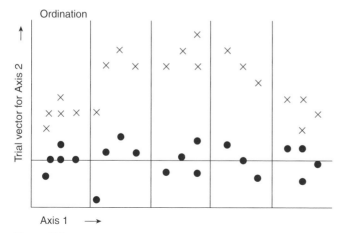

Figure 5-7.   Method of detrending used in detrended correspondence analysis. The (x) indicate sample scores before detrending, the (●) sample scores after detrending. The first axis is subdivided into five segments and a local mean is subtracted in each (source: Gauch, 1982).

Multivariate programs are listed in the Appendix. Some of the programs are available for use on microcomputers.

Examples of the Use of Multivariate Analyses in Marine Science

Figure 5-9 shows two examples of how one can use multivariate analysis in field studies.

Summaries of Ecological Multivariate Analyses

5.2   PRINCIPAL COMPONENT ANALYSIS

PCA is a data reduction method for normally distributed data (e.g., unimodal or Gaussian data such as bone lengths) or data with approximately linear relationships among variables (McCune et al., 2002). Although formerly the most common

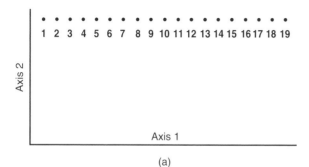

(a)

Figure 5-8a.   A theoretically desired ordination result. The straight line indicates no variance along the second axis.

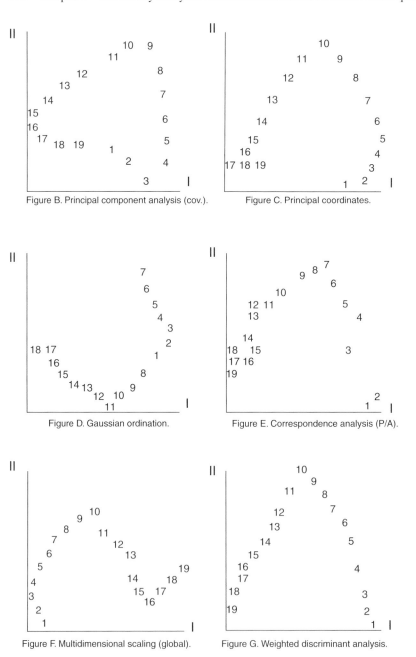

Figure B. Principal component analysis (cov.).

Figure C. Principal coordinates.

Figure D. Gaussian ordination.

Figure E. Correspondence analysis (P/A).

Figure F. Multidimensional scaling (global).

Figure G. Weighted discriminant analysis.

(b)

Figure 5-8b-i.   Comparison of 10 multivariate analyses by Ernie Iverson and Robert Smith. Note the strong "arch effect" along the second axis in Fig. 5-8 b-g.

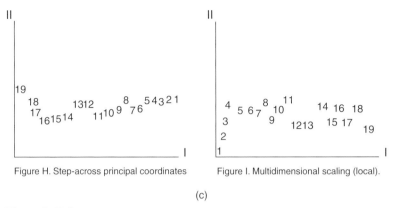

Figure H. Step-across principal coordinates   Figure I. Multidimensional scaling (local).

(c)

Figure 5-8b-i. *(Continued)*

ordination technique in ecology, it is not suitable for most field studies. However, the consequences of using PCA compared with other ordination techniques are not severe when beta diversity is low (McCune et al., 2002). True outliers need to be eliminated prior to the final analysis (McGarigal et al., 2000). PCA minimizes the total residual sum of squares often fitting straight lines to the species data. It reduces 20–30 variables (i.e., as dimensions in space) to 2 or 3 principal components. The variances of the axes are maximized. From the extensive experience of Robert Smith (pers. comm.) and my own limited studies, the first three axes often comprise 40–90% of the variance, but this may not hold true in other cases. The maximum number of principal components equals the number of variables in the original sample (Fowler et al., 1998). The axes are orthogonal and represent the linear combinations of species measurements. The first axis or principal component has the greatest variance (i.e., is the most important gradient). The axis length is expressed as standard deviation units of species scores. The axes are constructed by finding the eigenvalues and eigenvectors of a standardized sample covariance (correlation) matrix. During the calculations, the axes are rotated orthogonally to maximize the variances (often using Varimax rotation, which maximizes the variance of important variables and is generally considered to be the superior method of rotation for detecting clusters of sample units – McGarigal et al., 2000; McCune et al., 2002). The resulting eigenvalues represent the total amount of variance along the new (i.e., rotated) axis. The resulting eigenvectors show the position (i.e., specific variance) of individual measurements and their clusters. Tests of significance are often not appropriate for PCA (Green, 1979).

One major problem with principle component analysis is that correlation coefficients and variance-covariance matrices include data on conjoint absences. Conjoint absences tell nothing biological about an association and may distort the results. Therefore, **sampling that results in numerous quadrats with zeros or blanks for a species (i.e., no occurrence) would not be suitable for principal component analysis**. This phenomenon might be common in certain tropical habitats where diversity is high and many species are uncommon or rare. Another major problem

Here we show the use of DCA for two different situations:

1. A bay or estuary is polluted from industrial sources. Measurements of physical–chemical parameters are taken and animals are counted in the bay as well as offshore. **The environmental gradients of greatest importance involve pollutants.**

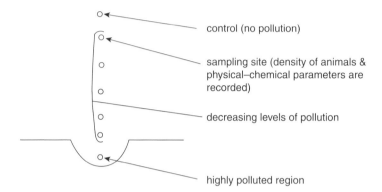

control (no pollution)

sampling site (density of animals & physical–chemical parameters are recorded)

decreasing levels of pollution

highly polluted region

2. A beach study. Animal densities and plant densities or percentage cover are recorded down and across the beach. **The environmental gradient of greatest impact to the organisms is the slope (= time of exposure to air).**

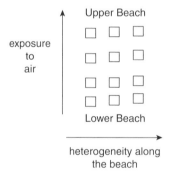

Upper Beach

exposure to air

Lower Beach

heterogeneity along the beach

Figure 5-9.   Two examples of the use of an ordination technique (e.g., MDS). (1) A bay or estuary is polluted from industrial sources. Measurements of the physical-chemical variables are taken and animals are counted in the bay as well as offshore. The environmental gradients of greatest importance involve pollutants. (2) A beach study. Animal and plant densities or percentage cover are recorded down and across the beach. **The environmental gradient of greatest impact to the organisms is the time of exposure to air, based on their positions along a slope.**

is that **principal component analysis is unsuitable when there is considerable environmental heterogeneity** (Swan, 1970). It assumes that variables respond along gradients in a linear fashion (McGarigal et al., 2000; Paukert and Wittig, 2002). McGarigal et al. (2000) and Clarke and Warwick (2001) describe further strengths and weaknesses of PCA.

5.3   FACTOR ANALYSIS (FA)

The major uses of factor analysis are to reduce the number of variables and to classify variables (Statsoft, Inc., 1995). The goal of factor analysis is to account for the covariance among descriptors (Legendre and Legendre, 1998). FA differs from PCA in that only the variability in an item that is common to the other items is used whereas in PCA all variability in an item is used (Statsoft, Inc., 1995). Varimax rotation can be orthogonal or oblique. Factor scores are calculated to detect structure. These are termed principal factors. Factor loadings are calculated. Factor analysis is seldom used in ecology. Several statisticians state that it should not be used because it is based on a special statistical model (e.g., see Clifford and Stephenson, 1975). Estimating the unique variance is the most difficult and ambiguous task in FA (McGarigal et al., 2000).

5.4   REDUNDANCY ANALYSIS

Redundancy analysis is the canonical form of PCA and has been generally neglected by ecologists (Jongman et al., 1995). It is based on a linear model of species responses in contrast to the unimodal model of CCA (McCune et al., 2002). RDA selects the linear combination of environmental variables that results in the smallest total residual sum of squares. RDA is PCA with a restriction on the site scores. See Legendre and Legendre (1998) for further details.

5.5   CORRESPONDENCE ANALYSIS (CA)
OR RECIPROCAL AVERAGING (RA)

Correspondence analysis is an extension of the method of weighted averages used in the direct gradient analysis of Whittaker (1967). This multivariate technique is similar to PCA. **It emphasizes weighted averages**. It can be applied to both presence–absence and abundance data (Jongman et al., 1995). Species ordination scores are averages of the sample ordination scores. The arch effect commonly occurs here, which can be improved with DCA (see below). The second axis introduces an arch into the first axis (McCune and Mefford, 1999). Points (e.g., species) are compressed toward the end of the first axis (Barbour et al, 1999). Outliers can have a strong effect on the results. CA or RA are seldom recommended for field studies. However, Gamito and Raffaelli (1992) concluded that correspondence analysis performed better than DCA, MDS and others in benthic studies.

5.6   DETRENDED CORRESPONDENCE ANALYSIS
(DCA OR DECORANA)

DCA is used for nonparametric distributions (e.g., field data). It is based on Reciprocal Averaging or Correspondence Analysis (McCune and Mefford, 1999). **Nonlinear rescaling and detrending by polynomials produces an ordination with uniform**

axis scaling and no arching (Fig. 5-7; Jongman et al., 1995). This technique has been effective with heterogeneous and difficult data sets, thus has been widely used in ecological studies in North America. Species points on the edge of the diagram are often rare species whereas those at the center may be unimodal, bimodal, or unrelated to the ordination axes (Jongman et al., 1995). CA and DCA are both sensitive to species that occur only in a few species-poor sites. Beware that in older computer programs, the order in which the numbers were entered affected the results! See Gauch (1982) and Peet et al. (1988). The computer program CANOCO detrends by up to the fourth-order polynomials (Ter Braak, 1987). The program TWINSPAN (in PC-ORD) produces a two-way table (species and sites) based on Reciprocal Averaging (Table 4-1). Perhaps the strongest criticism of DCA is that the results vary depending on the number of segments used to remove the arch effect. Moreover, DCA is sensitive to outliers and discontinuities in the data (McGarigal et al., 2000) and the behavior of an underlying χ^2 distance measure (Clarke, 1993). Some biologists recommend that DCA be avoided when analyzing data that represents complex ecological gradients (Legendre and Legendre, 1998). DCA is being replaced by Nonmetric Multidimensional Scaling and Canonical Correspondence Analysis.

5.7 NONMETRIC MULTIDIMENSIONAL SCALING (MDS, NMDS, NMS, NMMDS)

A nonparametric, ordinational technique that is widely used in Europe. **It arranges objects (e.g., sites) in multidimensional space based on ranked distances**. It is well suited for non-normal data (McCune et al., 2002). Any similarity or dissimilarity (distance) matrix (e.g., Jaccard, Bray–Curtis or Sorenson, etc.) can be used. An iterative procedure is used to minimize the stress or loss value (departure from monotonicity, a constant relationship between two or more sets of data) between the original distance and the reduced multidimensional space (i.e., how well the distances in the ordination compare in rank order with the dissimilarity values – or goodness of fit – see Barbour et al., 1999). Stress is defined in terms of total scatter of the data, with stress <0.05 = excellent representation, stress <0.01 = good ordination, stress <0.2 = usable picture with potential to mislead, and stress <0.2 = likely to produce plots that are dangerous to interpret (Clarke, 1993). Stress tends to increase with increasing numbers of samples. The program produces a stress output plot (Fig. 5-10) and plots sites, for example, between a given number of dimensions. It can plot sites (etc.) in 3-D. The axes scores depend on the number of axes selected. Local and global forms of MDS are available. Global forms of MDS are commonplace and use topology preserving methods. Local forms of MDS use shape preserving methods (see McCune et al. 2002, p. 137, for a discussion of local versus global, and the internet). **Most MDS programs require a priori the number of ordination axes and the initial ordination of sites**. The major strength of MDS is that it makes no assumption of linearity (McGarigal et al., 2000). Among the most important weaknesses of MDS are as follows: (1) convergence to the global minimum of stress is not guaranteed, and (2) the algorithm places most weight on the large distances (Clarke and Warwick, 2001). Other criticisms include that it

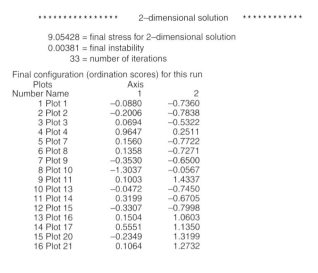

```
***************     2–dimensional solution     ************

  9.05428 = final stress for 2–dimensional solution
  0.00381 = final instability
       33 = number of iterations

Final configuration (ordination scores) for this run
    Plots                  Axis
Number Name                 1              2
    1 Plot 1            −0.0880        −0.7360
    2 Plot 2            −0.2006        −0.7838
    3 Plot 3             0.0694        −0.5322
    4 Plot 4             0.9647         0.2511
    5 Plot 7             0.1560        −0.7722
    6 Plot 8             0.1358        −0.7271
    7 Plot 9            −0.3530        −0.6500
    8 Plot 10           −1.3037        −0.0567
    9 Plot 11            0.1003         1.4337
   10 Plot 13           −0.0472        −0.7450
   11 Plot 14            0.3199        −0.6705
   12 Plot 15           −0.3307        −0.7998
   13 Plot 16            0.1504         1.0603
   14 Plot 17            0.5551         1.1350
   15 Plot 20           −0.2349         1.3199
   16 Plot 21            0.1064         1.2732
```

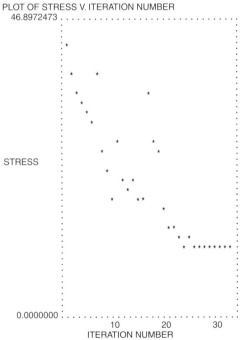

Figure 5-10 caption:

Figure 5-10. Nonmetric multidimensional scaling (MDS) ordination scores for plots (above) and a plot of stress (below). See text for explanation.

cannot be relied on to extract the correct configuration, even when applied to simulated data (McGarigal et al., 2000). Moreover, because the choice of initial configuration is somewhat arbitrary, the solution is never unique. See McCune and Mefford (1999), and especially Clarke and Warwick (2001) who have excellent programs and

Statistica: Multidimensional Scaling

Data file: lbmds.STA [29 cases with 5 variables]

0 CASE NAME	1 RALFSIA	2 LOTTIA	3 CHTHAMAL	4 PAGURUS	5 VAR5
	30	14	0	0	Plot 1
	40	10	0	9	Plot 2
	25	26	10	1	Plot 3
	0	10	50	0	Plot 4
	35	65	0	0	Plot 7
	30	54	0	0	Plot 8
	20	10	0	1	Plot 9
	1	4	0	1	Plot 10
	0	0	556	0	Plot 11
	27	22	0	0	Plot 13
	7	31	0	0	Plot 14
	35	5	0	70	Plot 15
	2	84	400	0	Plot 16
	0	2	200	0	Plot 17
	10	34	1000	2	Plot 20
	0	25	500	0	Plot 21

Figure 5-11. MDS raw data abundance file for four rocky intertidal species. See Fig. 5-4 for species names.

Data file: lbdist. STA [8 cases with 4 variables]

0 CASE	1 RALFSIA	2 LOTTIA	3 CHTHAMAL	4 PAGURUS
RALFSIA	.00000	316.0000	2934.000	248.0000
LOTTIA	316.0000	.00000	2782.000	442.0000
CHTHAMAL	2934.000	2782.000	.00000	2794.000
PAGURUS	248.0000	442.0000	2794.000	.00000
Means	16.37500	24.75000	169.7500	5.25000
Std. Dev.	15.20471	24.27482	294.4707	17.41072
No. Cases	16.00000			
Matrix	3.00000			

Figure 5-12. MDS initial distance matrix with four rocky intertidal species.

discussions of strengths and weaknesses of MDS. The procedure for running MDS with Statistica is presented in Figures 5-11 to 5-14. See the Appendix also.

5.8   MANOVA AND MANCOVA

These **parametric multivariate techniques** (Multivariate Analysis of Variance and Multivariate Analysis of Covariance) are similar to ANOVA and ANCOVA.

Stat. Multidim Scaling	Distances in Final Configuration (lbdist. sta) D–star: Raw stress = 0.000000; Alienation = 0.000000 D–hat: Raw stress = 0.000000; Stress = 0.000000			
	RALFSIA	LOTTIA	CHTHAMAL	PAGURUS
RALFSIA	0.000000	1.184507	1.985462	.578542
LOTTIA	1.184507	0.000000	1.913498	1.697217
CHTHAMAL	1.985462	1.913498	0.000000	1.943747
PAGURUS	.578542	1.697217	1.943747	0.000000

Figure 5-13. MDS final distance matrix with four rocky intertidal species.

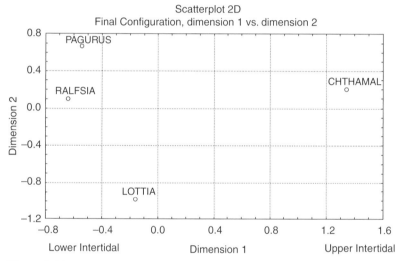

Figure 5-14. MDS scatterplot of four rocky intertidal species. Compare this with Fig. 5-4, which uses the same data for a DCA ordination. The DCA ordination indicates the zonation of the species more realistically (i.e., better spacing) than the MDS method in this example. Moreover, the more frequent occurrence of *Pagurus* below that of *Ralfsia* in the intertidal or littoral zone is the correct representation.

MANOVA (Wilks' Lambda) and ANCOVA are advantageous in that performing multiple univariate tests can inflate the α value (probability of Type I error), leading to false conclusions (Scheiner, 2001). They are more complex as they handle three or more variables simultaneously. **They are frequently used with the analysis of experimental studies, especially laboratory experiments**. MANOVA seeks differences in the dependent variables among the groups (McCune et al, 2002). Assumptions of MANOVA include multivariate normality (error effects included), independent observations, and equality of variance-covariance matrices (Paukert and Wittig, 2002). Because of these assumptions, among others, MANOVA is not often used in ecology although its use in increasing. The power of traditional MANOVA declines with an increase in the number of response variables (Scheiner, 2001). Unequal sample sizes are not a large problem for MANOVA, but may bias the results for factorial or nested designs. **Before ANCOVA is run, tests of the assumption of homogeneity of slopes need to be performed** (Petratis et al., 2001).

Early attempts to develop nonparametric multivariate analysis include those of Mantel and Valand (1970). More recently, the Analysis of Similarities was developed to compare communities or changes in communities because of pollution (Clarke, 1993). For typical species abundance matrices, an Analysis of Similarities (ANOSIM) permutation procedure is recommended over MANOVA (see p. 230 in Chapter 4 and Clarke and Warwick, 2001). This procedure is applied to a rank similarity matrix. Anderson (2001) has developed a new method for nonparametric MANOVA (NPMANOVA). It can be based on any measure of dissimilarity, can partition variation directly among individual terms in a multifactorial model of ANOVA, and provides a P-value using permutation methods (i.e., at least 1000 permutations for tests with an α value of 0.05 and 5000 permutations for $\alpha = 0.01$). The statistic used is analogous to Fisher's F-ratio and is constructed from sums of square distances (or dissimilarities) within and between groups. It does not have unrealistic assumptions and, like ANOSIM, is therefore useful for analyzing ecological data.

An example of how MANOVA is used is given in the Appendix under the filename MANOVA.

See Statistica for computer programs on ANOVA and ANCOVA, and Clarke and Gorley (2001) for ANOSIM. See Barker and Barker (1984), Sokal and Rohlf (1995), Zar (1999) and Scheiner and Gurevitch (2001) for further information.

5.9   DISCRIMINANT ANALYSIS (DA) (DISCRIMINANT FUNCTION ANALYSIS, CANONICAL VARIATES ANALYSIS)

DA is a powerful tool that can be used with both clusters of species data and environmental variables. It determines which variables discriminate between two or more groups, that is, independent variables are used as predictors of group membership (McCune et al., 2002). It is very similar to MANOVA and multiple regression analysis (Statsoft, Inc.,1995; McGarigal et al., 2000). Clusters can be identified by several methods using raw data: (1) constructing a dendrogram, (2) using PCA (even if you have field data) for initial visual identification of clusters, and (3) point rotation in space by rotating ordinations (i.e., rotating axes – see McCune and Mefford, 1999). If any method indicates groups or clusters of data then DA can be used. **However, the number of groups is set before the DA analysis**. DA finds a transform for the minimum ratio of difference between pairs of multivariate means and variances in which two clusters are separated the most and inflated the least (Fig. 5-15). DA produces two functions: (1) classification

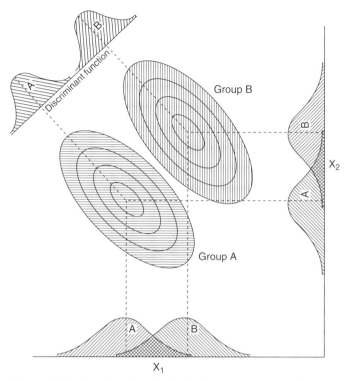

Figure 5-15.   Plot of two bivariate distributions, showing overlap between groups A and B along both variables X_1 and X_2. Groups can be distinguished (i.e., separated) by projecting members of the two groups onto the discriminant function line (source: Davis, 1986).

function consisting of 2+ groups or clusters of points (this information can be used for prediction with probabilities) (Fig. 5-16) and (2) discriminant function containing environmental variables that can be used to discriminate differences among the groups. DA differs from PCA and Factor Analysis in that no standardization of data is needed (PCA and FA need standardization because of scaling problems) and the position of the axes distinguishes the maximum distance between clusters. See Davis (1986). **DA assumes a multivariate normal distribution, homogeneity of variances, and independent samples** (Paukert and Wittig, 2002). Violations of normality are usually not fatal (i.e., somewhat non-normal data can be used). A description of the procedure to use DA with Statistica is given in the Appendix. Multiple Discriminant Analysis (MDA) is the term often used when three or more clusters of data are processed simultaneously. **MDA is particularly susceptible to rounding error**. Calculations in double precision for at least the eigenvalue-eigenvector routines are advisable (Green, 1979). Limitations of Discriminant Analysis are discussed by McGarigal et al. (2000). See Dytham (1999) for further information.

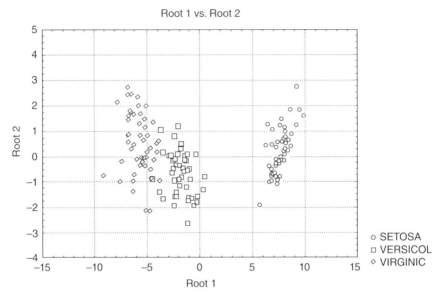

Figure 5-16. Scatterplot of canonical scores in a discriminant analysis of three species of *Iris* (source: Statistica 5.0).

5.10   PRINCIPAL COORDINATE ANALYSIS (PCOA) (METRIC MULTIDIMENSIONAL SCALING)

Principle coordinates are similar to principal components in concept. The advantage of PCoA is that it may be used with all types of variables (Legendre and Legendre, 1998). Because of this, PCoA is an ordination method of considerable interest to ecologists. Most metric (PCoA) and nonmetric MDS plots are very similar or even identical, provided that a similar distance measure is used. The occurrence of negative eigenvalues, lack of emphasis on distance preservation, and other problems are discussed in detail by Legendre and Legendre (1998) and Clarke and Warwick (2001).

5.11   CANONICAL CORRESPONDENCE ANALYSIS (CCA)

Canonical Correspondence Analysis was developed by ter Braak (1986) and implemented in the computer program CANOCO (see the Appendix and Legendre and Legendre, 1998). This is a combination of ordination with multiple regression to determine the most important environmental factors. **It assumes an approximate Gaussian distribution and emphasizes rare species. It is a dual ordination of species and sample sites that are constrained by environmental variables assessed by multiple regression** (McCune and Mefford, 1999; McGarigal et al., 2000). The math involved is basically the same as that of Redundancy Analysis (Legendre and Legendre, 1998).

Two matrices are produced as input into the analysis. The first matrix displays rows (representing sites) and columns (representing species). The second matrix shows rows (representing sites) and columns (representing environmental variables, which must be fewer in number than the number of sites). Two sets of correlations are presented: (1) weighted averages and (2) weighted averages regressed against environmental variables. **The plot shows the distribution of species along environmental gradients with a vector line or arrow, the length of which is proportional to the importance of the gradient** (Barbour et al., 1999; Fig. 5-17). The direction of the arrows indicate the direction of maximum change in each variable (McGarigal et al., 2000). CCA is replacing DCA used by plant ecologists in the United States. Arch effects can occur (Jongman et al., 1995). The advantage of CCA is that it is like discriminant analysis, in that it simultaneously handles species-sites and environmental factors (no separate multiple regression operation is needed). However, if any important environmental factors are not included, then the species and sample ordinations are constrained by more poorly correlated factors, possibly leading to some spurious results. Moreover, observations are required to be independent and randomly selected, seldom the case in ecological studies (McGarigal et al., 2000). On the contrary, CCA makes no assumption of linearity. **Because CCA uses a weighted multiple regression procedure, it will be strongly affected by high correlations among the independent variables.** CCA performs well with data from complex sampling designs, the data not requiring normal distributions. CCA is easily interpretable. It performs consistently better than

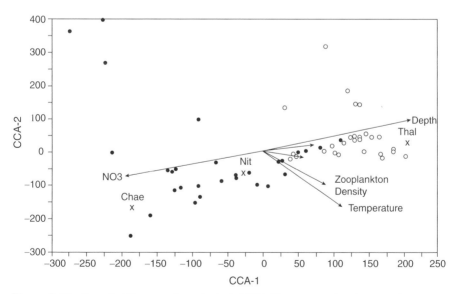

Figure 5-17. Canonical Correspondence Analysis (CCA). The genera and environmental variables are distributed along two axes (CCA-1 and CCA-2). The vector lines are proportional to the importance of the gradient. The variables influencing phytoplankton populations are indicated. Filled circles represent northern stations; open circles represent southern stations; Chae represents *Chaetoceros*; Nit represents *Nitschia*; and Thal represents *Thalassiothrix*.

DCA. McGarigal et al. (2000) consider CCA to be the best multivariate constrained ordination technique developed to date. However, because of the constraints, they recommend using DCA for a comparison of results with CCA.

5.12   MULTIPLE REGRESSION (MR) (MULTIPLE LINEAR REGRESSION)

Normally component loadings (e.g., scores in PCA) suggest which variables are most important. However, with species abundances the only variable in some ordinations, one must use other techniques to attempt to suggest what may have produced the gradients for Axis 1, 2, 3, and so forth. In order to show the relationship between biotic (principal component axes) and abiotic factors (environmental factors), a multiple regression type analysis is used. Univariate analyses such as the Spearman rank correlation coefficient are not as ecologically realistic as multivariate analyses such as multiple regression because some variables are correlated and there are interaction effects between variables (Jongman et al., 1995). MR is not considered by some statisticians as a multivariate procedure because it includes only one dependent variable (Paukert and Wittig, 2002).

The objective of multiple regression is to determine the influence of independent variables on a dependent variable, for example, the effect of depth, sediment grain size, salinity, temperature, and predator density on the population density of species X. The parameters are estimated by the least-squares method, that is, minimizing the sum of squares of the differences between the observed and expected response (Jongman et al., 1995). Multiple regression solves simultaneously normal equations and produces partial regression coefficients. Partial regression coefficients each give the rate of change or slope (i.e., in a regression line, see p. 107 in Chapter 2) in the dependent variable for a unit of change in a particular independent variable, assuming all other independent variables are held constant. The partial regression coefficients can be compared with each other by standardizing them, that is, converting them to units of standard deviation. The highest standardized partial regression coefficients (positive or negative) suggest the most important factors (e.g., sediment size, predator density, etc.) in controlling the population density of species X. Significance tests and standard errors then can be calculated from the data. An example of multiple regression is given on p. 259.

Multiple regression has many potential problems such as the type of response curve, error distribution, and outliers that may unduly influence the results (Jongman et al., 1995). **Multiple regression variables may be highly correlated, therefore, examine the correlation coefficients first (i.e., run a multiple correlation analysis between variables) to exclude some of them before doing multiple regressions. Multiple regression generally should employ a maximum of 6 variables**. Legendre and Legendre (1998) suggest a stepwise procedure for reducing numerous variables. This involves a process of alternating between forward selection and backward elimination. For an even more complete treatment of the subject see Kutner et al. (1996). Lee and Sampson (2000) took ordination scores, representing a gradient in fish communities, and regressed them against a group of environmental variables and time. See Davis (1986) and Sokal and Rohlf (1995) for further examples of

Table 5-5.   Intertidal rocky beach data for multiple regression. *Pagurus* = hermit crab and *Semicossyphus pulcher* = Sheephead (predaceous fish).

Quadrat	X1 Exposure to air (min/tide)	X2 Density of predators (# sheephead/h)	X3 Wave Height (x m/10 min)	X4 Density of *Pagurus* (No. of indiv./m)
1	120	13	1.6	10
2	118	12	2.1	9
3	110	6	1.4	7
4	119	15	0.8	9
5	116	20	1.2	8
6	110	19	1.3	7
7	125	32	0.7	12
8	105	3	1.1	6
9	115	13	2.0	8
10	117	14	1.5	9

multiple regression and computer programs. The computations involve rather complex matrix analyses.

The following is an example based on one species (as each species may be responding to somewhat different environmental factors).

We first decide on the dependent and independent variables.

X1, X2, X3 = independent variables (see Table 5-5)

X4 = dependent variable (i.e., density of a hermit crab)

You want to know which of the three independent variables (exposure to air, predators, wave height) is most important in controlling the density of the intertidal hermit crab *Pagurus sp.* **The highest positive and negative standardized regression coefficients (which you will calculate with these data) will be the most important. The answer you get is not proof, only correlative**. You will need to do experiments to prove that one variable is strongly affecting the density of hermit crabs.

The next step is to use multiple regression in a statistical package. See the Appendix for directions on using Stataq2 and Statistica 5 for multiple regression. See the Addendum for a newer technique (Hierarchical Partitioning).

5.13   PATH ANALYSIS

Path analysis is an extension of multiple linear regression, allowing interpretation of linear relationships among a small number of descriptors (see Legendre and Legendre, 1998, for details). Path analysis was originally developed by Sewall Wright in which he introduced the concept of a path diagram. **It handles more than one dependent variable and the effects of dependent variables on one another** (Mitchell, 2001). Path analysis assumes a normal distribution of residuals, additive and linear effects, inclusion of all important variables, that residual errors are uncorrelated, and that there is no measurement error.

5.14   CANONICAL CORRELATION ANALYSIS (CANCOR)

This is a type of multiple correlation that indicates relationships between two groups of variables. It can be viewed as a logical extension of multiple regression analysis. **CANCOR involves multiple dependent variables**. For example, coefficients for environmental variables (temperature, salinity, sediment grain size) and coefficients (scores) for species (species richness, diversity and evenness). The overall objective is to develop linear combinations of variables in each set (dependent and independent) such that the correlation between the composite variables is maximized (MacGarigal et al., 2000). The greatest limitation of CANCOR may be in the interpretation of the results. It is not used much in ecology these days (McCune and Mefford, 1999). See Gittins (1985), Manly (1986), and Jongman et al. (1995) for further details.

5.15   CANONICAL VARIATE ANALYSIS (CVA)

Canonical Variate Analysis is a special case of Canonical Correlation Analysis (Jongman et al., 1995). It is also described as a type of linear discriminant analysis (McGarigal et al., 2000). The set of environmental variables consists of a single nominal variable defining the classes. CVA is usable only if the number of sites is much greater than the number of species and the number of classes. Many ecological data sets cannot be analyzed by CVA without dropping many species, thus CVA is not used much in ecology. It has been successfully used in primate studies (Charles Oxnard, pers. comm.).

5.16   MULTI-RESPONSE PERMUTATION PROCEDURES (MRPP)

MRPP is a nonparametric procedure for testing the hypothesis of no difference between two or more groups (McCune and Mefford, 1999). It has the advantage of not requiring assumptions that are seldom met in ecological data (e.g., multivariate normality and homogeneity of variances) and can handle groups of varying sizes. For an introduction to this topic see Mielke et al. (1976), Biondini et al. (1985), Manly (1997), McCune and Mefford (1999), and McCune et al. (2002). The procedure is as follows: (1) calculate a distance matrix, (2) calculate the average distance in each group (x_i), (3) calculate delta or the weighted mean within-group distance by the following equation:

$$\text{delta} = \sum C_i x_i$$

where

$\Sigma = \text{sum}$

$C_i = n_i/N_i$

$n_i = \text{number of items in group } i$

$N_i = \text{total number of items}$

$x_i = \text{average distance in each group}$

(4) determine the probability of a delta (see McCune and Mefford, 1999 for details), and (5) calculate the effect-size (chance-corrected within-group agreement – see McCune and Mefford, 1999 for details).

Table 5-6 is an example of MRPP. In that example of 19 stands of plants comprising 49 species, the distance measure (Bray–Curtis) produces three groups of stands. These three groups are compared using MRPP. The value of $A = 0.05970947$ indicates that the groups are somewhat different from each other. If the value of $A = 1$ (maximum) then all items within each group are identical therefore maximum

Table 5-6. A multiresponse permutation procedure with data from a plant community. (Source: McCune and Mefford, 1999).

Differences in communities among levels of Brojap
    Groups were defined by values of: Brojap
    Input data has: 19 STANDS by 49 SPECIES
    Weighting option: $C(I) = n(I)/\text{sum}(n(I))$
    Distance measure: Sørensen (Bray–Curtis)

GROUP: 1
Code: 0
Size: 10   0.66267458 = Average distance
Members:

STAND1	STAND2	STAND3	STAND4	STAND6	STAND8	STAND9
STAN12	STAN13	STAN15				

GROUP: 2
Code: 2
Size: 4   0.61875399 = Average distance
Members:

STAND5	STAND7	STAN10	STAN19

GROUP: 3
Code: 1
Size: 5   0.54680432 = Average distance
Members:

STAN11	STAN14	STAN16	STAN17	STAN18

       Test statistic: $T = -2.5874500$
      Observed delta =   0.62293597
      Expected delta =   0.66249308
    Variance of delta =   0.23372499E-03
  Skewness of delta = -0.70854412

Chance-corrected within-group agreement. $A = 0.5970947$
    $A = 1 - (\text{observed delta/expected delta})$
    $A_{nax} = 1$ when all items are identical within groups (delta=0)
    $A = 0$ when heterogeneity within groups equals expectation by chance
    $A < 0$ with more heterogeneity within groups than expected by chance

Probability of a smaller or equal delta. $p = 0.01498651$

---

between-group differences occur. If A = 0 then the groups are neither similar nor different compared to expectations based on chance. If $A < 0$ then the groups are similar. Community ecology values for A are commonly < 0.1. An $A > 0.3$ is relatively high. It is very rare to have $A < -0.1$ (Bruce McCune, pers. comm.). Randomized block experiments or paired-sample data can be analyzed with a variant of MRPP called MRBP or blocked MRPP (seen McCune and Mefford, 1999 and McCune et al., 2002 for details).

5.17 OTHER MULTIVARIATE TECHNIQUES

McCune et al. (2002) discuss the history of and problems associated with Polar Ordination (Bray–Curtis Ordination). New methods of exploring differences among groups include the nonparametric, recursive classification, and regression tree (CART). It is used to classify habitats or vegetation types and their environmental variables (McCune et al., 2002). It produces a top-to-bottom visual classification tree that undergoes a "pruning" or optimization process. CART is used to generate community maps, wildlife habitats, and land cover types in conjunction with a GIS (see p. 195 in Chapter 3). Another multivariate technique is Structural Equation Modeling (unfortunately termed SEM), a merger of factor analysis and path analysis (McCune et al. (2002). It is a method of evaluating complex hypotheses (e.g., effects of abiotic factors on plant species richness) with multiple causal pathways among variables. It requires the initial development of a path diagram. It is an analysis of covariance relationships, effectively limited to about 10 variables. See Shipley (2000), McCune et al. (2002), and Pugesek et al. (2002).

SUMMARY

Green (1979) concluded long ago that Multiple Discriminant Analysis (or Canonical Variate Analysis) is best for environmental studies. Warwick and Clarke (1991) and Clarke and Warwick (2001) promote Nonmetric Multidimensional Scaling (MDS), which differs from the other multivariate methods because it is based on rank distances. Until recently, North American ecologists preferred to use Detrended Correspondence Analysis (DCA) for community studies whereas Europeans often preferred Nonmetric Multidimensional Scaling. I tested local intertidal data (the beach having a strong gradient of exposure time to air) with both techniques and found that DCA gave a slightly more realistic portrayal of community structure (i.e., intertidal zonation – compare Figures 5-4 and 5-14), but this may not be the case in other communities. For example, Cao et al. (1996) found that MDS performed better than DCA in the River Trent system of the United Kingdom when combined with the CY dissimilarity measure. Minchin (1987) found erratic performance of DCA in a simulation study. McCune et al. (2002) consider MDS the most generally effective ordination method for ecological community data. CCA is becoming very popular in part because it combines ordination with multiple regression (Barbour et al., 1999). CCA may be the best constrained ordination method to date. Because it is constrained, it should be coupled with DCA (McGarigal et al., 2000). Recommended by fishery biologists are PCA or MANOVA (ecology) and DA (morphology, Paukert and Wittig, 2002). For data clusters use DA. Different methods will give different results and thus the strengths and weaknesses of each method need to be considered before deciding on an appropriate method to use for a particular data set.

Chapter 6

Time Trend Analysis

A. INTRODUCTION

6.1 Introduction

Time trend analysis consists of a series of methods used to detect periodic events or patterns that reoccur over time. Time series are repeated measurements taken on the same experimental unit though time. There are two subdivisions of time trend analysis: (1) time domain analysis and (2) frequency domain analysis. Time trend analysis is used to detect temporal patterns within a variable or between variables. Chatfield (1996) presents an excellent review of time trend analysis. Statistica and Minitab are recommended software programs.

B. TIME SERIES ANALYSIS

6.2 Smoothing or Filtering Techniques

These techniques smooth sequences of data. They range from the Simple Moving Average or three-term moving average seen below to Spencer's 21-term equation (Davis, 1986). To clarify this technique, the Simple Moving Average is presented. See Davis (1986) for further information. The Maximum Liklihood method has also been used to smooth data (DeNoyer and Dodd, 1990). Other smoothing methods include LOESS, cubic splines, distance weighted least squares smoothing, and so forth.

Three-term moving or running average:

$$Y_i = \frac{\sum_{j=i-x}^{i+k} Xj}{m}$$

Quantitative Analysis of Marine Biology Communities: Field Biology and Environment
by Gerald J. Bakus
Copyright © 2007 John Wiley & Sons, Inc.

where

Y_i = moving average

X = original sequence

m = averaging interval

$$k = \frac{m-1}{2}$$

e.g. *Original sequence*   574   323   457   264

etc.

Example

$5+7+4 = 16$	$16/3 = 5.3$	
$7+4+3 = 14$	$14/3 = 4.7$	etc.

Three-Term moving or running average   5.3, 4.7, 3.0, 4.0, and so forth.

The results are presented in Fig. 6-1. Figure 6-2 shows further smoothing operations.

Mike Risk, a former student, was interested in the relationship between species diversity and substrate rugosity. Dental casting plaster was placed on a rocky intertidal substratum and allowed to firm up. The cast was cut into many thin slices. Each slice was positioned behind a light to cast a shadow

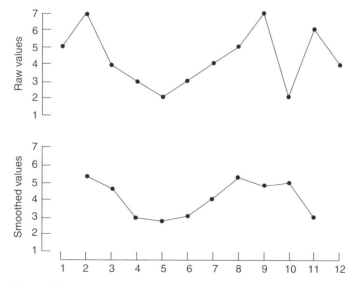

Figure 6-1.   Original data sequence and sequence smoothed by a three-term moving average. Note the shift in peak positions in the smoothed sequence (source: Davis 1973).

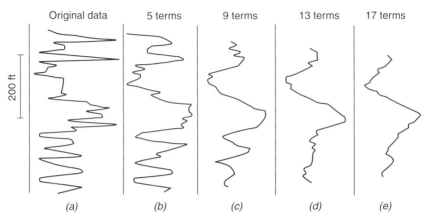

Figure 6-2. Digitized drilling time log smoothed by various equations that reflect an increase in the number of terms used (Davis, 1973). Beware that random data when smoothed can produce time trend patterns (Mosteller et al., 1973).

that was projected on a slide projector screen that had a vertical scale on it. The heights of all peaks and valleys of the shadow were measured and the data were smoothed with a Simple Moving Average. These data gave him a clearer quantitative visual assessment of rocky intertidal substrate micro-heterogeneity (Risk, 1971).

6.3 Serial Correlation (Auto- and Cross-Correlation)

Serial correlations are correlations between sets of time series data. Two major serial correlations are autocorrelation and cross-correlation. Data are plotted on a correlogram. Some individuals recommend that transforming the raw data be avoided prior to serial correlation (Chatfield, 1996). **They also recommend removing outliers before calculating serial correlations as outliers may severely effect time series** (e.g., you can use PC-ORD for removing outliers). Others may not agree with these alterations (e.g., Michael Fogarty, pers. comm.). In any event, you can plot a time series before your final analysis and remove any trend or seasonality by detrending, that is, subtract each data point from the monthly mean. This will help reveal other time series patterns in the data by eliminating the obscuring effects of seasonal trends.

(1) Autocorrelation

Correlation between a curve representing time period t (e.g., population curve) and time period $t + 1$ (lag). Parts of the same curve are compared with each other to detect cyclic or periodic patterns (e.g., plankton blooms).

To calculate the autocorrelation using the following equation
(George Moore, pers. comm.):

$$\phi(j) = \left[1/n \sum_{i=1}^{n} \left(X(i) - \bar{X} \right)\left(Y(1+j) - \bar{Y} \right) \right] 1/\phi 0$$

where

ϕ = phi
j = lag period
$\phi 0$ = reference point (see below)
n = number of observations (e.g., months)
Σ = sum
$X(i)$ = datum
$\bar{X}$ = mean

then:

1. Plot raw data.
2. Interpolate any missing points.
3. Find $\bar{X}$, which is equal to $\Sigma (Xi)/n$, where n = No. of observations.
4. Find $\bar{X}i - \bar{X}$
5. Calculate the autocorrelation coefficient at different lag periods
6. Plot the coefficients on a correlogram.

Example:

(1) See plot of raw data from Table 6-1 in Fig. 6-3. Month 1 data are not missing in Table 6-1. The data collection begins with Cruise 2 (month 2). Cruises with x were not undertaken. Note that monthly data were taken at different times of the month. **For time series analysis, the data should have been taken at equally spaced time intervals** (e.g., the 20th of each month) although this is not always possible with cruises by ship.

(2) Missing data points are interpolated.

(3) $\bar{X}(i) = \Sigma (Xi)/36 = $ **23.2 (mean temperature)**

(4) $\bar{X}(i) - X$ values are calculated (Table 6-2).

Just two examples are given here, that is, the calculation of Φ (phi) for 1 month lag and 5 months lag. Similar calculations would also be done for the remaining lag periods, that is, Φ (2) through Φ (4) and Φ (6) through Φ (12)

1 month lag:

$$\Phi(1) = \frac{20.13}{27.56 \text{ or } \Phi(0)} = 0.73$$

Table 6-1. Monthly temperature and salinity data from a ship cruise.

STATION : E-44
DEPTH : 7 M
LATITUDE : 29°15' N
LONGITUDE : 89°42' W

Cruise	Day	Month	Year	Temperature (°C) Depth (m)								Salinity (0/00) Depth (m)						
				X_t	3	11	24	43	70	107	8	Y_i	3	11	24	43	70	107
2	24	2	1963	14.3	14.3							32.40	32.32					
3	29	3	1963	19.9	19.9							21.15	21.50					
4	27	4	1963	27.1	26.8							19.91	19.77					
5	17	5	1963	28.1	28.1							28.61	29.04					
6	22	6	1963	27.3	27.3							25.78	25.94					
7	12	7	1963	28.9	28.9							31.53	32.02					
8	19	8	1963	29.9	29.4							27.71	33.08					
9	26	9	1963	27.1	27.0							31.10	31.13					
10	23	10	1963	25.1	25.1							32.51	32.48					
11	23	11	1963	21.8	21.8							32.29	32.13					
12	12	12	1963	16.0	16.0							32.31	33.19					
13	22	1	1964	13.8	13.7							32.93	33.09					
14	10	2	1964	15.6	15.6							34.44	34.40					
15	29	3	1964	14.7	14.8							32.41	31.88					
16	24	4	1964	27.3	27.2							15.64	17.83					
17	17	5	1964	25.1	25.0							13.26	13.22					
18	20	6	1964	30.9	30.5							14.60	14.71					
19	12	7	1964	29.1	29.1							26.55	33.16					

268

20	24	8	1964	31.7	30.3	24.75	28.29
21	20	9	1964	28.4	28.3	30.39	30.55
22	24	10	1964	22.2	22.3	32.89	32.92
23	13	11	1964	21.9	21.9	29.79	29.80
24	11	12	1964	17.6	17.6	32.28	32.26
25	18	1	1965	15.4	15.5	31.55	31.61
26	12	2	1965	18.7	18.7	30.28	30.51
27	15	3	1965	17.7	17.6	27.17	27.57
28	17	4	1965	23.5	23.5	20.55	20.47
29	22	5	1965	25.2	25.3	13.44	16.18
30	21	6	1965	28.4	28.4	23.34	23.37
x							
32	22	8	1965	28.4	28.3	00.00	00.00
x							
34	29	10	1965	22.5	22.5	31.96	31.94
35	4	12	1965	19.9	20.0	33.49	33.42

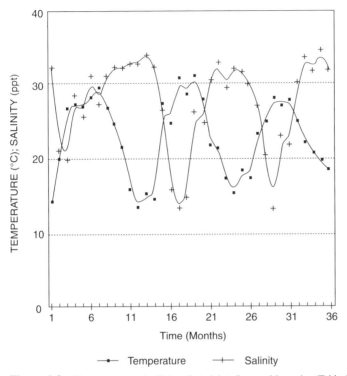

Temperature-salinity

Data Plot for 36 Months

Figure 6-3. Temperature and salinity plotted data from a ship cruise (Table 6-1).

5 months lag:

$$\Phi(5) = \frac{-20.28}{27.56 \,\text{or}\, \Phi(0)} = -0.74$$

$\Phi(0)$ = deviation from $\overline{X}$ of all raw data from month 1 times itself and divided by 36.

Phi of zero [$\Phi(0)$] is a reference point to which all other Phi (Φ) or lag periods are compared.

$$\Phi(0) = \frac{992.18}{36} = 27.56$$

(5) To calculate the autocorrelation, choose your lag period (j) (i.e., the time between observations). For a lag of one month we get:

$$\left| 1/n \left[(X(1) - \overline{X})(X(2) - \overline{X}) ... (X(36) - \overline{X}) \right] \right| 1/\Phi(0)$$

Table 6-2.   Values calculated from Table 6-1.

$X(i) - \overline{X} \phi(1)$ $X(1+j) - \overline{X}$

Lag 1 month	Lag 5 months
$1 \times 2 = (-8.9)(-3.3) = 29.37$	$1 \times 6 = (-8.9)(5.7) = -50.73$
$2 \times 3 = (-3.3)(3.9) = -12.87$	$2 \times 7 = (-3.3)(6.7) = -22.11$
$3 \times 4 = (3.9)(4.9) = 19.11$	$3 \times 8 = (3.9)(3.9) = 15.21$
$4 \times 5 = (4.9)(4.1) = 20.09$	$4 \times 9 = (4.9)(1.9) = 9.31$
$5 \times 6 = (4.1)(5.7) = 23.37$	$5 \times 10 = (4.1)(-1.4) = -5.74$
$6 \times 7 = (5.7)(6.7) = 38.19$	$6 \times 11 = (5.7)(-7.2) = -41.04$
$7 \times 8 = (6.7)(3.9) = 26.13$	$7 \times 12 = (6.7)(-9.4) = -62.98$
$8 \times 9 = (3.9)(1.9) = 7.41$	$8 \times 13 = (3.9)(-7.6) = -29.64$
$9 \times 10 = (1.9)(-1.4) = -2.66$	$9 \times 14 = (1.9)(-8.5) + -16.15$
$10 \times 11 = (-1.4)(-7.2) = 10.08$	$10 \times 15 = (-1.4)(4.1) = -5.74$
$11 \times 12 = (-7.2)(-9.4) = 67.68$	$11 \times 16 = (-7.2)(1.9) = 73.68$
$12 \times 13 = (-9.4)(-7.6) = 71.44$	$12 \times 17 = (-9.4)(7.7) = -72.38$
$13 \times 14 = (-7.6)(-8.5) = 64.6$	$13 \times 18 = (-7.6)(5.9) = -44.84$
$14 \times 15 = (-8.5)(4.1) = -34.85$	$14 \times 19 = (-8.5)(8.5) = -72.25$
$15 \times 16 = (4.1)(1.9) = 7.79$	$15 \times 20 = (4.1)(5.2) = 21.32$
$16 \times 17 = (1.9)(7.7) = 14.63$	$16 \times 21 = (1.9)(-1) = -1.9$
$17 \times 18 = (7.7)(5.9) = 45.43$	$17 \times 22 = (7.7)(-1.3) = -10.01$
$18 \times 19 = (5.9)(8.5) = 50.15$	$18 \times 23 = (5.9)(-5.6) = -33.04$
$19 \times 20 = (8.5)(5.2) = 44.2$	$19 \times 24 = (8.5)(-7.8) = -66.3$
$20 \times 21 = (5.2)(-1) = -5.2$	$20 \times 25 = (5.2)(-4.5) = -23.4$
$21 \times 22 = (-1)(-1.3) = 1.3$	$21 \times 26 = (-1)(-5.5) = 5.5$
$22 \times 23 = (-1.3)(-5.6) = 7.28$	$22 \times 27 = (-1.3)(0.3) = -0.39$
$23 \times 24 = (-5.6)(-7.8) = 43.68$	$23 \times 28 = (-5.6)(2) = -11.2$
$24 \times 25 = (-7.8)(-4.5) = 35.1$	$24 \times 29 = (-7.8)(5.2) = -40.56$
$25 \times 26 = (-4.5)(-5.5) = 24.75$	$25 \times 30 = (-4.5)(4.4) = -19.8$
$26 \times 27 = (-5.5)(0.3) = -1.65$	$26 \times 31 = (-5.5)(5.2) = -25.6$
$27 \times 28 = (0.3)(2) = 0.6$	$27 \times 32 = (0.3)(2) = 0.6$
$28 \times 29 = (2)(5.2) = 10.4$	$28 \times 33 = (2)(-0.7) = -1.4$
$29 \times 30 = (5.2)(4.4) = 22.88$	$29 \times 34 = (5.2)(-2.1) = -10.92$
$30 \times 31 = (4.4)(5.2) = 22.88$	$30 \times 35 = (4.4)(-3.3) = -14.52$
$31 \times 32 = (5.2)(2) = 10.4$	$31 \times 36 = (5.2)(-4.6) = -23.92$
$32 \times 33 = (2)(-0.7) = -1.4$	$*32 \times 1 = (2)(-8.9) = -17.8$
$33 \times 34 = (-0.7)(-2.1) = 1.47$	$33 \times 2 = (-0.7)(-3.3) = 2.31$
$34 \times 35 = (-2.1)(-3.3) = 6.93$	$34 \times 3 = (-2.1)(3.9) = -8.19$
$35 \times 36 = (-3.3)(-4.6) = 15.18$	$35 \times 4 = (-3.3)(4.9) = -16.17$
$36 \times \phi = (-4.6)(-8.9) = 40.94$	$36 \times 5 = (-4.6)(4.1) = -18.86$
$\Sigma = 783.46 - 58.63 = 724.83/36 = 20.13$	$\Sigma = 54.25 - 784.26 = -730.01/36 = -20.28$

(total + values)(total − values)

*indicates values that are not normally calculated. The interpolated numbers were included in this case due to lack of data points.

For the data set in Table 6-1, as mentioned above, **you do this for lag periods 1–12**. If dealing with data that contain a longer period, you must extend your lag number and vice versa. Note that 23.2 °C was the average temperature for column Xi in Table 6-1. This mean temperature (23.2 °C) minus the temperature under Xi (−14.3 °C) for Cruise 2 = −8.9. The mean temperature (23.2 °C) minus the temperature under Xi (19.9 °C) for Cruise 3 (or one month lag) = −3.3, and so forth.

(6) See plot of data (correlogram), Fig. 6-4. Note where the points for lag of one month [Φ (1)] and lag of 5 months [Φ (5)] are located on the correlogram.

At lags of near coincidence, we should see a peak of high correlation, or near +1. If there is a strong negative correlation, we would see values nearer −1, and if there is no correlation, we get 0. **The correlogram is a plot of the correlation coefficient and the time lag. It will disclose intervals of time or distance at which the series has a repetitive nature**. For some seasonal data the correlogram provides little extra information as the seasonal pattern is generally evident in the time plot. See also Figs. 6-5 to 6-7. Remember that detrending is necessary to remove the seasonal pattern when a full time series analysis is done (see p. 266).

CORRELOGRAM

Temperature and Salinity

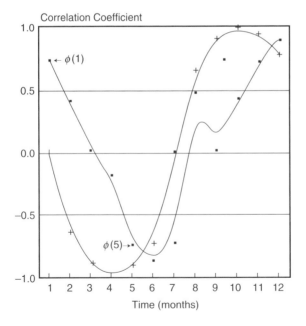

—•— Temperature (auto correlation)   —+— Temp. & Salinity (cross correlation)

Figure 6-4.   Correlogram of temperature and salinity from Figure 6-3.

Figure 6-5. Sequence of repeating values of Y along a traverse X through time or space (source: Davis, 1973).

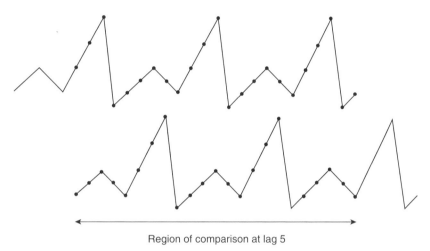

Region of comparison at lag 5

Figure 6-6. Sequence from Figure 6-5 compared to itself at lag 5 (source: Davis, 1973).

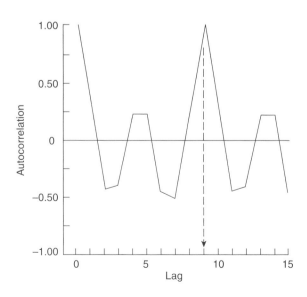

Figure 6-7. Correlogram of sequence from Figure 6-6. Autocorrelation of 1.00 at lag 9 indicates that the original sequence is periodic with a repetitive length of 9 units. (source: Davis, 1973).

273

(2) Cross Correlation

Correlation between two different variables (e.g., temperature and salinity). To calculate the cross correlation coefficient using the equation:

$$\phi(j) = \left[1/n \sum_{i=1}^{n} (X(i) - \bar{X})(Y(1+j) - \bar{Y}) \right] 1/\phi 0$$

See Serial Correlation above for definition of symbols.

(1) The procedure is exactly the same as for the autocorrelation, except this time you multiply your deviations between two sets of unlike features (e.g., temperature versus salinity) instead of temperature with itself.

(2) Interpretation of the correlogram is the same as for the autocorrelation. The equation above does not normalize (i.e., scale) the coefficients between $+1$ and -1. To do that (i.e., rescale the data), use the following equation:

$$\frac{\sum (X(i) - X)(Y(i+j) - Y)}{\sqrt{\sum (X(i) - \bar{X})^2 (Y(i) - \bar{Y})^2}}$$

(3) The raw data used in the calculation of the autocorrelation function (i.e., temperature) is combined with the data on salinity to calculate the cross correlation function.

Example

Cross correlation of two data sequences

A plot of the raw data and the correlogram are presented in Fig. 6-8.

Figure 6-9 shows different types of correlograms.

Legendre and Legendre (1998) present a detailed discussion of autocorrelation, cross correlation, and correlograms.

Most time series patterns can be described by trend and seasonality (Statsoft, Inc., 1995). Trend ideally comprises consistently increasing or decreasing (i.e., monotonous) data. If the time series contains considerable variations (e.g., from month to month or from year to year) then smoothing of data is required, the most common method being the moving average. When variations in data are very large then distance weighted least squares smoothing or negative exponentially weighted smoothing techniques can be used. Seasonal patterns of time series can be examined via correlograms. Serial dependency can be removed by differencing the series, that is, using time lags.

6.4 Autoregression

Prediction of the time series pattern (e.g., population density) at time $t + 1$ based on data from time t (example of a first order autoregression). Autoregression is

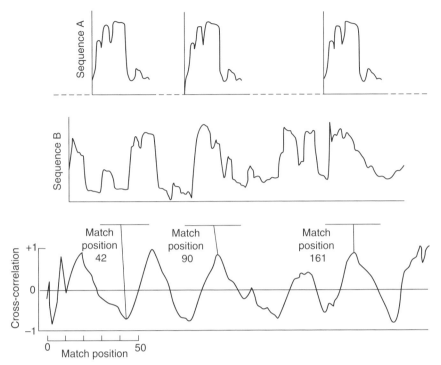

Figure 6-8.   Cross-correlation of two data sequences: A and B. Sequence A is shown at several positions of comparison. Cross-correlogram shows similarity of the two sequences at all match positions (source: Davis, 1973).

like a multiple regression with the dependent variable regressed on past values. This is a powerful statistical technique, occasionally used by ecologists. **A series of models (MA, AR, ARMA, ARIMA) have been developed by Box and Jenkins (1970), which combine autoregression and moving average terms** (a moving average different from that described previously). ARIMA, used most frequently in biology, is a complex modeling and forecasting technique requiring considerable experience, the results dependent on the level of expertise of the researcher (Statsoft, 1995). To determine if the ARIMA model fits the data, examine the residuals. Conduct a residual analysis for one step ahead forecast error (see Chatfield, 1996). ARIMA models can be fitted to a data series using a small number of parameters (Legendre and Legendre, 1998). The ARIMA autoregressive moving average model contains three parameters as follows: (1) autoregressive parameter (p), (2) the number of differencing passes or time lags (d), and (3) moving average parameters (q) (Statsoft, 1995). An ARIMA model described as (0, 1, 2) contains no autoregressive parameters (p) and two moving average parameters (q) which are computed for the time series after it is differenced once (d = one time lag).

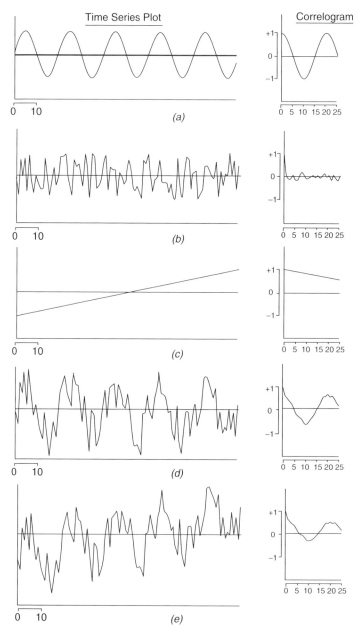

Figure 6-9. Some idealized time series and their autocorrelation functions. (a) Sine wave with wavelength of 20 units. (b) Sequence of random numbers or "noise". (c) Sequence of linearly increasing numbers or "trend". The trend can be removed by subtracting each number from the monthly average (a process called differencing). (d) Sine wave plus random noise (sequence a + sequence b). (e) Sine wave plus random noise plus linear trend (sequence a + sequence b + sequence c) (source: Wastler, 1969).

(a) Sine wave: the correlogram drops from +1 to 0 to –1. It then increases
  again until it reaches +1 when the signal is shifted one. Stationary, no
  trend exists.

(b) Random Noise: This signal was created by selecting points from a table
  of random numbers. The correlogram fluctuates around 0 . . . also
  stationary.

(c) Nonstationary: As observations increase in value over time, the
  correlogram shows decreasing correlation. Most time series are
  nonstationary.

(d) A + B sine wave + superimposed noise: Perfect correlation only occurs
  at lag 0, but periodic components are revealed in the correlogram. Pure
  noise is also known as white noise, chaos, random noise, and stochastic
  noise.

(e) A + B + C sine wave with trend and superimposed noise: The trend
  reduces the ability to distinguish the periodic component in the signal.

Figure 6-9. *(Continued)*

An example of the equations involved:

$$\ln z_i = \left(\varphi_1 z_{t-1} + \cdots + \varphi_p Z_{t-p}\right) + \\ + a_t - \left(\theta_1 a_{t-1} + \theta_q a_{t-q}\right) + \theta_0$$

where

  z_t is the time series variable, possible after transformation and/or differencing
  and corrected for an assumed stationary mean;

  a_t are independent random terms with zero mean and constant variance O_a^2;

  $\varphi_1, \varphi_2, ..., \varphi_p$ are auto-regressive parameters;

  $\theta_1, \theta_2, ... \theta_q$ are moving average parameters;

θ_0 is the possibly non-zero deterministic trend parameter if the data have been differenced.

where

  p = order of the auto-regressive segment of the model

  q = order of the moving average part

  d = the number of differencings used to remove non-stationarity of the mean
  function.

The a_t values are calculated recursively from the model and from the z_t values. The fitting of the ARIMA model to the observed data set involves three steps: (1) determination of how many and which z and a terms to use in the model, (2) estimation of Φ and θ parameters to make the model fit the data best, and (3) Diagnostic checking of the fitted model to detect any shortcomings.

The data need to be plotted in an autocorrelogram. Time series data usually need to be differenced until they become stationary (i.e., a relatively constant mean, variance and autocorrelation over time). This often first requires log transforming the data to stabilize the variance. Strong upward or downward changes usually require a lag = 1 differencing. Strong changes in slope usually require a lag = 2 differencing. Seasonal patterns or trends require seasonal differencing. The p and q parameters are rarely greater than 2. Next, the parameters are estimated so that the sum of squared residuals (error) is minimized. The estimates are used in forecasting new values beyond the original data set. The number of estimated parameters is based on the shape of the autocorrelogram and partial autocorrelogram and is almost never greater than 2. The data are standardized so that forecast data are expressed in values compatible with input data. ARIMA models may also include a constant. ARIMA operates in an iterative mode until the parameter estimation procedure converges. An example of the predictive utility of such models is given by Saila et al. (1980) who used an ARIMA model in predicting the catch of rock lobsters (Fig. 6-10). Automatic time series forecasting can be accomplished with a new program. The IMSL Auto_ARIMA (Autoregressive Integrated Moving Average) algorithm includes a number of techniques that reduce

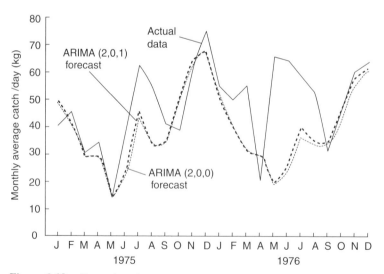

Figure 6-10. Comparisons between the actual monthly average catch per day fished of rock lobster from the Gisborne area (North Island, New Zealand) and the predicted catches determined from a harmonic regression model and from the monthly averages taking the trend into account. Solid line represents actual data; dotted line represents forecast by the ARIMA (2,0,0) simple auto-regressive model; and dashed line represents forecast by the ARIMA (2,0,1) model but includes a moving average term (source: Saila et al., 1980).

the complexity of the operation and time to analysis, such as estimation of missing values, identification and adjustment for the effects of outliers, seasonal adjustments, and selection of the best input parameters (p, d, q) and output of the forecast (FitzGerald, 2004). Also see Mendelssohn (1981), Parkratz (1983), Powell and Steele (1994), Chatfield (1996) and Parsons and Colbourne (2000).

Long-term forecasting requires 7–10 years of historical data. More than 50 observations are needed to use ARIMA. **Although the results of the Box-Jenkins models are mixed, they are recommended by numerous individuals**. For example, Michael Fogarty (pers. comm.) of the National Marine Fisheries Service uses Box-Jenkins models regularly with animal populations and with good results. **On the contrary the complexity of the models may not be worth the effort, according to Chatfield (1996), who recommends the widely used Holt-Winters forecasting method**. A number of business software programs include autoregression using ARIMA as part of their statistical package (e.g., Minitab). See Fogarty (1989) for information on forecasting yield and abundance of exploited invertebrates. See Rasmussen et al. (2001) for further information on ARIMA models. In order to understand more clearly how ARIMA models are used, see the Appendix and specific examples given in statistical packages (e.g., Statistica, Minitab).

C. FREQUENCY ANALYSIS

6.5 Frequency Analysis

The technique for analyzing frequencies is commonly known as spectral analysis or Fourier analysis. Spectral analysis is a method of estimating the spectral density or spectrum. Spectral analysis is a division of a signal into its harmonic components (described by sines and cosines), comparable to shining a beam of light through a prism, which transforms the light into a spectrum (Fig. 6-11).

A series of fluctuations (e.g., population fluctuations) may be composed of (1) a linear trend, (2) cyclic components (Fig. 6-12), and (3) a random component. Fourier analysis compares fluctuations in a time series with periodic functions such as a sine wave (one that repeats at regular intervals). The power spectrum is the Fourier transform of an autocovariance function (i.e., covariance over time). When using a power spectrum, indicate the bandwidth (analogous to a class interval with data). The figure showing the power spectrum (Fig. 6-13) is called a **periodogram** (see Legendre and Legendre, 1998, for a detailed discussion).

A simple Fourier series (Sine wave) is defined as:

$$Y_i = \sum_{n=1}^{x} \left(\alpha_n \cos \frac{2\pi n X i}{\lambda} + \beta \sin \frac{2\pi n X i}{\lambda} \right)$$

where

n = harmonic number

λ = wave length

α and β = phase angles

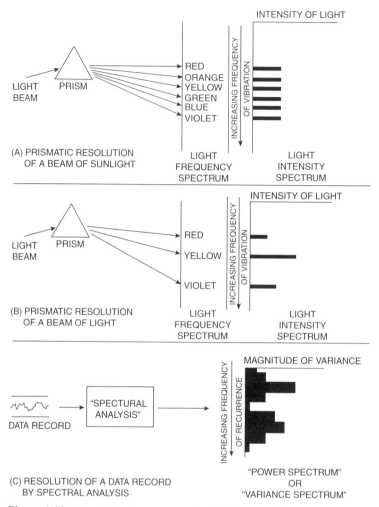

Figure 6-11.   Physical analogy to spectral analysis (source: Wastler, 1969).

This indicates that the amplitude Y_i at point X_i is determined by the sum of the component sine and cosine waves at a distance X from the origin of the series (Figure 6-12). The following figure (Fig. 6-13) illustrates the type of plots one may expect after transforming periodic and non-periodic data. The periodogram is a discontinuous frequency whereas the spectrum is a continuous function of frequencies (Legendre and Legendre, 1998).

Two algorithms are commonly used in spectral analysis: (1) Slow Fourier transform, requiring a minimum of about 100–200 data points for significant results (Chatfield, 1996), and (2) Fast Fourier Transform (FFT), usually requiring more than 1000 data points (Otnes and Enochson, 1978). Slow Fourier Transforms

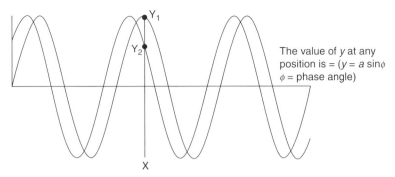

Figure 6-12. Two sinusoidal waves that are identical in form. The difference between Y_1 and Y_2 for a specified value of X is attributable to the difference in phase between the two wave forms (source: Davis, 1973).

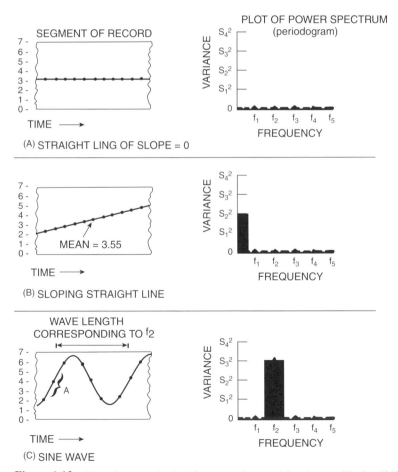

Figure 6-13. Typical spectra obtained from several types of data (source: Wastler, 1969).

are one-dimensional transforms requiring N^2 complex multiplications and divisions whereas FFTs are two-dimensional transforms requiring N^4 complex multiplications and divisions. This process can now be accelerated over 100-fold, reducing the number of computations to $N/\log_2 N$. Spectral analysis is seldom used in marine ecology because of the data requirement. For example, monthly population counts of a species would need to be taken for at least five years to represent an adequate sample. For this reason, serial correlation is of greater use to the marine ecologist. Another less serious problem with serial correlation, autoregression, and spectral analysis is that all data points should be complete. If data are missing, values must be interpolated. **Missing data can produce spurious time trend results if sampling is incomplete**. A smoothed spectrum is shown in Fig. 6-14.

Two power spectra can be compared by cross-spectral analysis (co-spectrurm or coherence spectrum – see Fig. 6-15). Cross-spectral analysis is used to uncover the correlations between two data series at different frequencies (e.g., weather patterns and changes in plankton populations over time). Multidimensional or multivariate spectral analysis can also be performed (Legendre and Legendre, 1998). See Platt and Denman (1975) for examples in marine ecology and Legendre and Legendre (1998) for information on the use of spectral analysis in ecology. Stephenson (1978) shows how correlograms, power curve spectra, and multiple regression are used in analyzing periodicities in the populations of marine macrobenthos. Emery

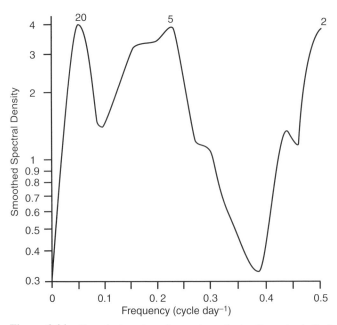

Figure 6-14.   Smoothed spectrum for numbers of migrating animals. Periods corresponding to the main peaks are indicated above the curve, that is, 20, 5, and 2. (source: Legendre and Legendre, 1998).

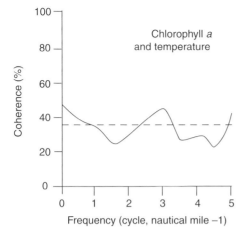

Figure 6-15. Coherence spectrum between chlorophyll a and water temperature. The dashed line represents 95% confidence limits. Close correspondence between the chlorophyll a-temperature plot and the 95% confidence limits line indicates significant coherence (source: Legendre and Legendre, 1998).

and Thomson (1998) discuss time series analysis in physical oceanography. Also see Shugart (1978), Jensen (1985), Diggle (1990), and Sugihara and May (1990).

For general computer programs on time series analysis see Statistica, Minitab, SAS and BMDP on the Internet.

Chapter 7

Modeling and Systems Analysis

7.1 INTRODUCTION

Systems analysis is an orderly and logical organization of data and information into models, followed by rigorous testing, validation, and improvement of those models. The purposes of environmental modeling, ecosystems analysis and simulation (imitation of real phenomena) are threefold: (1) to understand better how each of the components of an ecosystem contribute to its overall functioning, that is, to generate hypotheses for understanding the structure and behavior of nature (e.g., the Beverton-Holt stock-recruitment model in fisheries), (2) to predict future events, useful in environmental decision making (e.g., dynamic simulation models), and (3) to manipulate the system for maximization of some quantity (e.g., economic benefit by maximum economic yield in fisheries). Numerical models (modeling the environment largely with numbers and equations) in the study of marine ecosystems have been used for more than 60 years (Kremer and Nixon, 1978). Modeling can be relatively simple to perform, but before considering the process, it is useful to understand how different modelers view their specialty (their conceptions, approach, philosophy, mentality), one that is rapidly evolving.

There are different methods of categorizing the use of mathematics and modeling of ecosystems. This difference in approach creates considerable confusion for the beginner (see Table 7-1 and Figs. 7-1 to 7-3). Most theoretical models have been developed using analytical mathematical approaches but the number of problems that can be solved by routine analytic techniques is small (Hall, 1991). The limits of analytical methods in solving mathematical problems are presented in Table 7-2. A summary of the different types of models currently used in ecosystem analysis is presented in Table 7-3. See p. 138 in Chapter 3 for information on cellular automata models.

Perhaps the most commonly used models in ecological and environmental studies are dynamic simulation models (mechanistic or functional models). The full development of dynamic feedback models in ecology really began with the application of analog and digital computers in the 1960s (Kremer and Nixon, 1978). In

Quantitative Analysis of Marine Biology Communities: Field Biology and Environment
by Gerald J. Bakus
Copyright © 2007 John Wiley & Sons, Inc.

Table 7-1. Methods of categorizing the use of mathematics and modeling of ecosystems.

1. Kremer (pers. comm.)
   A. Statistical
      1. Advocate: statisticians, community ecologists
      2. Techniques: coefficient of correlation, covariance, multivariate analysis
   B. Donor-controlled
      1. Advocate: B.C. Patten
      2. Techniques: Donor (e.g., plants) producing energy for recipients (e.g., herbivores) without feedback mechanisms see Fig. 7-1.
   C. Phenomenological
      1. Advocate: H.T. Odum
      2. Techniques: Both donor (e.g., plants) and recipient (e.g., animals) control a sample process with feedbacks (e.g., nutrients). See Fig. 7-2.
   D. Mechanistic
      1. Advocate: modelers
      2. Techniques: Employ a series of functional relationship, interactions and feedbacks. See Fig. 7-3.
2. Hall and Day (1977)
   Ecological models and modelers
   A. Analytic models – attempt to find exact solutions to differential and other equations (e.g., the Lotka-Volterra population models).
   B. Simulation models – solve very many equations nearly simultaneously.
3. Jeffers (1978)
   A. Word models
   B. Mathematical models
      1. Deterministic – for a given set of initial conditions the model will always yield the same value.
      2. Stochastic – contain certain values that are independent and random.
      and
         a. matrix models – use matrices (e.g., Leslie model)
         b. stochastic models – incorporate probabilities
         c. multivariate models – multiple regression, principal component analysis, etc.
         d. optimization models – produce graphical solutions (i.e., intersecting curves)
         e. other models – game theory, catastrophe models
         f. dynamic models – simulation

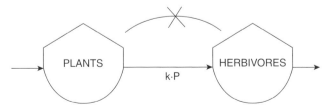

Figure 7-1. A donor (plants) controlled, simple linear model lacking feedback mechanisms. The flux of energy and associated matter from the donor (plants) to herbivores is a function only of the plants. $J = k \cdot P$ (source: Kremer, 1983).

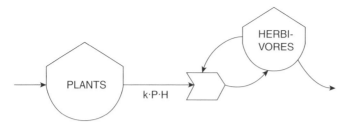

Figure 7-2. A phenomenological model showing donor (plants) and recipient (herbivores) with feedbacks. The flux of energy from plants to herbivores is a simple function of both compartments. J = k.P.H (source: Kremer, 1983).

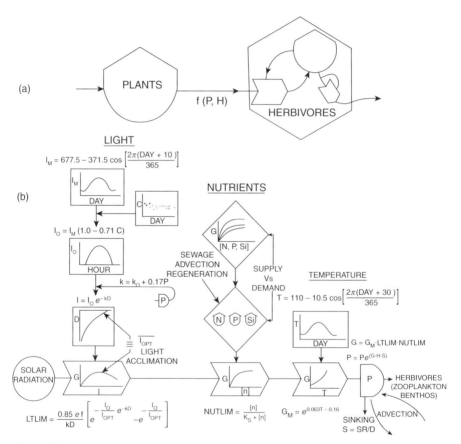

Figure 7-3. Simple (a) and detailed (b) mechanistic model. (Source: (a) Kremer, 1983); (b) Kremer and Nixon, 1978).

dynamic modeling, one identifies the temporal behavior of the system, creates a computer simulation of the system, and eventually uses the model for prediction (i.e., systems analysis). This process necessitates the use of computers and a variety of mathematics (such as differential equations and integration). Numerous assumptions

Table 7-2. Classification of mathematical problems and their ease of solution by analytical methods (source: Hall, 1988).

	Linear equations			Nonlinear equations		
Equation	One equation	Several equations	Many equations	One equation	Several equations	Many equations
Algebraic	Trivial	Easy	Essentially impossible	Very difficult	Very difficult	Impossible
Ordinary differential	Easy	Difficult	Essentially impossible	Very difficult	Impossible	Impossible
Partial differential	Difficult	Essentially impossible	Impossible	Impossible	Impossible	Impossible

Table 7-3. Descriptions of different types of models commonly used in ecosystem analysis (source: Shugart, 2000).

Type of Model	Description
Conceptual models	Diagrams or descriptions of the important connections among the components of an ecosystem
Microcosms	Small (usually small enough to fit on a laboratory bench) physical and biological analogs to a larger ecosystem of interest
Population models	Usually systems of differential or difference equations that compute the change in the numbers of individuals in a population
Community models	Often of similar structure to population models but including terms that involve interactions with other populations
Compartment models	Usually systems of differential or difference equations that follow the transfer of elements, energy, or other material through an ecosystem
Multiple commodity models	Compartment models that treat the interactive transfers of several different materials through an ecosystem
Individual-based models	Models in which the dynamic changes in the individuals in an ecosystem are used as a basis to understand larger system dynamics

must often be made since field data are frequently lacking. Once the variables have been identified, sensitivity analysis can be used to test whether changes in input variables and parameters produce large or small variations in the performance of the model, in other words, one simply changes the input numbers and determines which inputs cause the greatest effect on the model output.

Models need to be optimized for "realism," "generality," and "precision." **Realism is the degree to which the model mimics real world processes. Generality**

is the degree to which a model can be applied to other systems. Precision is the degree to which model-generated data match realworld data in a quantitative sense. Both a high level of precision and generality are difficult to develop together. One must remember that many dynamic models are but crude representations of real ecosystems and that poor or erroneous data will be reflected in poor predictive value. Moreover, some authors believe that only those who develop a model should use it because of the implicit assumptions and limitations inherent to it (James Kremer, pers. comm.).

7.2   PHILOSOPHY OF MODELING

Before we discuss the mechanics of modeling, it is worthwhile to expand somewhat on the philosophy of modeling. Models take many forms such as abstract or primitive art, a picture, a diagram, an abstract concept, or an equation. Modeling is an idealistic abstraction of the processes of nature. Marine ecology began with the biocoenosis model (i.e., mutual interdependence of species in a colony [Hedgpeth, 1977]). Modeling and predictive biology attracts those who are willing to sacrifice descriptive precision and detail for generality and application in prediction (Peters, 1986). The constructs used in limnology to predict the variables are (1) highly simplified models and (2) regressions utilizing only a few variables but resting on an extensive data base. Most complex living systems are nonlinear and exhibit discontinuities and chaotic behavior that can only be adequately represented with numerical methods and simulations (Costanza et al., 1993). It is inappropriate to think of ecosystem models as anything but crude, abstract representations of complex phenomena. Realistic ecosystem models would be orders of magnitude more complex than physical–chemical models in use today (e.g., in meteorology and oceanography). No single model can maximize all three modeling goals of realism, precision, and generality. Complex ecosystems are characterized by strong, usually nonlinear, interactions between the parts, complex feedback loops making it difficult to distinguish cause from effect, and significant time and space lags, discontinuities, thresholds, and limits. Even simple classical theoretical models in ecology have very little data from real populations to support their validity (e.g., the logistic population growth curve or S-shaped curve, the Lotka-Volterra predator-prey equation, the competition model of New Guinea birds by Diamond (1973), and Ricker's density-dependent stock recruitment model in fisheries, according to Hall, 1985, 1988, 1991). The track record of relating mathematical models to ecological data is not good, especially concerning the formulation of testable hypotheses and predictions (Cushing et al., 2003).

Some of the main criticisms of modeling are:

(1) Comprehensive ecosystem models have incomplete data bases (Hedgpeth, 1977, and many others).

(2) Ecosystem models must be based on a thorough understanding of the relevant natural history, and when possible, on some form of controlled manipulation (Hedgpeth, 1977, and many others).

(3) It may be practically impossible to distinguish data that have been generated by a rather simple deterministic process either from true stochastic noise or from "experimental error" in sampling or measurements (Hedgpeth, 1977).

(4) In modeling complex systems, the issues of scale and hierarchy are central. Information and measurements are generally collected at small scales then often used to build models at radically different scales (e.g., regional, national, global – Costanza et al., 1993).

(5) Needed are better measures of model correspondence with reality and long-term system performance with the conflicting criteria of realism, generality and precision (Costanza et al., 1993 and many others).

Hall (2003) states that simulation models have been criticized on two following counts: (1) they may get the right answer for the wrong reason and (2) their results are too sensitive to model parameters that may be poorly known. Peters (1986) maintains that predictive limnology offers the opportunity for rapid scientific development and has been instrumental in programs of eutrophication control in the Great Lakes and elsewhere. In any event, models have considerable heuristic value (i.e., speculation serving as a guide in scientific research) even if they generalize. Models are of considerable help in organizing large quantities of diverse information. As time progresses, ecosystem models will gradually become more realistic, accurate, and valuable as a predictive tool. Ecosystem models are destined to improve over time.

7.3   MODEL COMPONENTS AND MODEL DEVELOPMENT

The four major components of a model are (1) **compartments** (e.g., a box labeled "herbivores"), (2) **forcing functions** (processes coming from outside the system, e.g., solar energy), (3) **feedbacks** (e.g., predator-prey interactions), and (4) **transfer coefficients** (mathematical coefficients that express the rate of transfer from one compartment to another, e.g., prey [phytoplankton] to predator [copepod] in $g/m^3/day$).

The **steps in dynamic modeling** (Fig. 7-4) are as follows:

(a) **Develop a conceptual or mental model**. One needs to define the research or management question of interest that helps define which of the modeling techniques in Table 7-1 would be useful. (b) **Develop a diagrammatic model indicating forcing functions (e.g., solar energy input) and state variables** (i.e., boxes or compartments – see Fig. 7-5). Different authors use their own preferred symbols for functions and state variables; only a few are widely used. Odum (1994) presents a complex variety of symbols, an energy circuit language that has been used by some modelers.

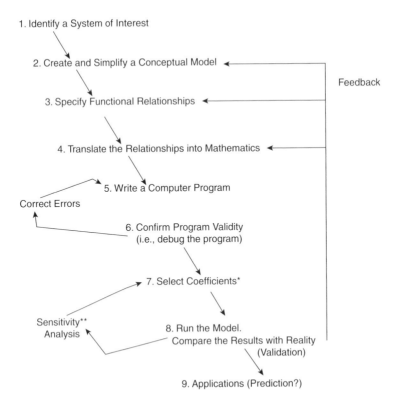

*This usually is an iterative process.
** This pathway is optional. Sensitivity analysis can be performed independent of altering or choosing coefficients.

Figure 7-4. The modeling process (Source: James Kremer and Joseph DeVita, pers. comm.).

For example,

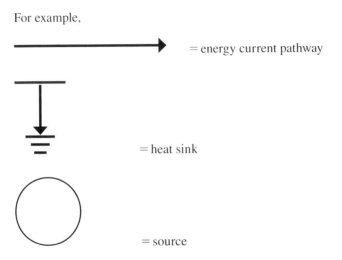

= energy current pathway

= heat sink

= source

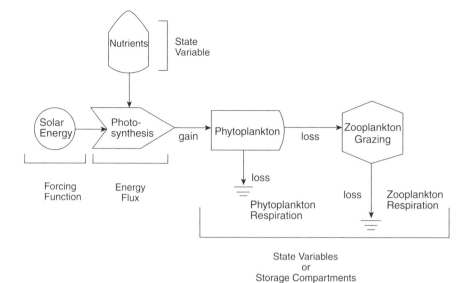

Figure 7-5.   Example of a simple compartmental model (source: Riley, 1946).

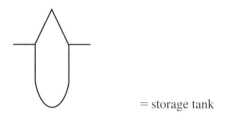

= storage tank

(c) **Develop a mathematical model by converting the diagrammatic model into equations**. A simple mathematical model to represent this compartmental model (Fig. 7-5) would be (Riley, 1946):

$$\frac{dp}{dt} = P(P_h - R - G)$$

where

$\dfrac{dp}{dt}$ = rate of change (d) of the phytoplankton population (p) with respect to time (t)

P = photosynthetic rate

P_h = photosynthetic rate per unit of population

R = rate of phytoplankton respiration

G = rate of grazing by zooplankton

A more detailed equation describing phytoplankton production might be (modified from Dugdale, 1982):

$$\frac{dp}{dt} = G_r - R - S - G + A + D - E - M$$

where

$\dfrac{dp}{dt}$ = rate of change (d) of phytoplankton (p) with respect to change (d) in time (t)

G_r = growth

R = respiration

S = sinking

G = grazing

A = advection

D = diffusion

E = excretion

M = natural mortality

The important thing to emphasize is that all parts of the equation be expressed in the same units (e.g., carbon units, calories, etc.). Expressions for energy exchange between compartments may be presented in the form of transfer coefficients (see p. 299 and Table 7-6 on p. 299). **The various equations can then be treated mathematically by a simple finite difference method, some more complex method of integration, or by successive summation of changes over small time increments** (see Fig. 7-6 and Kremer and Nixon, 1978).

Figure 7-6 shows methods of handling the math in modeling. The finite difference method is simply measurements taken over time then subtracted from one another (e.g., the human population is surveyed 10 or 25 years apart). The differential equation method shows an instantaneous change in population size described by:

$$\frac{dN}{dt} = rN$$

where r is the biotic potential or population growth described by $N_t = N_o e^{rt}$. See p. 21 in Chapter 3.

It should be emphasized that many differential equations (unlike the one shown above) cannot be solved analytically. One then needs to use numerical integration approximations for a solution (see Fig. 7-6).

Measuring changes over small time increments is the easiest method for beginning modelers. For example, one can simply measure phytoplankton production per

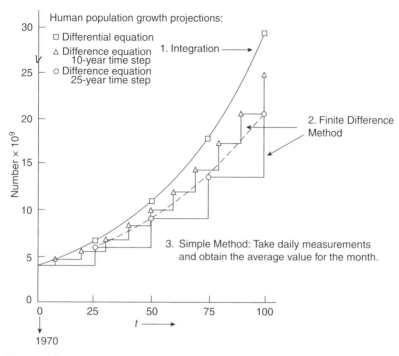

Figure 7-6. A comparison of predictions for the growth of human populations. Three solutions are offered: (a) analytical solution using integration, (b) Δ and o are numeric solutions with 10-year and 25-year time steps, and (c) a simple method of averaging the population increase each month.

unit time (day) and sum these values, divide by the number of day measurements, and arrive at mean monthly values. (d) **Develop a model output and validate the model**. Compare values of real or actual productivity to those generated by the model, using a different data set than used to develop the model (i.e., validation process – see Fig. 7-7). Other methods of validation include experimental prediction, exposing the model to critics, adaptive management, monitoring, feedback for model fine tuning, and sensitivity analysis (Hall and Day, 1977; David Dow, pers. comm.). (e) **Optimize the model**. Modify the model so that the predicted and actual data patterns are as similar as possible. In complicated interactions, the process of modification of variables by simulation and comparison with field data may need to be repeated numerous times, and (f) **Manipulate the model**. Ask questions; "feed" the model data to determine cause and effect, maximum efficiency, maximum profit, and so forth. An example of a generalized model of an estuarine bay ecosystem is presented in Fig. 7-8. A more detailed and complex estuarine bay ecosystem model is presented in Fig. 7-9. Fundamental steps in the modeling of the ecosystem in Figure 7-9 are given in Tables 7-4 to 7-6. A newer concern is stability analysis or the long-term response of the system to an environmental perturbation or change. This

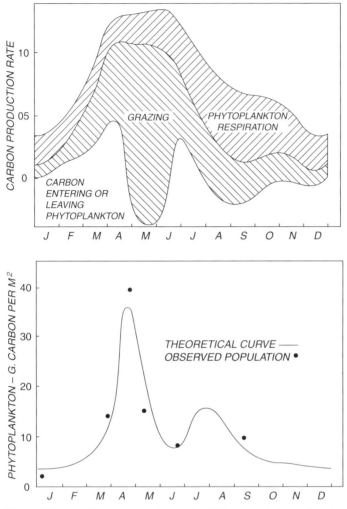

Figure 7-7.   A validation process in modeling. Using numeric integration, the rates of change are summed to estimate the time series of phytoplankton abundance. Annual calculated values are compared with observed values of phytoplankton populations. (source: Riley, 1946).

is accomplished by loop analysis (linear systems) or Lyapunov analysis (nonlinear or linear system; Shugart, 2000).

Three recent texts that emphasize the modeling of populations include Roughgarden (1998), Gotelli (2001), and Turchin (2003). For examples of using artificial neural networks in modeling, see the Special Issue on Marine Spatial Modelling (Høisaeter, 2001). For computer modeling of seas and coastal regions see Brebbia (2002). Modeling to guide policy and management decisions is discussed

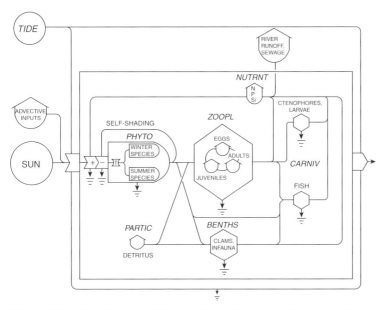

Figure 7-8.   Generalized model of energy flow in Narragansett Bay (source: Kremer and Nixon, 1978).

Table 7-4.   Estimated values for an estuarine model (source: McKellar, 1977).

Notation	Description	Value
External driving forces		
I_1	Sunlight	$6000\,\mathrm{kcal/m^2/day}$ (1)
I_2	Water temperature	$30°\mathrm{C}$ (1)
I_3	Tidal water exchange	$1.4 \times 10^6\,\mathrm{m^3/day}$ (1)
I_4	Advective water exchange	$1.2 \times 10^6\,\mathrm{m^3/day}$ (3)
I_{01}	Phytoplankton import due to water exchange	$1.4\,\mathrm{g/m^2/day}$ (2)
I_{02}	Zooplankton import due to water exchange	$0.21\,\mathrm{g/m^2/day}$ (2)
I_{03}	Macrofauna immigration	$0.74\,\mathrm{g/m^2/day}$ (3)
I_{04}	Detrital import due to water exchange	$32.5\,\mathrm{g/m^2/day}$ (3)
I_{05}	Phosphorus import due to water exchange	$0.11\,\mathrm{g/m^2/day}$ (2)
Internal standing stocks		
Q_1	Producer biomass	$45.9\,\mathrm{g/m^2}$ (1)
Q_2	Zooplankton biomass	$0.16\,\mathrm{g/m^2}$ (1)
Q_3	Macrofauna biomass	$20.7\,\mathrm{g/m^2}$ (2)
Q_4	Detritus stock (water and sediments)	$75\,\mathrm{g/m^2}$ (3)
Q_5	Total phosphorus in the water	$0.08\,\mathrm{g/m^2}$ (1)

(Continued)

Table 7-4. (*Continued*)

Notation	Description	Value
Exchange rates		
J_{GP}	Community gross primary production	$8.25 \, \text{g/m}^2/\text{day}$ (1)
J_R	Total community respiration	$8.40 \, \text{g/m}^2/\text{day}$ (1)
J_{R1}	Producer respiration	$4.13 \, \text{g/m}^2/\text{day}$ (1)
J_{R2}	Zooplankton respiration	$0.11 \, \text{g/m}^2 \, \text{day}$ (1)
J_{R3}	Macrofauna respiration	$1.38 \, \text{g/m}^2/\text{day}$ (2)
J_{R4}	Respiration of detritus and bacteria	$2.48 \, \text{g/m}^2/\text{day}$ (2)
J_N	Phosphorus recycle from total respiration	$0.08 \, \text{g/m}^2/\text{day}$ (2)
J_{10}	Phytoplankton export due to water exchange	$1.39 \, \text{g/m}^2/\text{day}$ (2)
J_{12}	Zooplankton grazing on phytoplankton	$0.08 \, \text{g/m}^2/\text{day}$ (2)
J_{13}	Macrofauna grazing on phytoplankton	$1.0 \, \text{g/m}^2/\text{day}$ (2)
J_{14}	Death rate of producers	$3.0 \, \text{g/m}^2/\text{day}$ (2)
J_{20}	Zooplankton export due to water exchange	$0.21 \, \text{g/m}^2/\text{day}$ (2)
J_{23}	Macrofauna consumption of zooplankton	$0.025 \, \text{g/m}^2/\text{day}$ (2)
J_{24}	Zooplankton death and faeces production	$0.025 \, \text{g/m}^2/\text{day}$ (2)
J_{30}	Macrofauna emigration	$0.74 \, \text{g/m}^2/\text{day}$ (3)
J_{34}	Macrofauna death and faeces production	$1.7 \, \text{g/m}^2/\text{day}$ (2)
J_F	Commercial and sport fishery harvest	$0.006 \, \text{g/m}^2/\text{day}$ (1)
J_{40}	Detritus export due to water exchange	$32.5 \, \text{g/m}^2/\text{day}$ (3)
J_{42}	Zooplankton consumption of detritus	$0.08 \, \text{g/m}^2/\text{day}$ (2)
J_{43}	Macrofauna consumption of detritus	$2.0 \, \text{g/m}^2/\text{day}$ (2)
J_{50}	Export of phosphorus due to water exchange	$0.11 \, \text{g/m}^2/\text{day}$ (2)
J_{51}	Phosphorus uptake by producers	$0.08 \, \text{g/m}^2/\text{day}$ (2)

(1) values which are based on direct measurements in the Crystal River area.
(2) values which involved a combination of direct measurements, literature values, and assumptions.
(3) values which were based on assumptions with little validation but which were reasonable considering an organic balance in the ecosystem

Estimated values for external driving forces, internal stocks, and exchange rates; summer conditions in the outer control bay, g = grams inorganic matter except for phosphorus exchanges in grams total phosphorus. Source: McKellar (1977).

(1) Note that only 10/36 are based on direct measurements.
(2) 20/36 are based on partial information.
(3) 6/36 are guesses

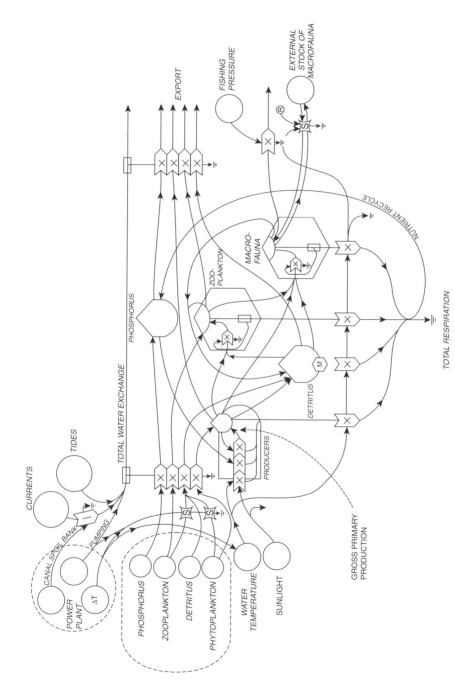

Figure 7-9. Compartmental model of an estuarine bay ecosystem. Crystal River area, about 90 miles (145 km) northwest of Tampa, west coast of Florida (source: McKellar, 1977).

Table 7-5. Differential equations for the estuarine model of McKellar (1977).

Rate of change	= trophic inputs	+ import − export	− losses to next trophic level	− organic recycle	− respiration
Producers:	$\dot{Q}_1 = \dfrac{K_{GP}I_1I_2Q_5Q_1}{1+K_1K_{GP}I_2Q_5Q_1}$	$+J_{01}$	$-K_{10}Q_1$	$-K_{12}Q_1(K_{R2}Q_2I_2)$ $-K_{13}Q_1(K_{R3}Q_3I_2)$	$-K_{14}Q_1$	$-K_{R1}Q_1I_2$
Zooplankton:	$\dot{Q}_2 = K_{12}Q_1(K_{R2}Q_2I_2)$ $+K_{42}Q_4(K_{R2}Q_2I_2)$	$+J_{02}$	$-K_{20}Q_2$	$-K_{23}Q_2(K_{R3}Q_3I_2)$	$-K_{24}Q_2$	$-K_{R2}Q_2I_2$
Macrofauna:	$\dot{Q}_3 = K_{13}Q_1(K_{R3}Q_3I_2)$ $+K_{23}Q_2(K_{R3}Q_3I_2)$ $+K_{43}Q_4(K_{R3}Q_3I_2)$	$+J_{03}$	$-K_{30}Q_3$ $-K_{30H}Q_3$ (when $I_2 > R_H$) $-K_{30C}Q_3$ (when $I_2 < R_C$)	$-K_FQ_3$	$-K_{34}Q_3$	$-K_{R3}Q_3I_2$
Detritus and microbes:	$\dot{Q}_4 = K_{14}Q_1$ $+K_{24}Q_2$ $+K_{34}Q_3$	$+J_{04}$	$-K_{40}Q_4$	$-K_{42}Q_4(K_{R2}Q_2I_2)$ $-K_{43}Q_4(K_{R3}Q_3I_2)$		$-K_{R4}Q_4I_2$
Phosphorus:	$\dot{Q}_5 = K_N\left(\sum J_R\right)$	$+J_{05}$	$-K_{50}Q_5$	$-K_N\,\dfrac{K_{GP}I_1I_2Q_3Q_1}{1+K_1K_{GP}I_2Q_5Q_1}$		

Notation:

Q = state variable
K = transfer coefficient
I_1 = sunlight
I_2 = temperature
J_0 = import forcing functions
Rh = upper reference temperature
Re = lower reference temperature
$(\Sigma J_R) = K_{R1}Q_1I_2 + K_{R2}Q_2I_2 + K_{R3}Q_3I_2 + K_{R4}Q_4I_2$

*Source: McKellar (1977)

Table 7-6. Transfer coefficients for the estuarine model of McKellar (1977).

Coefficient	Value
K_{GP}	4.99×10^{-5}
K_1	546.0
J_{01}	2.08
$K_{10} = J_{10}/Q_1$	0.0303
$K_{12} = J_{12}/Q_1 J_{R2}$	0.0158
$K_{13} = J_{13}/Q_1 J_{R3}$	0.0174
$K_{14} = J_{14}/Q_1$	0.0654
$K_{R1} = J_{R1}/Q_1 J_2$	0.0031
$K_{42} = J_{42}/Q_4 J_{R2}$	0.0073
J_{02}	0.2300
$K_{20} = J_{20}/Q_2$	1.3125
$K_{23} = J_{23}/Q_2 J_{R3}$	0.1132
$K_{24} = J_{24}/Q_2$	0.1563
$K_{R2} = J_{R2}/Q_2 J_2$	0.0230
$K_{42} = J_{43}/Q_4 J_{R3}$	0.0145
$K_F = J_F/Q_3$	0.0003
$K_{34} = J_{34}/Q_3$	0.0821
$K_{R3} = J_{R3}/Q_3 J_2$	0.0022
J_{04}	32.50
$K_{R4} = J_{R4}/Q_4 I_2$	0.0008
K_N	0.0100
J_{05}	0.1300
$K_{50} = J_{50}/Q_5$	1.3750
$K_{30} = J_{30}/Q_3$	0.0357
$K_{30H} = J_{30H}/Q_3$	0.0357
$K_{30C} = J_{30C}/Q_3$	0.0740

by van den Belt (2004). See the journal series Modeling Dynamic Systems and Ecological Modelling. See the Appendix for a tutorial on modeling, that is, modeling a simple food chain.

For further information on modeling see Nihoul (1975), Hedgpeth (1977), Jeffers (1978, 1982), Getz (1980), Nisbet and Gurney (1982), Spain (1982), Odum (1982), Huston et al. (1988), Jorgensen et al. (1996), Ruth and Lindholm (1996), Odum and Odum (2000), Sala et al. (2000), Haddon (2001), Hannon and Ruth (1997, 2001), May (2001), Shenk and Franklin (2001), Ruth and Lindholm (2002), Adam (2003), Canham et al. (2003), Costanza and Voinov (2003), Dale (2003), Lindsey (2003), Christensen et al. (2004), van den Belt (2004), Walters and Martell (2004), Fennel and Neumann (2005), Grace (2005), Ulanowiez (2005), and Stanciulescu (2005).

Future directions in modeling include multiple commodity models (combined models of change in structure and material flow) and individual-based models (Shugart, 2000). Huston et al. (1988) introduced the concept of "individual-based"

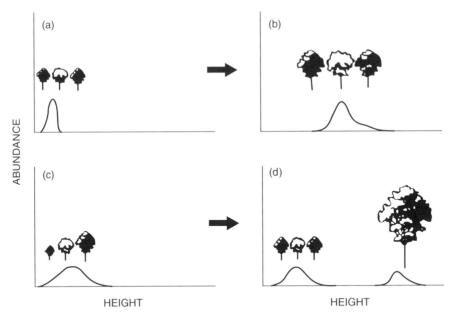

Figure 7-10.   The effect of initial variation in the height of plants on the consequences of competition for light. Initial height distribution with low variance (a) leads to a population composed of many stunted individuals of similar size (b). Initial height distribution with high variance (c) leads to a bimodal population structure with a few very large plants and many small suppressed individuals (d) (source: Huston et al., 1988).

models of populations. For example, instead of treating all plants as one group of primary producers in a model, they are subdivided into a few subclasses by size variance then modeled separately (Fig. 7-10). However, individual-based models would imply modeling on the species level by many systmatists. See Grimm and Railsback (2005).

Three of the most popular computer programs for modeling ecological systems are ModelMaker (PC), Stella (PC and Mac), and Ecobeaker (PC and Mac). Others can be downloaded from the internet (e.g., Ecopath with Ecosim, etc.). See the Appendix for information on these.

Chapter 8

Marine Sampling and Measuring Devices

8.1 INTRODUCTION

There are numerous sampling devices used in marine sciences. The choice of a sampler is based on the following: (1) size, density and behavior of the organism(s), including plankton, nekton, and benthon; water chemistry, suspended particulates, and sediments; (2) the collection of qualitative or quantitative results, or both types; (3) nature and area of the seafloor; (4) size of the vessel, if a ship or submersible is needed; (5) depth; (6) nature of the sample taken by a specific type of sampling device; and (7) equipment, facilities, and time available. Various authors have discussed the frequent problems associated with collecting marine organisms (such as patchiness, spot sampling versus continuous sampling, animals that escape nets, vertical migration, and fish that eat midwater animals in the net). These are discussed in numerous books and papers on biological oceanography and marine biology (see below and the References). McConnell (1982) describes the history of oceanographic instruments. The following summarizes many of the major sampling devices in use today.

8.2 OCEANOGRAPHIC DEVICES

There are numerous new developments in oceanographic devices. This information is taken in large part from Sea Technology (2001) and summarized in the following subsections.

1. Spatial Visualization

Spatial visualization software in 3-D, such as instruments supported by the Canadian Centre for Marine Communications (CCMC) of Saint Johns, Newfoundland (Fig. 8-1). The visualizations communicate the meaning of the data to nontechnical

Quantitative Analysis of Marine Biology Communities: Field Biology and Environment
by Gerald J. Bakus
Copyright © 2007 John Wiley & Sons, Inc.

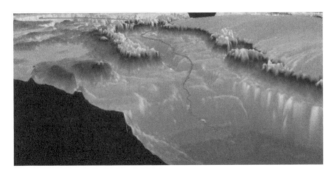

Figure 8-1. 3-D visualization of Baffin Bay (source: Safer, 2001).

people yet enable technicians to analyze, edit and run "what if" scenarios. High and low resolution data can be fused in a single image, providing details where needed. An example of a popular device is the Fledermaus, a 3-D tool that enables the viewer to fly through data and see it from all angles. See also Mayer et al. (2002).

2. *Moored Instruments*

Moored instruments typically measure physical-chemical properties such as current velocity and direction, conductivity, depth, temperature, and salinity (Fig. 8-2). Now they also measure turbidity, D.O., pH, organic material (e.g., by fluorescence),

Figure 8-2. Moored oceanographic instruments (source: Sea Technology February 2001 and Andrew Clark).

Figure 8-3.   Salmon mariculture south of Puerto Montt, Chile.

nutrients, redox potential, water level and waves (direction and height). Optical sensors measure the underwater light field and particle absorption and scattering (helpful for distinguishing types of particles, even differentiating major taxa of phytoplankton [Burt Jones, pers. comm.]). Other moored devices are used for offshore aquaculture (Fig. 8-3; Cates et al., 2001).

3. Weather

Moored and drifting buoy networks report near real time data primarily via satellites (Fig. 8-2). They measure wind speed and direction, barometric pressure, air and water temperature, relative humidity, solar radiation, and waves.

4. Current Meters

Current meters measure water velocity by acoustic frequency shift or electromagnetic techniques (Fig. 8-4). Low frequency, deep water (2000 m), multiplayer acoustic current profiles are being obtained.

5. Pressure

Pressure is measured by pressure tranducers whose sensors employ a silicon piezoresistive bridge.

6. Temperature

Temperature is measured by thermometers, reversing thermometers (in Nansen bottles) and thermocouples with platinum, gold, or palladium wire sensors. Most oceanographers use CTDs with a highly sensitive thermistor for measuring

Figure 8-4. Current meter (source: http://www.interoceansystems.com/html).

temperatures. Satellite sea surface temperatures are obtained with scanning radiometers (see p. 56 in Chapter 1).

7. *Conductivity*

Conductivity is measured by the following techniques: (a) an electrode method where the current flow between the electrodes depend on the conductivity of the water and (b) an inductive method using conductivity between two coils in water. CTDs measure conductivity, temperature, depth, and other variables (Fig. 8-5). The record

Figure 8-5. CTD surrounded by Niskin bottles (source: http://www.pmel.noaa.gov./.../ bot-over-the-side.html).

is transferred to a computer and plotted in real time as the CTD profiles through the water column.

8. *Turbidity*

Turbidity is measured gravimetrically, with a Secchi disk (Fig. 8-6), optically (by transmissometer; Fig. 8-7), and acoustically. The transmissometer reading is the amount of light reaching the photon sensor at the end of a seawater path. The light that is lost along the path is caused by scattering and absorption. The transmissometer does not measure the scattering of light by particulate matter. Light backscatter and forescatter are measured with a scattering meter or scatterometer.

9. *Oxygen*

Electrochemical oxygen sensors are calibrated against Winkler titrations and by optical oxygen sensors. The most recent technique is that of employing optodes. Measurements are based on the ability of oxygen to act as dynamic fluorescence quenchers (Tengberg et al., 2003).

Figure 8-6. The author deploying a Secchi disk in the Bering Sea in June.

Figure 8-7. Transmissometer used in measuring ocean turbidity.

10. Nutrients

Real-time nutrient profiling is accomplished by in-situ chemical analyzers. They measure concentrations of nitrate, nitrite, phosphate, iron, trace metals, and so forth. The analyzers can also be mounted on moorings for time series measurements. See Millero (2006) for information on chemical oceanography.

11. Underwater Vehicles

There are four major classes of underwater vehicles: (1) Towed undulating vehicles (TUVs), (2) Remote operated vehicles (ROVs), (3) Autonomous underwater vehicles (AUVs), and (4) Manned vehicles/Submersibles.

TUVs are equipped with CTDs, optical sensors, automated chemical detectors, optical plankton counters, videoplankton recorders, and so forth (Fig. 8-8). They can also be equipped with pumps so that seawater can be pumped aboard a ship for additional measurements.

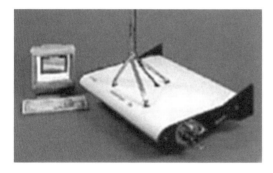

Figure 8-8. A towed undulating vehicle (TUV). The Scanfish MK1 (source: http://www.chelsea.co.uk/Vehicles.htm).

Figure 8-9. The Max Rover remote operated vehicle (ROV) (source: http://www.deepseasystems .com/.../maxrover/thrust2.html).

ROVs have become smaller (600–900 kg operating to 6500 m depth) with the availability of computer controllers for nearly all vehicle functions. The Max Rover and Mini-Max (Deep Sea Systems) are virtually all electric with high powered thrusters for operations in strong currents (Fig. 8-9). The major uses for ROVs include undersea cable route surveying, pipe line tracking, precision control operations, and submarine rescue. See OPL Staff (2002).

Autonomous underwater vehicles (AUVs such as Ocean Voyager II) are used to obtain ground truth data by satellites. AUVs now include a range of vehicles (Fig. 8-10).

Figure 8-10. An autonomous underwater vehicle (AUV). The EAVE-III (source: http://www.cdps .umcs.maine.edu/html).

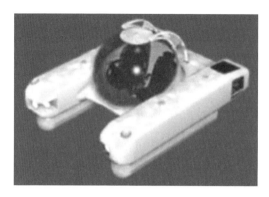

Figure 8-11.   Triton 650 luxury submarine (source: http://www .brownmarine.com/ss/submarine.html).

Some are gliders which require low power because they simply change buoyancy and travel long distances by using buoyancy to propel them through the water (Tomaszeski, 2004). Also see Wood et al. (2004) and Okamoto et al. (2005). Autonomous surface vehicles are described by Lammons (2005).

Submersible categories include the following: (1) Luxury Submarines, (2) Tourist Submarines, (3) Underwater Habitats, (4) Deep Submersibles, (5) Small Military Submarines, and (6) Floating Activity Centers.

Luxury submarines for individual ownership range from small units (e.g., Triton 650, 200 m depth, 2 passengers) to larger units and are costly (Fig. 8-11). Tourist submarines (e.g., ST-100 Argos, 100 m depth, 16 passengers) are developed for commercial entertainment purposes and are found in many parts of the world. Underwater habitats with underwater research began in ernest with the 1969 Tektite project in the U.S. Virgin Islands. This has been continued with NOAA Hydrolab and Aquarius projects in various places (e.g., Alaska, Florida). One of the earliest deep submersibles was the bathyscaph Trieste in which Donald Walsh and Jacques Piccard on January 3, 1960, reached a depth of 11,524 m (37,800 ft) in the Challenger deep. Modern deep-sea vessels include the ROV (remote operated vehicle) Jason/Medea that can reach a depth of 6000 m. ALVIN dives up to 200 times a year for 6–10 h per dive to a maximum depth of 4000 m and holds 3 people (Fig. 8-12). The deep-sea hydrothermal vents were discovered by ALVIN. Many of these submersible programs have been supported by ONR and NOAA over the years. The Ocean Explorer series (e.g., Explorer 1000, depth 305 m, 2 passengers) typically contain a variety of core sensors and operate in the epipelagic zone (0–200 m), even under Arctic ice. Floating activity centers operate on the Great Barrier Reef of Australia where people go offshore to live and dive for short periods of time (Fig. 8-13). See Burgess (2000) for further information and *Sea Technology* magazine and the internet for current information.

12. *Bottom Topography*

Bottom topography is recorded by multibeam echosounders (penetration into sediments to 150 m), towed magnetometers (Fig. 8-14), and towed streamers (e.g.,

Figure 8-12. ALVIN submersible
(source: http://www.geokem.com/
oceanic-atlantic.html).

Thomson Marconi Sonar) for oil and gas exploration. See OPL Staff (2002) for
further information. Sidescan sonar (Fig. 8-15) is being used in small unmanned
underwater vehicles (UUVs) to produce ultra-high resolution imagery of the seafloor
(Wilcox and Fletcher, 2004).

Figure 8-13.   Reefworld platform on the Great Barrier Reef, eastern Australia (source: http://www.
whitsundays.com.au/fantasea.html).

Figure 8-14.   Magnetometer (source: http://www.
resa.net/nasa/images/science/magnetometer.html).

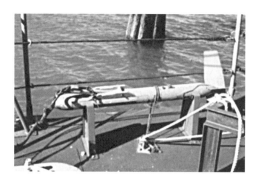

Figure 8-15.   A "fish" towed from a ship
in sidescan sonar studies of the benthos.
Sidescan sonar has been modified to
operate in unmanned underwater vehicles.

13. Cameras

TV cameras and still cameras are mounted on a metal frame that sweeps across the
ocean floor (Fig. 8-16 and Holme and Barret, 1977). The images are recorded in the
"doghouse" on the deck of the ship (Fig. 8-17). ROBIO and DOBO landers freefall to
the seabed at 30 m/minute to depths of 6000 m (Jamieson and Bagley, 2005). They are
recovered by shedding ballast weights. They are equipped with time-lapse still cameras, current meter-CTD, and a microprocessor. Motor-powered paragliders can be
used to photograph coastal areas with digital SLR and videocameras (Asami, 2005).

14. Remote Sensing

Remote sensing is used to measure the following: (1) sea surface temperature,
(2) surface wind, (3) sea surface elevation (i.e., dynamic height), (4) ocean color
(e.g., chlorophyll, colored dissolved organic matter, suspended particulate matter,
and (5) sea surface roughness. See Large Scale Sampling (p. 56 in Chapter 1) and
Landscape Ecology (p. 195 in Chapter 3) for further information.

15. Diving Instruments

WetPC is an underwater computer developed at the Australian Institute of Marine Sciences and now sold commercially. It consists of a computer mounted in a waterproof
housing on the diver's air tank. A cable connects it to a waterproof virtual display
attached to the diver's mask. A second cable connects to a waterproof pod. Five keys
on the pod can be pressed to record information, even while swimming.

Figure 8-16.   Underwater TV and still camera in a frame protected by old tires.

Figure 8-17.   Images from the TV and still cameras are recorded on tape in the "doghouse" on the deck of a ship.

SeaSlate is a pen computer with a LCD screen. Position information is obtained with an underwater GPS (DGPS). The diver can log the positions of organisms on the seafloor. Data can be transferred to a desktop PC. For further information contact the following internet sites: (1) wetpc.com.au/html/ and (2) www.aims.gov.au/wetpc/index.html/.

A diver detection system (Cerberus360) has been developed that can detect a diver up to 800 m away (Sea Technology, November 2004). Although developed as an anti-terrorist detection system, it could be used to locate lost divers (e.g., drift divers). For further information visit www.security.qinetiq.com on the Internet.

Closed-circuit rebreather technology is discussed by Lombardi (2004).

16. Oceanographic Data with Animal Tags

Data Storage Tags (DST-CTD) are used for tagging fish such as tuna. They measure conductivity (salinity), temperature and depth. Pop-up satellite archival tags (PSAT) measure ambient light (to determine latitude and longitude), temperature and depth. They are used with a variety of large game and commercial fishes, seabirds, and marine mammals. See Sea Technology (2006) and p. 44 in Chapter 1 (Mark-Recapture Techniques) for further details.

17. Data Analysis Methods

Data acquisition and data processing methods in physical oceanography are discussed by Emery and Thomson (1998). See Mann (2005).

8.3   MARINE BOTTOM SAMPLING DEVICES

Eleftheriou and McIntyre (2006) discuss methods for the study of marine benthos.

Intertidal and Shallow Subtidal

(a) Rock and Reef

Frequently used are transect lines and quadrat methods (Fig. 8-18). Some divers prefer round (sometimes called a circlet) samples. Large kelps require various techniques (for epibiota, frond, or holdfast organisms). Large benthic algae may be collected in plastic bags. Video and photography may be used (Fig. 8-19). Infrared aerial photography is used to indicate the distribution and abundance of benthic algae along coasts or in lagoons. Remote sensing by satellite or multispectral analysis by aircraft make large scale data collecting possible (see p. 56 in Chapter 1). See Stephenson and Stephenson (1977), Knox (2001), and Murray et al. (2006).

(b) Coral

Transect and quadrat methods are used on coral reefs (Loya, 1978; Unesco, 1984). The heterogeneity of a reef is determined by using chain draped over the reef in a

Figure 8-18. Quadrat frame placed in the rocky intertidal (littoral) zone.

Figure 8-19. Underwater housing with videocamera (source: http://www. underwaterphotography.com/html).

line, to give a ratio of chain length to straight length (Fig. 8-20). For obtaining all the organisms from a coral head, place a plastic bag around the coral then break the coral near the base and cinch the bag (Fig. 8-21). In the laboratory, slowly break the coral using a geopick and/or hammer and chisel. Treat the sample with acetic acid. This will dissolve away the calcareous material leaving the organisms (Hutchings, 1986). Samples can be air-lifted to the surface by attaching an inflatable plastic hood to a bucket then using compressed air from a scuba tank to fill the hood with air sufficient to raise the load to the surface. Aerial photography is very important for coral reef studies (Stoddart and Johannes, 1978) as is remote sensing (Bour et al., 1986). Manta tows give a rapid assessment of coral reef diversity but the surveyor must be able to sight identify corals from a distance (Fig. 8-22). A coral reef survey sheet helps in the recording of data (Fig. 8-22). Divers use a clipboard

Figure 8-20. Chain length or contour heterogeneity measurement on a rocky intertidal (littoral) reef.

Figure 8-21. Collecting coral infauna and cryptofauna.

and waterproof paper (polypaper). The sheets are held down with two large rubber bands (See Fig. 1-5 on p. 5). A pencil is tied to the clipboard, attached to a brass link on the diver's belt. Alternatively, small polypaper notebooks with pencils are now available. Underwater voice recorder systems are available for divers but they are expensive (Fig. 8-23). A special but simple system of scuba tanks, tubing, housing, and sep-paks (Fig. 8-24a,b) are used to trap chemicals (e.g., toxins) secreted by marine organisms that are competing chemically for space (Schulte et al., 1991).

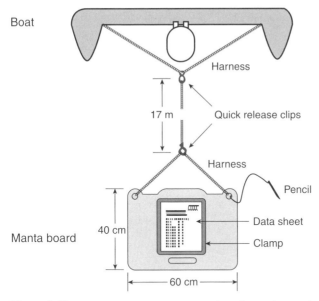

Figure 8-22.   Towing a Manta board (source: http://www.aims.gov/pages/research/reefmonitoring/ltm/mon-sop1/monsop1-06.html).

Figure 8-23.   Underwater communications system (source: http://www.oceantechnologysystems.com).

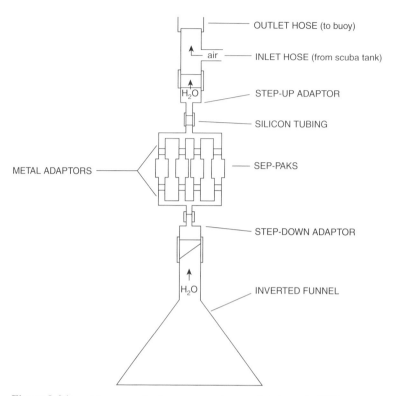

Figure 8-24a. Allomone collecting apparatus (source: Schulte et al., 1991).

Figure 8-24b. Allomone collector in the field, placed over sponges.

(c) *Sand and Mud*

Use a volumetric clam shovel or a hand corer (e.g., 8 cm in inner diameter and 80 cm long) in the intertidal (Fig. 8-25a,b). A shovel can also be used with a meter stick and quadrat frame. Divers may use 3 lb coffee cans or a small box corer for sediments. Sometimes samples are vacuumed off the bottom. Collections may be airlifted to the surface, easing the physical strain on the diver. Sieves are used to separate organisms from sediment (Figure 8-26a–e). These may be shaken manually or with a mechanical shaker. Sediment sizes can be determined by placing a sediment sample in a column of water. The sediments fall onto a sensor in the bottom of the water column. The sediment sizes are plotted on paper, the heavier sediments falling first and the lightest sediments (e.g., clays) last (grarimetric method).

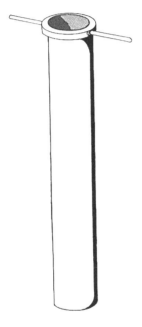

Figure 8-25a. Hand corer for sand and mud flats.

Figure 8-25b. Bill Jessee coring mud in a salt marsh at Ballona Lagoon, southern California.

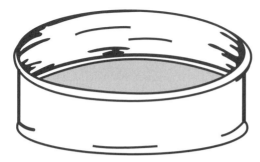

Figure 8-26a. Sieve for separating organisms (often polychaete worms and small clams and snails) from sediments.

Figure 8-26b. Shipboard sieve used in the Gulf of Alaska.

Figure 8-26c. Washing the sediment in a shipboard sieve.

Figure 8-26d. Organisms (polychaete worms, snails) retained on the shipboard sieve after washing.

Figure 8-26e.   The author sorting organisms aboard ship in the Gulf of Alaska.

See Eltringham (1971), Grussendorf (1981), Mudroch and MacKnight (1994), Little (2000), Knox (2001), and Bale and Kenny (2005).

Bottom sediments are often studied with instrumented tripods and landers. Tripods often have current meters, videocameras, water samplers, and optical sensors for plankton and suspended sediments (Fig. 8-27). Landers may be equipped with sensors to analyze fluxes of oxygen, nutrients, gases, and so forth across the seawater/sediment interface (Fig. 8-28). Deep-sea benthic sampling is discussed by Gage and Bett (2005).

Figure 8-27. A Tripod with oceanographic instrumentation (source: http://woodshole.er.usgs.gov/operations/stg/gear/lgtripod.html).

Figure 8-28. A Lander with oceanographic instrumentation (source: http://www. geomar.de/zd/deep-sea/deslandercomplete.html).

Either a plastic dome or a conical plankton net are used to collect demersal zooplankton that move from the infauna or epifauna of sediments into the water column (Fig. 8-29) (Hamner, 1981). Seagrass beds require various techniques, similar to those for kelps, including the use of pushnets (Fig. 8-30) (Ott and Losert, 1979;

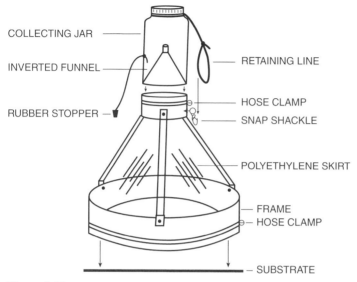

COLLECTING JAR

INVERTED FUNNEL

RUBBER STOPPER

RETAINING LINE

HOSE CLAMP

SNAP SHACKLE

POLYETHYLENE SKIRT

FRAME
HOSE CLAMP

SUBSTRATE

Figure 8-29.   Trap for demersal zooplankton (source: Hamner, 1980).

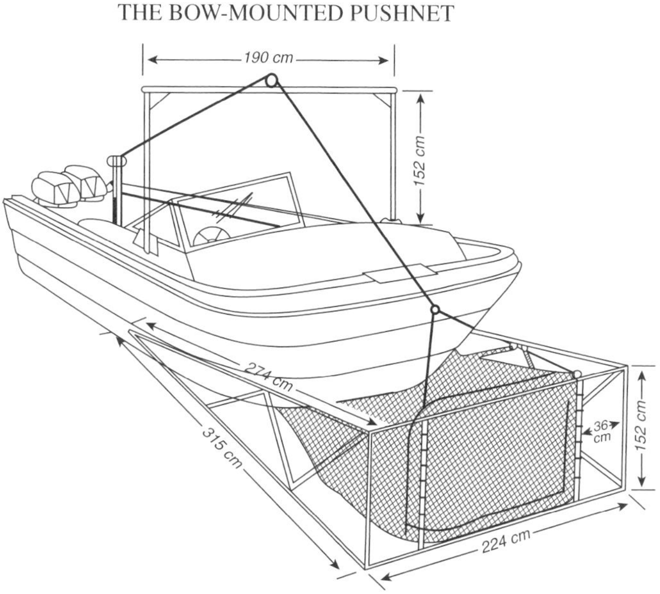

THE BOW-MOUNTED PUSHNET

190 cm

152 cm

274 cm

315 cm

36 cm

152 cm

224 cm

Figure 8-30.   The bow-mounted pushnet, used in seagrass beds (source: http://www. fisheries.vims. edu/anadromous/pushnet.html).

Phillips and McRay, 1979; Knight et al., 1980; Short and Coles, 2001). See Baker and Wolff (1987) and Blonquist (1991) for further information.

(d) Interstitial Fauna

Organisms living between sand grains (i.e., meiofauna) require special techniques (Fig. 8-31). See Hulings and Gray (1971), Zool. J. Linnean Society 96: 213–280, Heip et al. (1988), Giere (1993), and Somerfield et al. (2005).

(e) Pilings

A large hand corer (inner diameter 25 cm, 40 cm long) can be fitted with a sharp toothed edge. The corer is then placed against a mussel or barnacle patch and rotated to bite into the encrusting piling organisms, giving a quantitative sample (Fig. 8-32).

(f) Commercial Organisms

Clams are frequently taken with a hydraulic harvester, consisting of a steel frame with leading blade, cage, and high velocity water jets (Fig. 8-33a,b) (Kauwling and

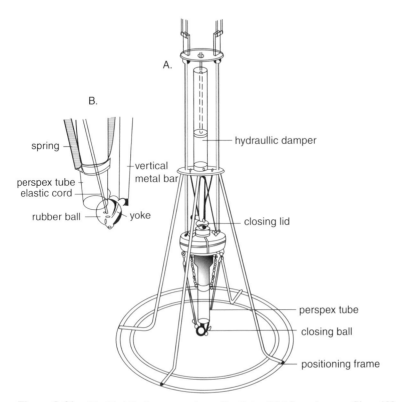

Figure 8-31.   Modified Craib corer used to collect interstitial fauna (source: Giere, 1993).

Bakus, 1979). Magnell et al. (2005) describe a data acquisition and telemetry system for shellfish dredge operations. Crustaceans and fishes are collected with commercial trawls (von Brandt, 1985).

(g) Fungi

One of the easiest ways to see marine fungi is to look for them inside the gas-filled bladders (pneumatocysts) of kelp drift (e.g., *Macrocystis*, *Nereocystis*,

Figure 8-32.   Piling corer to sample piling organisms. Note the fine teeth on the lower edge.

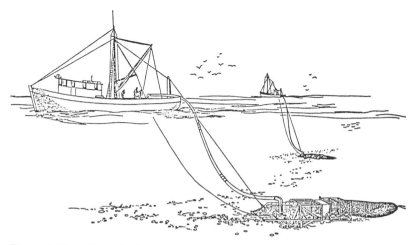

Figure 8-33a.   Hydraulic clam dredging (source: Kauwling and Bakus, 1979).

Figure 8-33b.   Leading edge (below) of a hydraulic clam dredger used in the Bering Sea.

Pelagophycus) on the beach. They often look like a spiderweb inside the bulbous bladders.

Acetate to Ergosterol Method

Collect marshgrass leaf blades, store them in bacteria-free seawater, and incubate. Add 1-^{14}C acetate and regular sodium acetate. Extract ergosterol by methanol refluxing and pentane partitioning. Run the samples through HPLC and collect the ergosterol sample peak. Obtain radioactive and background scintillation counts (CPM) and convert these to fungal biomass. See Newell (2001) for details.

Offshore

(a) Dredges

Dredges are steel-frame devices that scrape along the bottom, collecting organisms. Commonly used devices are the biological dredge (rock dredge) (Figs. 8-34 and 8-35), the beam trawl (Fig. 8-36) and the epibenthic dredge (Fig. 8-37). The otter trawl (actually a functional dredge) collects epibenthic and nektobenthic organisms (Fig. 8-38a,b). The efficiency of a bottom dredge at a variety of depths can be determined by sending a diver down to watch the dredge in operation (Green, 1979). See Holme and McIntyre (1984) and Eleftheriou and Moore (2005) for further information.

(b) Grabs

Grabs are quantitative bottom samplers. Among the most frequently used grabs are the Peterson grab, Campbell grab, Van Veen (Fig. 8-39a,b), and Smith-Mcintyre grab. Each grab has a specific biting profile and successful biting record. Among the best in this regard is the Soutar-modified Van Veen grab (see Kauwling and

Figure 8-34. Rock dredge with a chain mesh.

Figure 8-35. Dredges are deployed by winch and cable in the Gulf of Alaska. Left to right: the author, Ed Noda, and Greg Smith.

Bakus, 1979) and the improved SEABOSS (Fig. 8-40) (Sea Technology, 2001). The SEABOSS has a frame that holds a van Veen grab, depth sensor, lasers for scale and height off the seafloor, two videocameras, a 35 mm camera, strobe and lights. It gives high quality images and seabed samples in a timely manner. Saila et al. (1976)

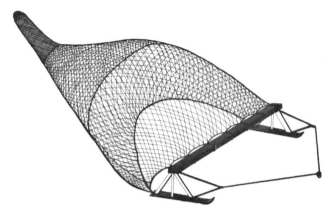

Figure 8-36. Beam trawl, used on soft bottoms.

Figure 8-37. Epibenthic dredge, used on hard bottoms.

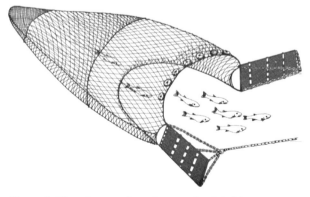

Figure 8-38a. Otter trawl, used on gravel and shell bottoms.

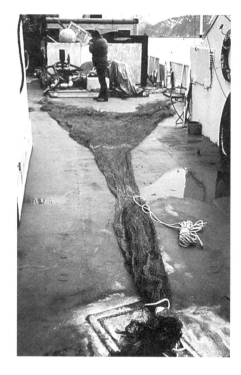

Figure 8-38b. Otter trawl on the deck of a ship in the Gulf of Alaska.

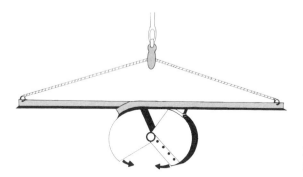

Figure 8-39a. Van Veen grab for soft bottoms.

stated that a single marine grab sample cost $25 to collect and $300 to sort – the total cost today is nearly $1000 per grab sample.

(c) Corers

Corers provide samples of undisturbed sediments and are frequently used by the geologist in offshore work. They consist of a metal tube with a plastic liner. Corers are either of the gravity (free fall) or impact type (weighted, pneumatic, or explosive charge). Among the most widely used devices are the Phleger, (Fig. 8-41), Ewing

Figure 8-39b. Dee Chamberlain holding the Van Veen grab with sediment.

piston (Fig. 8-42), and dart corers (Fig. 8-43). The box corer (Fig. 8-44a,b) is now commonly used by both biologist and geologist. It provides an undisturbed sample and is preferred over grab samples. The only disadvantage is that a ship is required to deploy the apparatus, which is relatively bulky and heavy. Macrofaunal techniques are discussed by Eleftheriou and Moore (2005) and deep-sea benthic sampling by Gage and Bett (2005).

(d) Other Aspects of Sampling

Further methods used to sample benthic marine organisms include collections by hand, scuba and speargun (Fig. 8-45), traps (Fig. 8-46), slurp gun (for small fish – Fig. 8-47), cameras (still, motion and videotape; Fig. 8-16), and by submersible (Fig. 8-12). Microfauna are often collected by subsampling box cores with narrow cores (2–3 cm inner diameter). Paleontologists use rose bengal or alizarin red stain to detect living foraminiferans from these subsamples. Flotation techniques (bubbling air in water or altering the density of seawater) may be used for collecting animals passing through sieves.

Figure 8-40. SEABOSS grab sampler system (source: http://www.woodshole .er.usgs.gov/operations/seaboss.html).

Figure 8-41. Phleger corer for sediments. Often used by biologists.

329

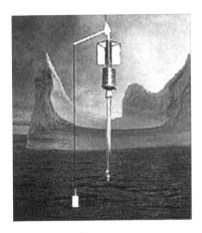

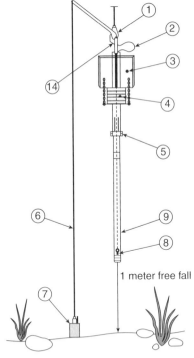

1 meter free fall

Figure 8-42. Piston corer for sediments. Used by geologists (source: http://www.kc-denmark .dk/public_html/p4.html).

The standing crop of animals collected is expressed in various ways as follow: numbers, settled volume, live weight, wet weight, fresh weight (with the body surface towel-dried), dry weight (dried to constant weight in an oven), body weight less skeletal weight or body weight less tube weight, body weight less alimentary tract contents (e.g., sand), formalin weight, alcohol weight, ash-free dry weight or organic weight (using a muffle furnace – Fig. 8-48), carbon weight, nitrogen weight, calories per unit weight, ATP, and RNA or DNA. Fresh or wet weight is often used

Figure 8-43. Dart corer for sediments. Used by geologists.

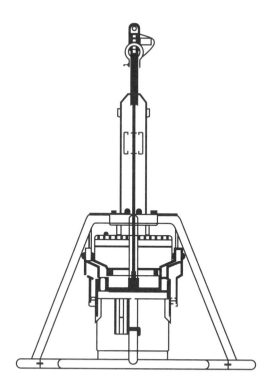

Figure 8-44a. Box corer for undisturbed sediments. Used by both biologists and geologists.

for describing the catch of commercial fishes or kelps because it saves time. Dry weight is much better than wet weight, especially for small organisms (e.g., meio-fauna), because the rate of evaporation of body surface water often is very rapid. Calorimetry is best for energetics studies and ATP values for plankton. The loss

Figure 8-44b. Box corer on the deck of a ship (source: http://www.geopubs. wr.usgs.gov/open-file/of01-107.html).

Figure 8-45. Spear gun used for collecting fish.

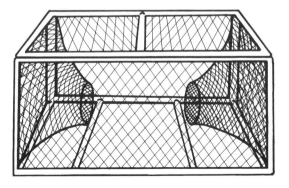

Figure 8-46. Fish trap or crab trap.

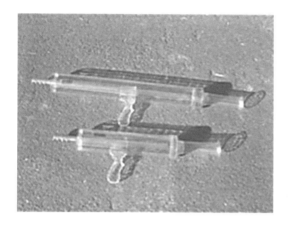

Figure 8-47.   Slurp gun for collecting small fish (source: http://www.deep-six.com/PAGE 45.HTM).

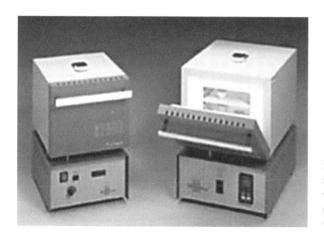

Figure 8-48.   Muffle furnace for ash-free dry weight (organic content) measurements (source: http://www.shamphavimpex.com).

of weight by storage in formalin can be as high as about 5% whereas in alcohol it can reach approximately 35%. Conversion factors (fresh weight to formalin weight) have been published (Vinogradov, 1953) but it is wise to determine your own if time permits.

8.4   MARINE WATER SAMPLING DEVICES

Bottles

Nansen and Knudsen bottles were used for many years by oceanographers (Fig. 8-49a,b). They are metallic and carry reversing thermometers. Plastic bottles (lacking contaminating metallic ions) are now in vogue. Among the most popular

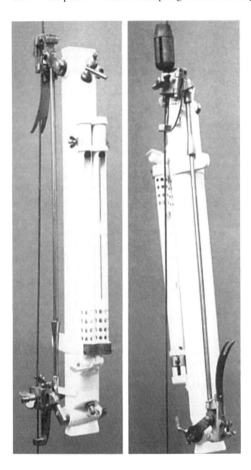

Figure 8-49a. Nansen bottle with two reversing thermometers (source: http://www.oceanworld.tamu.edu/.../satellite2.htm).

are the Van Dorn, Frautschey, and Niskin bottles (Fig. 8-50), the lattermost available in large volumes (50 liters) for seawater hydrocarbon samples. These devices are used for measuring physical–chemical and biological properties of seawater and plankton. There is also a "GO-FLO" bottle that prevents the possibility of metal contamination in samples (Fig. 8-51). This is especially important for studies of phytoplankton productivity and growth. Perhaps the major problem in collecting plankton is its patchiness and temporal variability and the relationships between these factors and physical–chemical characteristics of the waters in which they occur.

A water sampling device called POPEIE and a tube are launched from an aircraft and descended by parachute into the ocean (Chase and Sanders II, 2005). The sampling chamber is opened by a remotely operated switch and this exposes a tetrafluoroethylene (TFE) mesh to the oil/water interface after an oil spill. The sampling chamber closes after one hour and the sample is returned to the laboratory for analysis, used as forensic evidence.

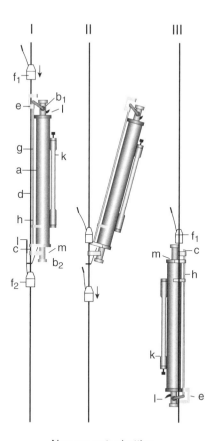

Nansen water bottles
before (I), during (II),and
after (III) reversing.
(From Dietrich et al. 1980)

Figure 8-49b.   Nansen bottle in operation
for water samples at depth (source: http://
www.140.11268.243/Chap4/Chap.html).

Plankton Nets

(a) Simple

Standard silk mesh sizes are $\geq 64\,\mu$ and follow a conventional numbering system
(Sverdrup et al., 1942). Teflon nets (Nitex) are now popular, with mesh sizes of $\geq 20\,\mu$
(Fig. 8-52a–c).

(b) Complex

Some earlier plankton nets had opening-closing devices (Clark-Bumpus) activated
by a messenger or weight dropped from the ship. Now they are operated by electronic
signals (CalCoFi net; Fig. 8-53). A bongo net consists of two large (70 cm diameter)

Figure 8-50.   Niskin bottle (non-metallic) for water samples (source: http://www.porifera. org/a/niskin/niskin.html).

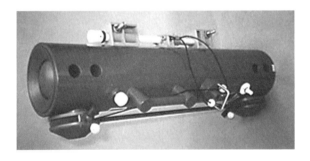

Figure 8-51.   GO-FLO non-metallic water sampling bottle (source: http://www.mumm .ac.be/EN/Monitoring/Belgica/ niskin.php).

plankton nets arranged side by side (Fig. 8-54). It takes replicate samples to determine patchiness.

MOCNESS is a modern version of multiple open/closing nets complete with CTDs and acoustical and optical sensors (Fig. 8-55). These nets are being replaced by video plankton recorder-type systems that elaborate realtime taxonomy through image processing (e.g., BIOMAPPER II – see Fig. 8-56 and the following Internet sites:

http://globec.whoi.edu/images/gb/bmapcombo.html

http://www.whoi.edu/institutes/oli/facilities/index.html

A holographic camera for subsea imaging of plankton is discussed by Watson (2004). Bioluminescence measurements are made with a bathyphotometer (HIDEX-BP GEN I; Widder et al., 2005).

Figure 8-52a. Deploying a plankton net from a ship (source: http://www.hys.geol.uib .no/.../plankton%20net.jpg).

Figure 8-52b. Towing a plankton net off the coast of southern California.

Figure 8-52c. A plankton sample from a 10 minute tow in the Gulf of Alaska in June. The sample is the color of pea soup because of a plankton bloom.

Figure 8-53. CalCoFi plankton net, used in California waters for many years (source: http://www.swsfc.ucsd.edu/ paulsmith/images/SCAN209_ small.jpg).

(c) Continuous

The Hardy-Longhurst continuous plankton recorder is sometimes towed long distances by ocean liners and the plankton are preserved (Fig. 8-57 and Hardy, 1958). An undulating recorder (UOR) operates at depths of 5–70 m, collecting continuous plankton samples and recording data on temperature, pressure, salinity, and chlorophyll a.

Figure 8-54.   Bongo net. The paired nets test for patchiness in plankton (source: http://www
.ucsdnews.ucsd.edu/newsrel/science/CalCOFI.htm).

Figure 8-55. MOCHNESS,
a multiple open/closing net
(source: http://www
.swfsc.nmfs.noaa.gov/frd/
images%20-%20ours/
coastalfisheriesphtotoarchive/
moc42.jpg).

(d) Pump

Seawater is pumped through plastic piping to a plankton net or to a holding tank then
through a plankton net on the deck of a ship. Alternatively, the pump and net are sub-
mersed to the desired depth for sampling. Both techniques are used at depths down
to 100 m (Figs. 8-58 and 8-59). Krebs (2000a) has a computer program to determine
mean plankton density with subsamples.

Figure 8-56. BIOMAPPER, a video plankton recording system (source: http://www.zooplankton
.lsu.edu/survey-system.htm).

(e) Pleuston and Neuston

Special nets are used to capture organisms living at or just below the air-sea interface
(Fig. 8-60; Banse, 1975). See also van Vleet and Williams (1980) and Schram et al.
(1981).

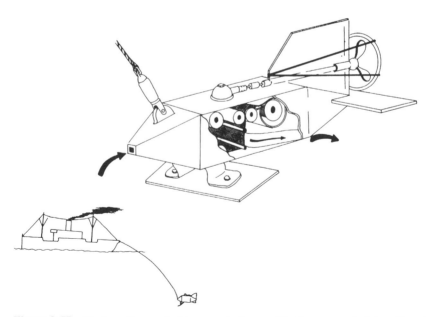

Figure 8-57. Hardy continuous plankton recorder (source: http://www.gso.uri.edu/ maritimes/
Back_Issues/99Spring/Images/jossi_fig. 1.gif).

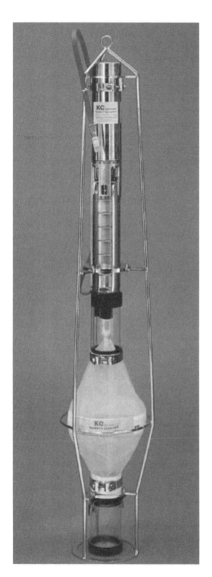

Figure 8-58. Submersible plankton pump
(source: http://www.biobull.org/cgi/content/
abstract/186/2/168).

(f) Plankton Subsamples

A rotometer or flowmeter (the irreversible type is preferable) is added to the mouth of the plankton net and calibrated to measure the volume of water passed (Fig. 8-61a,b). Plankton splitters are used to take subsamples of plankton samples since it requires so much time to identify and count plankton. Two major types of plankton splitters are commonly used, the Folsom radial splitter (Fig. 8-62) and the Motoda rectangular splitter (Fig. 8-63). Plankton splitters are useful in splitting plankton samples taken by plankton nets but problems occur involving

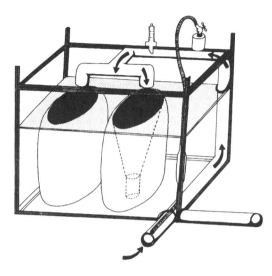

Figure 8-59. Plankton pumped to the deck of a ship. The water then flows downward through fixed plankton nets.

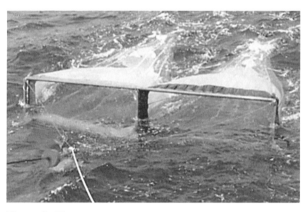

Figure 8-60. Neuston nets for ocean surface plankton (source: http://www.moc.noaa.gov/ot/visitor/otneuston.htm).

Figure 8-61a. Flow meter or rotometer. This is placed in the mouth of a plankton net then calibrated before use.

Figure 8-61b.   A large flow meter showing dials set to zero before towing. The dials indicate the number of revolutions the propeller has turned.

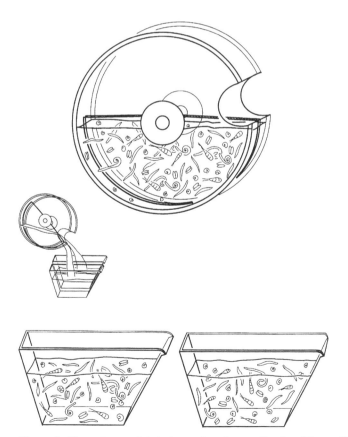

Figure 8-62.   Folsom radial plankton splitter (source: Hayek and Buzas, 1997).

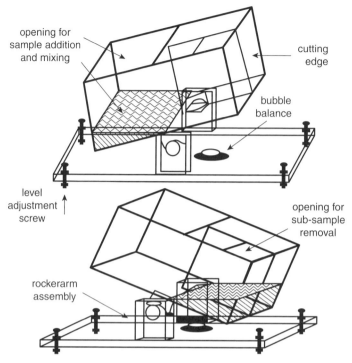

opening for
sample addition
and mixing

cutting
edge

bubble
balance

level
adjustment
screw

opening for
sub-sample
removal

rockerarm
assembly

Figure 8-63. Motoda rectangular plankton splitter (source: http://www.eman-reso.ca/eman/
ecotools/marine/zooplankton/intro.html).

patchiness (therefore, it is best to collect larger initial samples and examine relatively small aliquots – see Venrick, 1971), clumping, proper swirling, and so forth.
(see Longhurst and Siebert, 1967). Plankton splitters often cannot be used with
plankton taken by trawls (midwater plankton) since the organisms tend to be relatively large and adhere to one another. For a summary of zooplankton methodology see Omori and Ikeda (1984) and Harris et al. (2000). The Russian Federal
Research Institute of Fisheries and Oceanography in Moscow recently developed
a laser plankton meter (called TRAP-7A) that measures the size of mesoplankton
simultaneously with measurements made by an accompanying CTD (Levashov
et al., 2004).

Trawls

Trawls are frequently used to sample midwaters (mesopelagic or bathypelagic
zone; Fig. 8-64). Among the most useful of the current trawls is the Tucker trawl
(Fig. 8-65). The trawl should have a flowmeter and an opening–closing device for
quantitative, at-depth samples. Organisms collected by trawling frequently clump
and must be separated on the deck or deck table by hand.

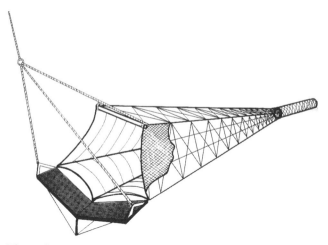

Figure 8-64.   Isaacs-Kidd midwater trawl.

Figure 8-65.   Tucker midwater trawl (source: caribou .mi.mun.ca).

Indirect Sampling Techniques

A variety of indirect sampling techniques have been developed in recent years. These include the use of fluorometry (Fig. 8-66), free-fall particle counters (Fig. 8-67), photography, video, acoustics, and aerial techniques (remote sensing).

Figure 8-66.   Shipboard fluorometer (F) (source: http://www.aol.wiff.nasa.gov/html/slf.html).

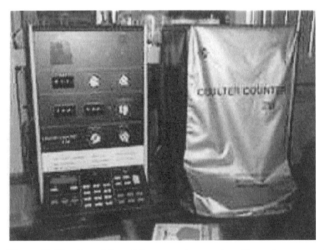

Figure 8-67.   Free-fall particle counter (source: http://www.jcsmr.anu.edu/.../images/Coulter_Counter.GIF).

8.5 SAMPLING PLANKTON, BACTERIA AND VIRUSES

A number of special techniques are required for determining standing crops of plankton and bacteria, as follows:

1) General Phytoplankton

Water samples undergo millipore filtration (Fig. 8-68). The pigments on the filter paper (chlorophyll a or chlorophyll a-b-c) are extracted in acetone and their amounts

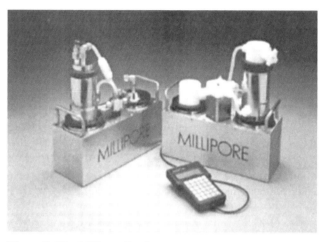

Figure 8-68. Millipore filtration system.

determined with a spectrophotometer (Fig. 8-69). The absorption values are fed into an equation to give the chlorophyll concentration (Holm-Hansen and Riemann, 1978). HPLC analysis of phytoplankton allows investigators to identify organisms to major taxa. A newer technique widely used is to continuously measure phytoplankton biomass with a fluorometer (Fig. 8-70) while underway on research vessels or on towed vehicles, AUVs, and so forth. A Coulter counter (Fig. 8-67) is used to count individual cells (see p. 346).

The Environmental Sample Processor (ESP) detects whole cells, nucleic acids, toxins (e.g., Domoic acid)) and has real time molecular probes for phytoplankton (e.g., the dinoflagellate *Gonyaulaux*) and invertebrate larvae (e.g., cypris larvae of barnacles). It was developed by Chris Scholin (pers. comm.) and associates at the

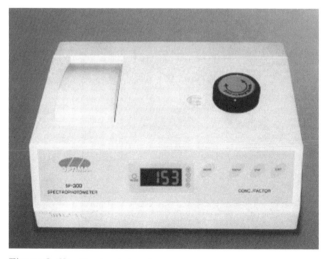

Figure 8-69. Spectrophotometer.

Monterey Bay Research Institute, Monterey, California, and has been successfully tested in Monterey Bay and the Gulf of Maine.

Phytoplankton primary productivity is measured using dark and light bottles (Fig. 8-71) that are incubated in ambient surface seawater temperature aboard a ship. Bottles can also be lowered by cable to desired depths for water column measurements. Earlier studies measured changes in oxygen concentration over time (oxygen technique). More accurate measurements are made with $NaH^{14}CO_3$ (^{14}C technique). The uptake of $^{14}CO_2$ is measured with a beta particle counter and the results entered into an equation to produce primary production rates. Primary productivity can also be measured using a Fast Repetition Rate Fluorometer (Fig. 8-70) (e.g., Chelsea Instruments FASTtracka). See Oquist et al. (1982).

For information on photosynthesis in aquatic ecosystems see Kirk (1994).

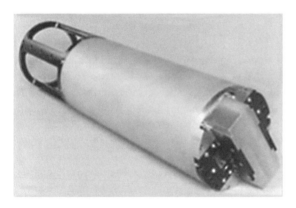

Figure 8-70. Fast repetition rate flourometer (source: http://www.marine.rutgers.edu/.../staff/zkolber.html).

Figure 8-71. Light and dark bottles for plankton primary production measurements.

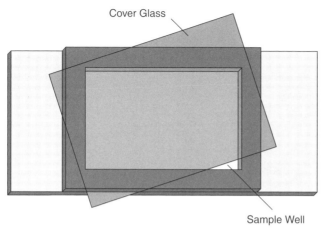

Cover Glass

Sample Well

Figure 8-72. Sedgwick-Rafter cell for counting plankton (source: http://www.vic.waterwatch.org .au/.../manual/sect3e.htm).

2) Small Phytoplankton

Nannoplankton (10 μm) and μ-flagellates are centrifuged and counted alive in a counting chamber (Sedgwick-Rafter cell [Fig. 8-72]), haemocytometer, and they may be counted with an inverted microscope (Untermöhl technique) (Fig. 8-73). An automatic cell counter (Coulter counter – Fig. 8-67) may also be used for single-cell phytoplankton. The Coulter counter can also distinguish size categories of plankton. However, some phytoplankton will stick together, resulting in spurious results.

Flow cytometry is used for picoplankton (<2 μm in diameter). This consists of an analysis of light scatter and fluorescence of individual cells through an intensely focused light source. See Campbell (2001) for details.

3) Gelatinous Zooplankton

Gelatinous zooplankton have been studied in open ocean waters using special dry suits. See Hamner et al. (1975).

4) Bacteria

A sterilized water bottle or plastic bag is used to obtain a water sample at desired depth. Alternatively, water is pumped through a filter, the filter is frozen then returned to the laboratory for analysis. Bacteria may be counted directly under a microscope (ordinary or fluorescence [Fig. 8-74]) or they are plated on nutrient media, incubated, and counted with a colony counter (Fig. 8-75).

The current preferred technique is to collect water samples using Niskin bottles, triple acid-rinsed (5% HCl) bottles, or by bucket and transfer the water

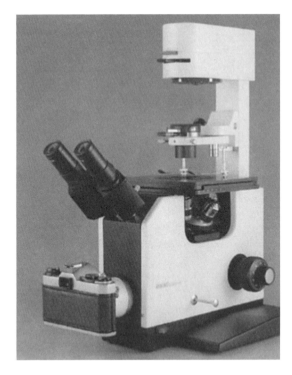

Figure 8-73. Inverted microscope for counting plankton (source: http://www .microscopeworld.com/high/hplu .htm).

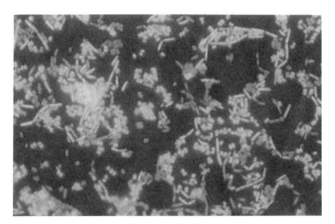

Figure 8-74. Fluorescence microscopy showing live (red) and dead bacteria (source: http://www .nobelprize.org/.../fluoresencegallery/5html).

into acid-rinsed and sample-rinsed polypropylene tubes. Maintain the samples in a freezer or fix them with 1% formalin. The seawater is filtered with an ultra-fine pore size filter (Anodisc, 0.02 μm), stained with SYBR Green 1, and the bacteria counted by epifluorescence microscopy. See Noble and Fuhrman (1998) and Noble (2001) for details. Also see Bernhard (1974), Tabor et al. (1981), and Colwell (1979).

Figure 8-75. Colony counter for bacteria (source: http://www .labequip.com/itemcatalog/ stkno/15023/FISHER-SCIENTIFIC.html).

5) Viruses

Viral particles outnumber microbial cells in seawater ten-fold (Colwell, 2004). Archaea comprise 20–30 percent of microbes in the sea. Seawater samples (2–10 ml) are filtered (25 mm Anodisc, 0.02 µm pore size,) then preserved with 1% 0.02 µm filtered formalin (Noble and Fuhrman, 1998; Noble, 2001). Viruses are examined under epifluoresence microscopy and counted. The sequencing of ribosomal RNA (or DNA) can be used to identify virus particles (virions) without having to culture them, although this is not an easy process.

8.6   SAMPLING FISHES

Fish are killed and collected at fish poison stations. Rotenone may be added to a tidepool, coastal mangrove, or placed in plastic bags and broken at desired depths (Fig. 8-76). Fish are collected with dip nets (Fig. 8-77) and goody bags (mesh dive bags). This operation involves using 10 or more people to be most effective. Concentrated rotenone products (Noxfish, Chemfish) can be used in a similar manner. Dynamite (legal only by permit in the United States) is used to collect fish in deeper water (38–213 m) since divers can spend only a very short period of time at the deeper depths without extensive decompression. Fish can be narcotized with quinaldene (some collectors suffer neurological effects). However, it is expensive and is used successfully only in relatively calm water. Fishes are also collected by hook and line (Fig. 8-78), slurp gun (Fig. 8-47), spear, dip net, beach seine (Fig. 8-79), and

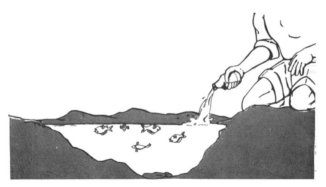

Figure 8-76.   Collecting fishes in a tidepool by poisoning with rotenone.

traps (Fig. 8-46). Commercial operations include collecting by trawling, purse sein-ing (Fig. 8-80), gill netting (Fig. 8-81a,b), lining (Fig. 8-82), trapping, and spearing (see Everhart et al., 1980). Fish movements are being monitored with the use of Mini GPS tags (Gudbjornsson et al., 2004).

Figure 8-77.   Dip net for collecting fish.

8.7   SAMPLING REPTILES, BIRDS AND MAMMALS

These organisms are sampled by hand, gun, nets (Fig. 8-83), traps, harpoon (Fig. 8-84) or they may be narcotized (Fig. 8-76).

Freshwater Biology

For sampling methods in freshwater biology see Welch (1948), Hauer and Lamberti (1998), and Wetzel and Likens (2000). Sidescan sonar used in lakes and reservoirs is discussed by Song (2005).

Figure 8-78. Fishing by hook and line in an estuary (source: leahy.senate.gov/../mercury/fishin1.jpg).

Figure 8-79. Beach seine (source: htto://www.absc.usgs .gov).

For further information on sampling devices and techniques see the following:

(1) Aquatic Sciences and Fisheries Abstracts on the Internet.

(2) Sheppard, C. 2000. Seas at the Millennium: An Environmental Evaluation (see references).

(3) Steele, J.H. et al. 2001. Encyclopedia of Ocean Sciences (see references).

(4) Large Marine Ecosystem Series. 13+ volumes. Elsevier, St. Louis

Also see Longhurst and Pauly (1987), Firestone (1976), Mann and Lazier (2005), and Stow (2006).

Figure 8-80.   A salmon purse seiner in Alaska hauling in the net after pursing it. The seine skiff (behind the fishing boat) is pulling the seiner away from the net (source: http://www.fishingnj.org/techps.htm).

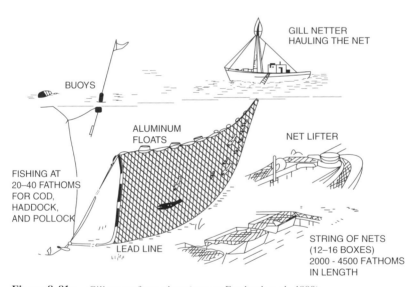

Figure 8-81a.   Gill net set from a boat (source: Everhardt et al., 1980).

8.8   NATURAL HISTORY OBSERVATIONS

Binoculars:

7×35 for bird watching – get central focus rocker for fast focusing

Nikon has a 10×50 binocular, a bit heavy but excellent for birding.

lightweight – get a 8×23 quality Bushnell (etc.)

Figure 8-81b.   Gill net arranged on a sandy beach.

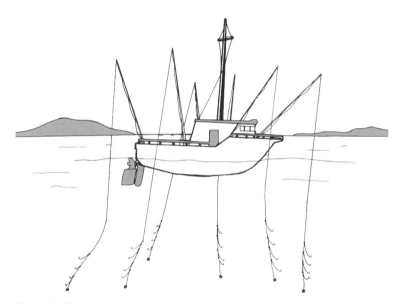

Figure 8-82.   Lining (hook and line) for salmon and albacore.

   shorebirds, waterfowl, brown bears – get a 20X to 60X spotting scope with tripod
      (Nikon has a waterproof model)

   Do not use self-focusing binoculars for natural history observations

Cameras:

Get a quality brand 35 mm camera (e.g., Nikon, Minolta, Leica, etc.). Digital SLR's
(Single Lens Reflex) cameras are preferred but expensive. See Popular Photography
magazine for information and reviews.

   For scenery – wide angle lens (e.g., 20 mm)

   For general purpose – 50 mm lens

Figure 8-83.   Cannon net for trapping ducks, geese, and seabirds.

Figure 8-84.   Primitive whale harpoon (source: heritage .elizaga.net).

For close-up – macro lens (focuses down to 2–4")

For distance shots – telephoto lens (300–600 mm) and tripod

Combination – Zoom lens – 35 to 200 mm (etc.)

Videocameras are now out with high resolution (>5 megapixels) and internal hard disks. Choose a name brand. If you plan to take underwater videos determine which cameras will fit into the housing you want then purchase the camera and housing. See Bartholomew (1986) for comments on natural history and Smith and Rumohr (2005) for imaging techniques.

For the best camera buys see the latest Popular Photography magazine and look in the back. I have had good service and reasonable prices from B & H in New York (800-254-0996) and Camera World of Oregon (800-222-1557). Check the Internet also.

Appendices

Note: The majority of computer programs mentioned here and placed on the accompanying CD were developed on a PC computer using Windows XP. The use of PC programs on a Mac computer can be accomplished by purchasing a software program called Virtual PC at a cost of less than $100 with an academic discount. Check the Internet under Virtual PC. Alternatively, Apple Computer is now offering software ($200) that will allow PC programs to operate on an Intel-based Mac computer. This software will be included in the next version of the Mac operating system OS X.

CHAPTER 1

Pointcount 99 is a computer program developed by Phil Dustan and associates that assigns random points to still digital photos or videotape frames for counting organisms. Contact Dr. Dustan for further information: Department of Biology, College of Charleston, Charleston, S.C. 29429. Telephone: 843-953-8086. Fax: 843-953-5453. E-mail: dustanp@cofc.edu

CHAPTER 2

1. A brief summary or introduction to statistics is included on the CD-ROM associated with this book. The file is a PC only file: *Stats4BN*. This is an attempt to make statistics interesting to beginners. Stats4BN is located in a folder called STATISTICS REVIEW.

2. Ezstat and other programs for PCs. Ezstat, developed by Howard Towner, Department of Biology, Loyola Marymount University, is the best statistical program for beginning biologists because of its ease of use and broad coverage. Other associated programs include Ecostat (community analyses) and Ecosim (simulation). These and other ecological computer programs are available from the following distributors:

Exeter Software 47 Route 25A, Suite 2
East Setauket, NY 11733-2870
Contact: Telephone 1-800-842-5892 or 1-631-689-7838
    Fax 1-631-689-0103
    Internet: www.exetersoftware.com
    E-mail: sales@Exetersoftware.com

Quantitative Analysis of Marine Biology Communities: Field Biology and Environment
by Gerald J. Bakus
Copyright © 2007 John Wiley & Sons, Inc.

Trinity Software. 607 Tenny Mountain Hwy, STE 215
Plymouth, NH 03264-3156.
Contact: Telephone 1-800-352-1282
    Fax 1-603-536-9951
    Internet: www.trinitysoftware.com

3. PASS 2002+ is a software program that deals with power analysis and sample size. It calculates sample size for t-tests, regression, correlation, ANOVA, and many other analyses. Contact NCSS Statistical Software, 329 N. 1000 East, Kaysville, UT 84037. Telephone: 800-898-6109 or 801-546-0445. Fax 801-546-3907. E-mail: sales@ncss.com. Internet: www.ncss.com

4. Statistical Tables

Chi square table

χ^2 **table**

Degrees of freedom	Level of significance		Degrees of freedom	Level of significance	
	0.05	0.01		0.05	0.01
1	3.84	6.63	20	31.41	37.57
2	5.99	9.21	21	32.67	38.93
3	7.81	11.34	22	33.92	40.29
4	9.49	13.28	23	35.17	41.64
5	11.07	15.09	24	36.42	42.98
6	12.59	16.81	25	37.65	44.31
7	14.07	18.48	26	38.89	45.64
8	15.51	20.09	27	40.11	46.96
9	16.92	21.67	28	41.34	48.28
10	18.31	23.21	29	42.56	49.59
11	19.68	24.72	30	43.77	50.89
12	21.03	26.22	40	55.76	63.69
13	22.36	27.69	50	67.50	76.15
14	23.68	29.14	60	79.08	88.38
15	25.00	30.58	70	90.53	100.43
16	26.30	32.00	80	101.88	112.33
17	27.59	33.41	90	113.15	124.12
18	28.87	34.81	100	124.34	135.81
19	30.14	36.19			

Coefficient of Correlation (r) table

Degrees of freedom	Level of significance		Degrees of freedom	Level of significance	
	0.05	0.01		0.05	0.01
1	0.997	0.9999	32	0.339	0.436
2	0.950	0.990	34	0.329	0.424
3	0.878	0.959	35	0.325	0.418
4	0.811	0.917	36	0.320	0.413
5	0.754	0.874	38	0.312	0.403
6	0.707	0.834	40	0.304	0.393
7	0.666	0.798	42	0.297	0.384
8	0.632	0.765	44	0.291	0.376
9	0.602	0.735	45	0.288	0.372
10	0.576	0.708	46	0.284	0.368
11	0.553	0.684	48	0.279	0.361
12	0.532	0.661	50	0.273	0.354
13	0.514	0.641	55	0.261	0.338
14	0.497	0.623	60	0.250	0.325
15	0.482	0.606	65	0.241	0.313
16	0.468	0.590	70	0.232	0.302
17	0.456	0.575	75	0.224	0.292
18	0.444	0.561	80	0.217	0.283
19	0.433	0.549	85	0.211	0.275
20	0.423	0.537	90	0.205	0.267
21	0.413	0.526	95	0.200	0.260
22	0.404	0.515	100	0.195	0.254
23	0.396	0.505	125	0.174	0.228
24	0.388	0.496	150	0.159	0.208
25	0.381	0.487	175	0.148	0.193
26	0.374	0.479	200	0.138	0.181
27	0.367	0.471	300	0.113	0.148
28	0.361	0.463	400	0.098	0.128
29	0.355	0.456	500	0.088	0.115
30	0.349	0.449	1,000	0.062	0.081

F table

F table

Level of significance = 0.05

df1 = df for the greater variance
df2 = df for the lesser variance

df2 \ df1	1	2	3	4	5	6	7	8	9
1	161.45	199.50	215.71	224.58	230.16	233.99	236.77	238.88	240.54
2	18.513	19.000	19.164	19.247	19.296	19.330	19.353	19.371	19.385
3	10.128	9.5521	9.2766	9.1172	9.0135	8.9406	8.8867	8.8452	8.8323
4	7.7086	6.9443	6.5914	6.3882	6.2561	6.1631	6.0942	6.0410	5.9938
5	6.6079	5.7861	5.4095	5.1922	5.0503	4.9503	4.8759	4.8183	4.7725
6	5.9874	5.1433	4.7571	4.5337	4.3874	4.2839	4.2067	4.1468	4.0990
7	5.5914	4.7374	4.3468	4.1203	3.9715	3.8660	3.7870	3.7257	3.6767
8	5.3177	4.4590	4.0662	3.8379	3.6875	3.5806	3.5005	3.4381	3.3881
9	5.1174	4.2565	3.8625	3.6331	3.4817	3.3738	3.2927	3.2296	3.1789
10	4.9646	4.1028	3.7083	3.4780	3.3258	3.2172	3.1355	3.0717	3.0204
11	4.8443	3.9823	3.5874	3.3567	3.2039	3.0946	3.0123	2.9480	2.8962
12	4.7472	3.8853	3.4903	3.2592	3.1059	2.9961	2.9134	2.8486	2.7964
13	4.6672	3.8056	3.4105	3.1791	3.0254	2.9153	2.8321	2.7669	2.7444
14	4.6001	3.7389	3.3439	3.1122	2.9582	2.8477	2.7642	2.6987	2.6458
15	4.5431	3.6823	3.2874	3.0556	2.9013	2.7905	2.7066	2.6408	2.5876
16	4.4940	3.6337	3.2389	3.0069	2.8524	2.7413	2.6572	2.5911	2.5377
17	4.4513	3.5915	3.1968	2.9647	2.8100	2.6987	2.6143	2.5480	2.4443
18	4.4139	3.5546	3.1599	2.9277	2.7729	2.6613	2.5767	2.5102	2.4563
19	4.3807	3.5219	3.1274	2.8951	2.7401	2.6283	2.5435	2.4768	2.4227
20	4.3512	3.4928	3.0984	2.8661	2.7109	2.5990	2.5140	2.4471	2.3928
21	4.3248	3.4668	3.0725	2.8401	2.6848	2.5727	2.4876	2.4205	2.3660
22	4.3009	3.4434	3.0491	2.8167	2.6613	2.5491	2.4638	2.3965	2.3219
23	4.2793	3.4221	3.0280	2.7955	2.6400	2.5277	2.4422	2.3748	2.3201
24	4.2597	3.4028	3.0088	2.7763	2.6207	2.5082	2.4226	2.3551	2.3002
25	4.2417	3.3852	2.9912	2.7587	2.6030	2.4904	2.4047	2.3371	2.2821
26	4.2252	3.3690	2.9752	2.7426	2.5868	2.4741	2.3883	2.3205	2.2655
27	4.2100	3.3541	2.9604	2.7278	2.5719	2.4591	2.3732	2.3053	2.2501
28	4.1960	3.3404	2.9467	2.7141	2.5581	2.4453	2.3593	2.2913	2.2360
29	4.1830	3.3277	2.9340	2.7014	2.5454	2.4324	2.3463	2.2783	2.2329
30	4.1709	3.3158	2.9223	2.6896	2.5336	2.4205	2.3343	2.2662	2.2507
40	4.0847	3.2317	2.8387	2.6060	2.4495	2.3359	2.2490	2.1802	2.1240
60	4.0012	3.1504	2.7581	2.5252	2.3683	2.2541	2.1665	2.0970	2.0401
120	3.9201	3.0718	2.6802	2.4472	2.2899	2.1750	2.0868	2.0164	1.9688
∞	3.8415	2.9957	2.6049	2.3719	2.2141	2.0986	2.0096	1.9384	1.8799

10	12	15	20	24	30	40	60	120	∞
241.88	243.91	245.95	248.01	249.05	250.10	251.14	252.20	253.25	254.31
19.396	19.413	19.429	19.446	19.454	19.462	19.471	19.479	19.487	19.496
8.7855	8.7446	8.7029	8.6602	8.6385	8.6166	8.5944	8.5720	8.5594	8.5264
5.9644	5.9117	5.8578	5.8025	5.7744	5.7459	5.7170	5.6877	5.6381	5.6281
4.7351	4.6777	4.6188	4.5581	4.5272	4.4957	4.4638	4.4314	4.3085	4.3650
4.0600	3.9999	3.9381	3.8742	3.8415	3.8082	3.7743	3.7398	3.7047	3.6689
3.6365	3.5747	3.5107	3.4445	3.4105	3.3758	3.3404	3.3043	3.2674	3.2298
3.3472	3.2839	3.2184	3.1503	3.1152	3.0794	3.0428	3.0053	2.9669	2.9276
3.1373	3.0729	3.0061	2.9365	2.9005	2.8637	2.8259	2.7872	2.7475	2.7067
2.9782	2.9130	2.8450	2.7740	2.7372	2.6996	2.6609	2.6211	2.5801	2.5379
2.8536	2.7876	2.7186	2.6464	2.6090	2.5705	2.5309	2.4901	2.4480	2.4045
2.7534	2.6866	2.6169	2.5436	2.5055	2.4663	2.4259	2.3842	2.3410	2.2962
2.6710	2.6037	2.5331	2.4589	2.4202	2.3803	2.3392	2.2966	2.2524	2.2064
2.6022	2.5342	2.4630	2.3879	2.3487	2.3082	2.2664	2.2229	2.1778	2.1307
2.5437	2.4753	2.4034	2.3275	2.2878	2.2468	2.2043	2.1601	2.1141	2.0658
2.4935	2.4247	2.3522	2.2756	2.2354	2.1938	2.1507	2.1058	2.0589	2.0096
2.4499	2.3807	2.3077	2.2304	2.1898	2.1477	2.1040	2.0584	2.0107	1.9604
2.4117	2.3421	2.2686	2.1906	2.1497	2.1071	2.0629	2.0166	1.9681	1.9168
2.3779	2.3080	2.2341	2.1555	2.1141	2.0712	2.0264	1.9795	1.9302	1.8780
2.3479	2.2776	2.2033	2.1242	2.0825	2.0391	1.9938	1.9464	1.8963	1.8432
2.3210	2.2504	2.1757	2.0960	2.0540	2.0102	1.9645	1.9165	1.8657	1.8117
2.2967	2.2258	2.1508	2.0707	2.0283	1.9842	1.9380	1.8894	1.8380	1.7831
2.2747	2.2036	2.1282	2.0476	2.0050	1.9605	1.9139	1.8648	1.8128	1.7570
2.2547	2.1834	2.1077	2.0267	1.9838	1.9390	1.8920	1.8424	1.7896	1.7330
2.2365	2.1649	2.0889	2.0075	1.9643	1.9192	1.8718	1.8217	1.7684	1.7110
2.2197	2.1479	2.0716	1.9898	1.9464	1.9010	1.8533	1.8027	1.7488	1.6906
2.2043	2.1323	2.0558	1.9736	1.9299	1.8842	1.8361	1.7851	1.7306	1.6717
2.1900	2.1179	2.0411	1.9586	1.9147	1.8687	1.8203	1.7689	1.7138	1.6541
2.1768	2.1045	2.0275	1.9446	1.9005	1.8543	1.8055	1.7537	1.6981	1.6376
2.1646	2.0921	2.0148	1.9317	1.8874	1.8409	1.7918	1.7396	1.6835	1.6223
2.0772	2.0035	1.9245	1.8389	1.7929	1.7444	1.6928	1.6373	1.5766	1.5089
1.9926	1.9174	1.8364	1.7480	1.7001	1.6491	1.5943	1.5343	1.4673	1.3893
1.9105	1.8337	1.7505	1.6587	1.6084	1.5543	1.4952	1.4290	1.3519	1.2539
1.8307	1.7522	1.6664	1.5705	1.5173	1.4591	1.3940	1.3180	1.0214	1.0000

F table

Level of significance = 0.01

df2 \ df1	1	2	3	4	5	6	7	8	9
1	4052.2	4999.5	5403.4	5624.6	5763.6	5859.0	5928.4	5981.1	6022.5
2	98.503	99.000	99.166	99.249	99.299	99.333	99.356	99.374	99.388
3	34.116	30.817	29.457	28.710	28.237	27.911	27.672	27.489	27.345
4	21.198	18.000	16.694	15.977	15.522	15.207	14.976	14.799	14.659
5	16.258	13.274	12.060	11.392	10.967	10.672	10.456	10.289	10.158
6	13.745	10.925	9.7795	9.1483	8.7459	8.4661	8.2600	8.1017	7.9761
7	12.246	9.5466	8.4513	7.8466	7.4604	7.1914	6.9928	6.8400	6.7188
8	11.259	8.6491	7.5910	7.0061	6.6318	6.3707	6.1776	6.0289	5.9106
9	10.561	8.0215	6.9919	6.4221	6.0569	5.8018	5.6129	5.4671	5.3511
10	10.044	7.5594	6.5523	5.9943	5.6363	5.3858	5.2001	5.0567	4.9424
11	9.6460	7.2057	6.2167	5.6683	5.3160	5.0692	4.8861	4.7445	4.6315
12	9.3302	6.9266	5.9525	5.4120	5.0643	4.8206	4.6395	4.4994	4.3875
13	9.0738	6.7010	5.7394	5.2053	4.8616	4.6204	4.4410	4.3021	4.1911
14	8.8616	6.5149	5.5639	5.0354	4.6950	4.4558	4.2779	4.1399	4.0297
15	8.6831	6.3589	5.4170	4.8932	4.5556	4.3183	4.1415	4.0045	3.8948
16	8.5310	6.2262	5.2922	4.7726	4.4374	4.2016	4.0259	3.8896	3.7804
17	8.3997	6.1121	5.1850	4.6690	4.3359	4.1015	3.9267	3.7910	3.6822
18	8.2854	6.0129	5.0919	4.5790	4.2479	4.0146	3.8406	3.7054	3.5971
19	8.1849	5.9259	5.0103	4.5003	4.1708	3.9386	3.7653	3.6305	3.5225
20	8.0960	5.8489	4.9382	4.4307	4.1027	3.8714	3.6987	3.5644	3.4567
21	8.0166	5.7804	4.8740	4.3688	4.0421	3.8117	3.6396	3.5056	3.3981
22	7.9454	5.7190	4.8166	4.3134	3.9880	3.7583	3.5867	3.4530	3.3458
23	7.8811	5.6637	4.7649	4.2636	3.9392	3.7102	3.5390	3.4057	3.2986
24	7.8229	5.6136	4.7181	4.2184	3.8951	3.6667	3.4959	3.3629	3.2560
25	7.7698	5.5680	4.6755	4.1774	3.8550	3.6272	3.4568	3.3239	3.2172
26	7.7213	5.5263	4.6366	4.1400	3.8183	3.5911	3.4210	3.2884	3.1818
27	7.6767	5.4881	4.6009	4.1056	3.7848	3.5580	3.3882	3.2558	3.1494
28	7.6356	5.4529	4.5681	4.0740	3.7539	3.5276	3.3581	3.2259	3.1195
29	7.5977	5.4204	4.5378	4.0449	3.7254	3.4995	3.3303	3.1982	3.0920
30	7.5625	5.3903	4.5097	4.0179	3.6990	3.4735	3.3045	3.1726	3.0665
40	7.3141	5.1785	4.3126	3.8283	3.5138	3.2910	3.1238	2.9930	2.8876
60	7.0771	4.9774	4.1259	3.6490	3.3389	3.1187	2.9530	2.8233	2.7185
120	6.8509	4.7865	3.9491	3.4795	3.1735	2.9559	2.7918	2.6629	2.5586
∞	6.6349	4.6052	3.7816	3.3192	3.0173	2.8020	2.6393	2.5113	2.4073

10	12	15	20	24	30	40	60	120	∞
6055.8	6106.3	6157.3	6208.7	6234.6	6260.6	6286.8	6313.0	6339.4	6365.9
99.399	99.416	99.433	99.449	99.458	99.466	99.474	99.482	99.491	99.499
27.229	27.052	26.872	26.690	26.598	26.505	26.411	26.316	26.221	26.125
14.546	14.374	14.198	14.020	13.929	13.838	13.745	13.652	13.558	13.463
10.051	9.8883	9.7222	9.5526	9.4665	9.3793	9.2912	9.2020	9.1118	9.0204
7.8741	7.7183	7.5590	7.3958	7.3127	7.2285	7.1432	7.0567	6.9690	6.8800
6.6201	6.4691	6.3143	6.1554	6.0743	5.9920	5.9084	5.8236	5.7373	5.6495
5.8143	5.6667	5.5151	5.3591	5.2793	5.1981	5.1156	5.0316	4.9461	4.8588
5.2565	5.1114	4.9621	4.8080	4.7290	4.6486	4.5666	4.4831	4.3978	4.3105
4.8491	4.7059	4.5581	4.4054	4.3269	4.2469	4.1653	4.0819	3.9965	3.9090
4.5393	4.3974	4.2509	4.0990	4.0209	3.9411	3.8596	3.7761	3.6904	3.6024
4.2961	4.1553	4.0096	3.8584	3.7805	3.7008	3.6192	3.5355	3.4494	3.3608
4.1003	3.9603	3.8154	3.6646	3.5868	3.5070	3.4253	3.3413	3.2548	3.1654
3.9394	3.8001	3.6557	3.5052	3.4274	3.3476	3.2656	3.1813	3.0942	3.0040
3.8049	3.6662	3.5222	3.3719	3.2940	3.2141	3.1319	3.0471	2.9595	2.8684
3.6909	3.5527	3.4089	3.2587	3.1808	3.1007	3.0182	2.9330	2.8447	2.7528
3.5931	3.4552	3.3117	3.1615	3.0835	3.0032	2.9205	2.8348	2.7459	2.6530
3.5082	3.3706	3.2273	3.0771	2.9990	2.9185	2.8354	2.7493	2.6597	2.5660
3.4338	3.2965	3.1533	3.0031	2.9249	2.8442	2.7608	2.6742	2.5839	2.4893
3.3682	3.2311	3.0880	2.9377	2.8594	2.7785	2.6947	2.6077	2.5168	2.4212
3.3098	3.1730	3.0300	2.8796	2.8010	2.7200	2.6359	2.5484	2.4568	2.3603
3.2576	3.1209	2.9779	2.8274	2.7488	2.6675	2.5831	2.4951	2.4029	2.3055
3.2106	3.0740	2.9311	2.7805	2.7017	2.6202	2.5355	2.4471	2.3542	2.2558
3.1681	3.0316	2.8887	2.7380	2.6591	2.5773	2.4923	2.4035	2.3100	2.2107
3.1294	2.9931	2.8502	2.6993	2.6203	2.5383	2.4530	2.3637	2.2696	2.1694
3.0941	2.9578	2.8150	2.6640	2.5848	2.5026	2.4170	2.3273	2.2325	2.1315
3.0618	2.9256	2.7827	2.6316	2.5522	2.4699	2.3840	2.2938	2.1985	2.0965
3.0320	2.8959	2.7530	2.6017	2.5223	2.4397	2.3535	2.2629	2.1670	2.0642
3.0045	2.8685	2.7256	2.5742	2.4946	2.4118	2.3253	2.2344	2.1379	2.0342
2.9791	2.8431	2.7002	2.5487	2.4689	2.3860	2.2992	2.2079	2.1108	2.0062
2.8005	2.6648	2.5216	2.3689	2.2880	2.2034	2.1142	2.0194	1.9172	1.8047
2.6318	2.4961	2.3523	2.1978	2.1154	2.0285	1.9360	1.8363	1.7263	1.6006
2.4721	2.3363	2.1915	2.0346	1.9500	1.8600	1.7628	1.6557	1.5330	1.3805
2.3209	2.1847	2.0385	1.8783	1.7908	1.6964	1.5923	1.4730	1.3246	1.0000

Fmax table

Fmax table Level of significance $= 0.05$

df	n	2	3	4	5	6	7	8	9	10	11	12
2		39.0	87.5	142	202	266	333	403	475	550	626	704
3		15.4	27.8	39.2	50.7	62.0	72.9	83.5	93.9	104	114	124
4		9.60	15.5	20.6	25.2	29.5	33.6	37.5	41.1	44.6	48.0	51.4
5		7.15	10.8	13.7	16.3	18.7	20.8	22.9	24.7	26.5	28.2	29.9
6		5.82	8.38	10.4	12.1	13.7	15.0	16.3	17.5	18.6	19.7	20.7
7		4.99	6.94	8.44	9.70	10.8	11.8	12.7	13.5	14.3	15.1	15.8
8		4.43	6.00	7.18	8.12	9.03	9.78	10.5	11.1	11.7	12.2	12.7
9		4.03	5.34	6.31	7.11	7.80	8.41	8.95	9.45	9.91	10.3	10.7
10		3.72	4.85	5.67	6.34	6.92	7.42	7.87	8.28	8.66	9.01	9.34
12		3.28	4.16	4.79	5.30	5.72	6.09	6.42	6.72	7.00	7.25	7.48
15		2.86	3.54	4.01	4.37	4.68	4.95	5.19	5.40	5.59	5.77	5.93
20		2.46	2.95	3.29	3.54	3.76	3.94	4.10	4.24	4.37	4.49	4.59
30		2.07	2.40	2.61	2.78	2.91	3.02	3.12	3.21	3.29	3.36	3.39
60		1.67	1.85	1.96	2.04	2.11	2.17	2.22	2.26	2.30	2.33	2.36

n is the number of samples being compared. **df** is the degrees of freedom of each sample (if samples do not have equal numbers of observations then use the degrees of freedom of the sample with the smaller number of observations). If n is different between samples, use the largest value of **df**.

Kolmogorov–Smirnov (K–S) two sample table

K–S 2-Sample table

Degrees of freedom	Level of significance	
	0.05	0.01
4	4	
5	4	5
6	5	6
7	5	6
8	5	6
9	6	7
10	6	7
11	6	8
12	6	8
13	7	8
14	7	8
15	7	9
16	7	9
17	8	9
18	8	10
19	8	10
20	8	10
21	8	10
22	9	11
23	9	11
24	9	11
25	9	11
26	9	11
27	9	12
28	10	12
29	10	12
30	10	12
35	11	13
40	11	14

Rayleigh's z table

Degrees of freedom	Levels of significance								
	0.50	0.20	0.10	0.05	0.02	0.01	0.005	0.002	0.001
6	0.734	1.639	2.274	2.865	3.576	4.058	4.491	4.985	5.297
7	0.727	1.634	2.278	2.885	3.627	4.143	4.617	5.181	5.556
8	0.723	1.631	2.281	2.899	3.665	4.205	4.710	5.322	5.743
9	0.719	1.628	2.283	2.910	3.694	4.252	4.780	5.430	5.885
10	0.717	1.626	2.285	2.919	3.716	4.289	4.835	5.514	5.996
11	0.715	1.625	2.287	2.926	3.735	4.319	4.879	5.582	6.085
12	0.713	1.623	2.288	2.932	3.750	4.344	4.916	5.638	6.158
13	0.711	1.622	2.289	2.937	3.763	4.365	4.947	5.685	6.219
14	0.710	1.621	2.290	2.941	3.774	4.383	4.973	5.725	6.271
15	0.709	1.620	2.291	2.945	3.784	4.398	4.996	5.759	6.316
16	0.708	1.620	2.292	2.948	3.792	4.412	5.015	5.789	6.354
17	0.707	1.619	2.292	2.951	3.799	4.423	5.033	5.815	6.388
18	0.706	1.619	2.293	2.954	3.806	4.434	5.048	5.838	6.418
19	0.705	1.618	2.293	2.956	3.811	4.443	5.061	5.858	6.445
20	0.705	1.618	2.294	2.958	3.816	4.451	5.074	5.877	6.469
21	0.704	1.617	2.294	2.960	3.821	4.459	5.085	5.893	6.491
22	0.704	1.617	2.295	2.961	3.825	4.466	5.095	5.908	6.510
23	0.703	1.616	2.295	2.963	3.829	4.472	5.104	5.922	6.528
24	0.703	1.616	2.295	2.964	3.833	4.478	5.112	5.935	6.544
25	0.702	1.616	2.296	2.966	3.836	4.483	5.120	5.946	6.559
26	0.702	1.616	2.296	2.967	3.839	4.488	5.127	5.957	6.573
27	0.702	1.615	2.296	2.968	3.842	4.492	5.133	5.966	6.586
28	0.701	1.615	2.296	2.969	3.844	4.496	5.139	5.975	6.598
29	0.701	1.615	2.297	2.970	3.847	4.500	5.145	5.984	6.609
30	0.701	1.615	2.297	2.971	3.849	4.504	5.150	5.992	6.619
32	0.700	1.614	2.297	2.972	3.853	4.510	5.159	6.006	6.637
34	0.700	1.614	2.297	2.974	3.856	4.516	5.168	6.018	6.654
36	0.700	1.614	2.298	2.975	3.859	4.521	5.175	6.030	6.668
38	0.699	1.614	2.298	2.976	3.862	4.525	5.182	6.039	6.681
40	0.699	1.613	2.298	2.977	3.865	4.529	5.188	6.048	6.692
42	0.699	1.613	2.298	2.978	3.867	4.533	5.193	6.056	6.703
44	0.698	1.613	2.299	2.979	3.869	4.536	5.198	6.064	6.712
46	0.698	1.613	2.299	2.979	3.871	4.539	5.202	6.070	6.721
48	0.698	1.613	2.299	2.980	3.873	4.542	5.206	6.076	6.729
50	0.698	1.613	2.299	2.981	3.874	4.545	5.210	6.082	6.736

Degrees of freedom	Levels of significance								
	0.50	0.20	0.10	0.05	0.02	0.01	0.005	0.002	0.001
55	0.697	1.612	2.299	2.982	3.878	4.550	5.218	6.094	6.752
60	0.697	1.612	2.300	2.983	3.881	4.555	5.225	6.104	6.765
65	0.697	1.612	2.300	2.984	3.883	4.559	5.231	6.113	6.776
70	0.696	1.612	2.300	2.985	3.885	4.562	5.235	6.120	6.786
75	0.696	1.612	2.300	2.986	3.887	4.565	5.240	6.127	6.794
80	0.696	1.611	2.300	2.986	3.889	4.567	5.243	6.132	6.801
90	0.696	1.611	2.301	2.987	3.891	4.572	5.249	6.141	6.813
100	0.695	1.611	2.301	2.988	3.893	4.575	5.254	6.149	6.822
120	0.695	1.611	2.301	2.990	3.896	4.580	5.262	6.160	6.837
140	0.695	1.611	2.301	2.990	3.899	4.584	5.267	6.168	6.847
160	0.695	1.610	2.301	2.991	3.900	4.586	5.271	6.174	6.855
180	0.694	1.610	2.302	2.992	3.902	4.588	5.274	6.178	6.861
200	0.694	1.610	2.302	2.992	3.903	4.590	5.276	6.182	6.865
300	0.694	1.610	2.302	2.993	3.906	4.595	5.284	6.193	6.879
500	0.694	1.610	2.302	2.994	3.908	4.599	5.290	6.201	6.891
∞	0.6931	1.6094	2.3026	2.9957	3.9120	4.6052	5.2983	6.2146	6.9078

Spearman rank correlation (rho) table

Degrees of freedom	Level of significance for two-tailed test			
	0.10	0.05	0.02	0.01
5	0.900	—	—	
6	0.829	0.886	0.943	—
7	0.714	0.786	0.893	—
8	0.643	0.738	0.833	0.881
9	0.600	0.683	0.783	0.833
10	0.564	0.648	0.745	0.794
11	0.523	0.623	0.736	0.818
12	0.497	0.591	0.703	0.780
13	0.475	0.566	0.673	0.745
14	0.457	0.545	0.646	0.716
15	0.441	0.525	0.623	0.689
16	0.425	0.507	0.601	0.666
17	0.412	0.490	0.582	0.645
18	0.399	0.476	0.564	0.625
19	0.388	0.462	0.549	0.608
20	0.377	0.450	0.534	0.591
21	0.368	0.438	0.521	0.576
22	0.359	0.428	0.508	0.562
23	0.351	0.418	0.496	0.549
24	0.343	0.409	0.485	0.537
25	0.336	0.400	0.475	0.526
26	0.329	0.392	0.465	0.515
27	0.323	0.385	0.456	0.505
28	0.317	0.377	0.448	0.496
29	0.311	0.370	0.440	0.487
30	0.305	0.364	0.432	0.478

t table

Degrees of freedom	Level of significance for two-tailed test			
	0.10	0.05	0.02	0.01
1	6.314	12.706	31.821	63.657
2	2.920	4.303	6.965	9.925
3	2.353	3.182	4.541	5.841
4	2.132	2.776	3.747	4.604
5	2.015	2.571	3.365	4.032
6	1.943	2.447	3.143	3.707
7	1.895	2.365	2.998	3.499
8	1.860	2.306	2.896	3.355
9	1.833	2.262	2.821	3.250
10	1.812	2.228	2.764	3.169
11	1.796	2.201	2.718	3.106
12	1.782	2.179	2.681	3.055
13	1.771	2.160	2.650	3.012
14	1.761	2.145	2.624	2.977
15	1.753	2.131	2.602	2.947
16	1.746	2.120	2.583	2.921
17	1.740	2.110	2.567	2.898
18	1.734	2.101	2.552	2.878
19	1.729	2.093	2.539	2.861
20	1.725	2.086	2.528	2.845
21	1.721	2.080	2.518	2.831
22	1.717	2.074	2.508	2.819
23	1.714	2.069	2.500	2.807
24	1.711	2.064	2.492	2.797
25	1.708	2.060	2.485	2.787
26	1.706	2.056	2.479	2.779
27	1.703	2.052	2.473	2.771
28	1.701	2.048	2.467	2.763
29	1.699	2.045	2.462	2.756
30	1.697	2.042	2.457	2.750
40	1.684	2.021	2.423	2.704
60	1.671	2.000	2.390	2.660
120	1.658	1.980	2.358	2.617
∞	1.645	1.960	2.326	2.576

CHAPTER 3

1. Pointcount 99. See information in Appendix above under Chapter 1.

2. A computer program of cellular automata is located on the enclosed CD-ROM. The folder is AUTOMATA and the program: *AUTOMT90*. The files to be called up from the program: *100PERC*, etc. This program was developed by Wilfredo Licuanan, Marine Science Institute, University of the Philippines, Diliman, Quezon City 1101, Philippines. A description of the program is found in the file: Licuanan.

3. Baev and Penev (1993) present a software program on biodiversity. The manual accompanying the program has a detailed discussion of biodiversity indices.

4. Idrisi is the most popular university and college GIS software program in the United States. It is a geographic information and image processing software system developed by the Graduate School at Clark University. It was introduced in 1987 and is used in more than 130 countries. Contact Clark Labs, Cartographic Technology and Geographic Analysis, Clark University, 950 Main St., Worcester, MA 01610-1477. Telephone: 508-793-7526. Fax: 508-793-8842. E-mail: idrisi@clark.edu Internet: www.idrisi.clarku.edu

5. The most recent version of Idrisi is Version 15—Andes Edition, including a tutorial. See the internet address above for further information.

6. An example of the use of expert systems in taxonomy involves two computer programs, that is, *editxs* and *exsys*, developed by Exsys. Inc. Dustin Huntington used *editxs to create* an expert system that would identify shorebirds. Use the program *exsys* to call up the file: *BIRDDEMO* – located in the folder EXPERT SYSTEMS on the CD-ROM.

7. An updated version of this program, called Corvid, can incorporate drawings, photos, etc. to make the expert system both visual, more interesting, and more accurate. Corvid is available from Exsys, Inc., 144 Louisiana Blvd. NE, Albuquerque, NM 87108. Telephone: 505-888-9494. Fax: 505-888-9509. E-mail: info@exsys.com Internet: www.exsys.com

8. A tutorial explaining the use of SIGMASCAN was written by Dr. Gregory Nishiyama. SIGMASCAN is used for counting and measuring items or organisms, either manually or automatically. The tutorial is included in the enclosed CD-ROM under the PC only file: *SigmaScanNish*. The file is located in a folder called SIGMASCAN.

9. A detailed program (PC and Mac) on Environmental Impact Assessments was developed by the present author and the programmer Phil Kim. It is located in the enclosed CD-ROM in the folder called EIA. Click on the red diamond with *eia _ pc* next to it. The program contains two mathematical techniques used to calculate utility values (i.e., to give the final arithmetic results for decision making), that is, the older SMART method and the more recent NATURAL RANGES method. A separate PC file with the NATURAL RANGES method is located in the folder called EIA: *Utility*—for EIAs.

CHAPTER 4

1. PRIMER is the major European software program for community analysis. It was developed by K. Robert Clarke and R.N. Corley. PRIMER 5 for Windows is now available. A companion booklet is Change in Marine Communities by K.R. Clarke and Richard M. Warwick (2001). PRIMER 5 discusses similarities–dissimilarities, nonmetric multidimensional scaling (MDS), data manipulation, linkage to environmental variables, analysis of similarities, species curves, and bio-diversity measures, among other topics. Contact: Dr. K. Robert Clarke, Plymouth Marine Laboratory, Prospect Place, West Hoe, Plymouth PL1 3DH, U.K. Internet: primer-e.com   E-mail: admin@primer-e.com

2. PC-ORD is the major North American software program for community analysis. It was developed by Bruce McCune and M.J. Mefford. Version 4 for Windows is now available. It discusses matrix operations, ordination, cluster analysis, indicator and outlier species analysis, diversity indices, species-area curves, and multiple-response permutation procedures, among other topics. Contact: Dr. Bruce McCune, Department of Botany and Plant Pathology, Oregon State University, Corvallis, Oregon 97331, USA. Internet: PC-ORD.com E-mail: mjm@centurytel.net Telephone: 1-800-690-4499.

3. ClustanGraphics is a software program developed by David Wishart. It conducts cluster analyses and presents graphical displays of the results. Contact: Clustan Limited, 16 Kingsburgh Road, Edinburgh EH12 6DZ, Scotland. Telephone: (44) 131-337-1448. E-mail: sales@clustan.com   Internet: www.clustan.com

CHAPTER 5

1. CANOCO is an extension of DECORANA, a Fortran program developed by ter Braak. The program is a canonical community ordination by correspondence analysis, principal components analysis, and redundancy analysis. The current version (3.1) operates on a PC or Macintosh and the data are input as ASCII files. Monte Carlo techniques are used in some of the procedures. Order from Dr. Richard E. Furnas. Telephone: 607-272-2188   Fax: 607-272-0782. See CANOCO on the Internet for further information.

The following files are located in a folder called MULTIVARIATE.

1. Multiple regression using Statistica (see files: 469mr1 and 469mr2)
2. Multiple regression using Stataq 2 (see files 369mr3 and 369mr4)
3. MDS using PC-ORD (see files: MDS1 and MDS2)
4. MDS using Statistica (see files: MDS3 and MDS4)
5. Discriminant analysis using Statistica (see file: Discrim)
6. Example of the use of MANOVA (see the files MANOVA and Figures MANOVA3 and MANOVA4).

CHAPTER 6

Example of a Time Series Forecasting Study Using ARIMA

A time series study by Parsons and Colbourne (2000) incorporated environmental data to construct predictive models for standardized catch rates in a northern shrimp (*Pandalus borealis*) fishing area off the mid-Labrador coast. An autoregressive, integrated, moving-average (ARIMA) procedure was used to devise prediction models for shrimp catch per unit effort (CPUE).

The three steps in the ARIMA procedure were: (1) Identification, (2) Estimation, and (3) Forecast. The identification step for CPUE data produced autocorrelations with no trend (i.e., stationarity) therefore differencing was unnecessary. There was a significant autocorrelation at a lag of one year (Figure 4). Similarly, winter ice input data (i.e., the major environmental data) also produced a strong autocorrelation at a lag of one year (Figure 4). The estimation step showed that cross-correlations between the CPUE and winter ice was highest at shifts of 0 and 6 years (Figure 5). The final or forecast step indicated a 6-year forecast of shrimp CPUE (Figure 6).

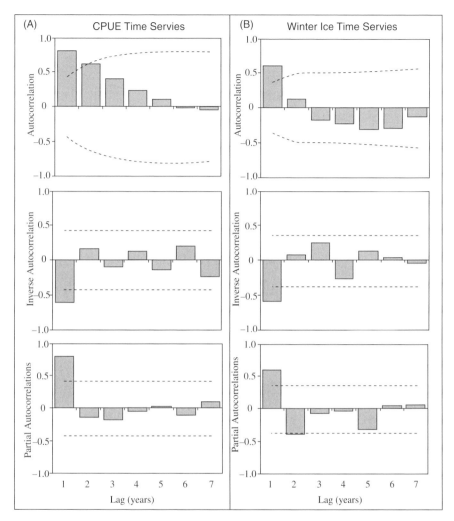

Fig. 4.   (A) Autocorrelations ± 2 standard errors (dotted line) for variable *In* CPUE from ARIMA modeling; mean of series = 6.266818, standard deviation = 0.403669, number of observations = 22. (B) Autocorrelations ± 2 standard errors (dotted line) for winter ice input variable from ARIMA modeling; mean of working series = 2.292069, standard deviation = 0.694036, number of observations = 29. Source: Parsons and Colbourne (2000)

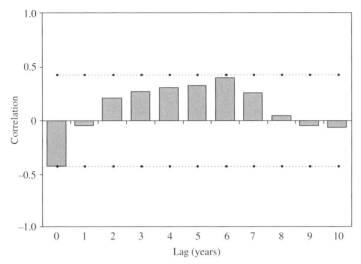

Fig. 5. Crosscorrelations (±2 standard errors − dotted line) for *In* CPUE and winter ice from ARIMA modeling. Both variable were prewhitened by the following autoregressive filter: 1−0.69304 B** (1). Variance of transformed series = 0.041778 and 0.276015; number of observations = 22. Source: Parsons and Colbourne (2000)

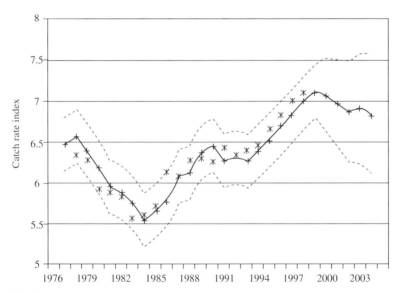

Fig. 6. Transfer function model for *In* CPUE and winter ice input series with six-year forecast. Observed = '*', predicted = '+', and dotted lines = 95% confidence intervals. Source: Parsons and Colbourne (2000)

The large statistical programs that use ARIMA include Statistica, Minitab, SAS, and BMDP. Detailed descriptions of these programs can be found on the Internet.

CHAPTER 7

A tutorial on modeling (i.e., a food chain study) using the software program Model-maker is located on the enclosed CD-ROM. The file is: *mmmodel4*. Contact Model-maker 4 on the Internet and download the demo software program. Call up the file: mmmodel4. The file is located in a folder called MODELING.

Addendum

BINARY REGRESSION

Binary regression is a type of regression analysis that is becoming popular in ecology (Trexler and Travis, 1993). Binary regression involves situations where the outcome variable takes one of two possible values (e.g., 1 or 0, such as presence/absence, small/large, and so forth). The two most popular models used in binary regression are logistic and probit. Logistic is favored for its mathematical tractability, interpretation of model coefficients, and usefulness in analysis of data from retrospective studies (Todd Alonzo, personal commun). The dependent variable for the logistic model is the logit of the probability of the outcome occurring, for example, p where logit is log $(p/(1-p))$. The goal of logistic regression is to correctly predict the category of outcomes for individual cases using the most parsimonious model. A model is created that initially includes all predictor variables that are useful in predicting the response variable. The dependent variable is usually dichotomous and is assumed to have a binomial distribution. Log-likelihood (maximum likelihood) is used to fit the model to the data and estimate the model parameters. Computer programs for binary regression are available in SPSS, BMDP, and so forth. R^2 values can be used to measure the strength of association between the independent variables and dependent variable, ranging from 0 to 1. Example: Are small and large sea urchins (i.e., size) related to food resources, wave surge, or predation? The discrete dependent variable is small/large and the continuous independent variables are quantity of available food resources, the degree of wave surge, and the incidence of predation in the environment.

REFERENCES

Agresti, A. 1996. An Introduction to Categorical Data Analysis. John Wiley & Sons, New York.
Hosmer, D. and S. Lemeshow. 1989. Applied Logistic Regression. John Wiley & Sons, New York.
Trexler, J.C. and J. Travis. 1993. Nontraditional regression analysis. Ecology 74(6):1629–1637.

COMMUNITY ANALYSES

Anderson et al. (2006a) propose that beta diversity for a group of units sampled from a given area can be measured as the average distance or dissimilarity from an individual unit to the group centroid, using an appropriate dissimilarity measure (see Chapter 4, p. 209). For multivariate analysis of species abundance data, the data should be transformed (by square roots or logs) in order to reduce the influence of numerically dominant species. They recommend using a variety of dissimilarity measures that range from emphasizing compositional changes (e.g., species) to changes in abundances of species. Distance-based tests for measuring the homogeneity of multivariate dispersion are discussed by Anderson (2006b). The tests are multivariate extensions of Levene's test. Computer programs are available from Marti Anderson (Internet: mja@stat.auckland.ac.nz).

REFERENCES

Anderson, M.J., K.E. Ellingsen, and B.H. McArdle. 2006a. Multivariate dispersion as a measure of beta diversity. Ecological Letters 9:683–693.
Anderson, M.J. 2006b. Distance-based tests for homogeneity of multivariate dispersions. Biometrics 62:245–253.

HIERARCHICAL PARTITIONING (DOMINANCE ANALYSIS)

Multiple regression is a multivariate technique used for decades by statisticians and others to determine the rank importance of independent variables in regard to one dependent variable. Care must be taken to be certain that the independent variables are not highly correlated (see Chapter 5, p. 237). An improvement on this technique is the use of hierarchical partitioning. Hierarchical partitioning is a method in which all orders of variables are used and *an average independent contribution of each variable is obtained* (Chevan and Sutherland, 1991) (see Table 1 below). The method is applicable to all regression models including ordinary least squares, logistic, probit, and log-linear regression. If the explanatory (i.e., independent) variables are truly independent from each other, then hierarchical partitioning and

Table 1. Eight possible models with one response (dependent) variable and three predictor (independent) variables. [Source: Quinn and Keough (2002:141).]

Label	Model	Level of Hierarchy
1	No predictors, $r^2 = 0$	0
2	X_1	1
3	X_2	1
4	X_3	1
5	$X_1 + X_2$	2
6	$X_1 + X_3$	2
7	$X_2 + X_3$	2
8	$X_1 + X_2 + X_3$	3

multiple regression produce identical results (Kim Berger, personal commun). An R^2 value provides an estimate of the goodness of fit. The analyses can be done with statistical program R within the hier.part package (Walsh and MacNally, 2004) available free from: http://cran.r-project.org/src/contrib/Descriptions/hier.part. html.

A SAS macro under the name of Dominance Analysis is located at: http://www.uwm.edu/~azen/damacro.html.

Although the current hier.part package is designed for a maximum of 12 variables, Yao et al. (2006) found that it worked well with only nine variables. See also MacNally (2002) and Berger (2006).

REFERENCES

Berger, K.M. 2006. Carnivore-Livestock Conflicts: Effects of Subsidized Predator Control and Economic Correlates on the Sheep Industry. Conservation Biology 20(3):751–761.

Chevan, A. and M. Sutherland. 1991. Hierarchical Partitioning. The American Statistician 45(2):90–96.

MacNally, R. 2002. Multiple regression and inference in ecology and conservation biology: further comments on identifying important predictor variables. Biodiversity and Conservation 11:1397–1401.

Quinn, G.P. and M.J. Keough. 2002. Experimental Design and Data Analysis for Biologists. Cambridge University Press, New York.

Walsh, C.J. and R. MacNally. 2004. The hier-part package version 1.0 Hierarchical Partitioning. Internet: http://cran.r-project.org/

Yao, J., D.P.C. Peters, K.M. Havstad, R.P. Gibbens, and J.E. Herrick. 2006. Multi-scale factors and long-term responses of Chihuahuan Desert grasses to drought. Landscape Ecology. In press.

RECENT PUBLICATIONS

Sampling

Murray et al. (2006) produced an informative text on techniques of monitoring rocky shores with emphasis on macro-algae and macro-invertebrates. They discuss site selection, biological units, sampling design, transects and quadrats, plot and plotless sampling, photographic sampling, biomass estimation, growth rates, and reproduction.

Conklin, A.R., Jr. 2004. Field Sampling: Principles and Practices in Environmental Analysis. CRC Press, Boca Raton.

Murray, S.N., R.F. Ambrose, and M.N. Dethier. 2006. Monitoring Rocky Shores. University of California Press, Berkeley.

Metapopulations

Kritzer, J.P. and P.F. Sale (eds.). 2006. Marine Metapopulations. Elsevier Academic Press, Burlington, MA.

Remote Sensing

Elachi, C. and J. van Zyl. 2006. Introduction to the Physics and Techniques of Remote Sensing. 2nd Edition. John Wiley & Sons, New York.

Gower, J.F.R. (ed.). 2006. Remote Sensing of the Marine Environment. American Society of Photogrammetry and Remote Sensing, Bethesda, MD. Internet: www.asprs.org

Time Series Analysis

Latini, G. and G. Passerini (eds). 2004. Handling Missing Data: Applications to Environmental Analysis. WIT Press, Billerica, MA.

Modeling

Clark, J.S. and A.E. Gelfand. 2006. Hierarchical Modelling for the Environmental Sciences. Oxford University Press, Oxford.

Radach, G. and A. Moll. 2006. Review of Three-Dimensional Ecological Modelling Related to the North Sea Shelf System. Part 2 – Model validation and Data needs. Oceanog. Mar. Biol. Ann. Rev. 44:11–60.

Marine Chemistry

Crompton, T.R. 2006. Analysis of Seawater: A Guide for the Analytical and Environmental Chemist. Springer, Berlin.

Underwater Photography

Aw, M. and M. Meur. 2006. An Essential Guide to Digital Underwater Photography. 2nd Edition. Best Publishing Co., Flagstaff.

Marine Ecological Processes

Valiela, I. 1995. Marine Ecological Processes. Springer, New York.

Augustus Vogel (Dissertation Research) (see p. 202)

Love, M. (1996). Probably More than You Want to Know About the Fishes of the Pacific Coast. Really Big Press, Santa Barbara, CA,

Vogel, A. (2006). Population Genetics and Recruitment of the Kelp Bass, *Paralabrax clathratus*. Department of Biological Sciences. Los Angeles, CA, University of Southern California.

References

Adam, J.A. 2003. Mathematics in Nature: Modeling Patterns in the Natural World. Princeton University Press, Ewing NJ.

Allen, T.K.H. and S. Skagen. 1973. Multivariate geometry as an approach to algal community analysis. Br. Phycol. J. 8:267–287.

Amstrup, S.C., McDonald, T.L., and B.F.J. Manly. 2006. Handbook of Capture–Recapture Analysis. Princeton University Press, Ewing NJ.

Anderberg, M.R. 1973. Cluster Analysis for Applications. Academic Press, New York.

Anderson, C.W., Barnett, V., Chatwin, P.C., and A.H. El-Shaarawi. 2002. Quantitative Methods for Current Environmental Issues. Springer-Verlag, New York.

Anderson, M.J. 2001. A new method for non-parametric multivariate analysis of variance. Austral. Ecol. 26:32–46.

Andrew, N.L. and B.D. Mapstone. 1987. Sampling and the description of spatial pattern in marine ecology. Oceanography Mar. Biol. Ann. Rev. 25:39–90.

Arnason, A.N. 1973. The estimation of population size, migration rates and survival in a stratified population. Res. Popul. Ecol. 15:1–8.

Arnqvist, G. and D. Wooster. 1995. Meta-analysis—synthesizing research findings in ecology and evolution. Tree 10:236–240.

Artiola, J.F., Pepper, I.L., and M.L. Brusseau (eds.). 2004. Environmental Monitoring and Characterization. Elsevier Academic Press, St. Louis.

Arvantis, L. and K.M. Portier. 2005. Natural Resources Sampling. CRC Press, Boca Raton, FL.

Asami, S. 2005. Applications of motor-powered paragliders. Sea Technol. 46(1):57–58.

Avery, T.E. and G.L. Berlin. 2004. Fundamentals of Remote Sensing and Airphoto Interpretation. Prentice-Hall, New York.

Avise, J.C. 1994. Molecular Markers, Natural History and Evolution. Chapman and Hall, New York.

Aquatic Sciences and Fisheries Abstracts. See the internet.

Baev, P.V. and L.D. Penev. 1993. BIODIV. Pensoft, Sofia, Bulgaria. Available from Exeter Software – see the Appendix.

Bailey. 1961. Statistical Methods in Biology. English Universities Press, London.

Bailey, N.T.J. 1981. Statistical Methods in Biology. Wiley, Somerset.

Bakanov, A.I. 1984. A nomogram for estimating the required number of samples of an aggregated distribution. J. Hydrobiol. 6:85–88.

Baker, J.M. and W.J. Wolff. 1987. Biological Surveys of Estuaries and Coasts. Cambridge University Press, Cambridge.

Bakun, A. 1996. Patterns in the Ocean, Ocean Processes and Marine Population Dynamics. California Sea Grant College System – NOAA – CIBN, La Paz, BCS, Mexico.

Quantitative Analysis of Marine Biology Communities: Field Biology and Environment
by Gerald J. Bakus
Copyright © 2007 John Wiley & Sons, Inc.

Bakus, G.J. 1959a. Observations on the life history of the Dipper in Montana. Auk 76:190–207.

Bakus, G.J. 1959b. Territoriality, movements and population density of the Dipper in Montana. Condor 61:410–452.

Bakus, G.J. 1964. The effects of fish-grazing on invertebrate evolution in shallow tropical waters. Allan Hancock Foundation Occasional Paper No. 27:1–29.

Bakus, G.J. 1966. Marine poecilascleridan sponges of the San Juan Archipeligo, Washington. J. Zoolog., London 149:415–531.

Bakus, G.J. 1969. Energetics and feeding in shallow marine waters. Internat. Revue Gen. Exper. Zoolog. 4:275–369.

Bakus, G.J. 1973. The biology and ecology of tropical holothurians. pp. 325–367 in: Jones O.A. and R. Endean (eds.). Biology and Geology of Coral Reefs. V. 2. Biology 1. Academic Press, New York.

Bakus, G.J. 1983. The selection and management of coral reef preserves. Ocean Manag. 8(1982/83):305–316.

Bakus, G.J. 1988. Practical and theoretical problems in the use of fouling panels. In: Marine Biodeterioration, (eds.) Mary-Frances Thompson et al., Oxford & IBH Publishing Co. Pvt. Ltd., New Delhi, pp. 619–630.

Bakus, G.J. 1990. Quantitative Ecology and Marine Biology. A.A. Balkema, Rotterdam.

Bakus, G.J. and G.K. Nishiyama. 1999. Sponge distribution and coral reef community structure off Mactan Island, Cebu, Philippines. Proceedings of 5th International Sponge Symposium, Brisbane, Australia. Mem. Queensland Mus. 44:45–50.

Bakus, G.J., G. Nishiyama, E. Hajdu, H. Mehta, M. Mohammad, U.S. Pinheiro, S.A. Sohn, T.K. Pham, Zulfigar bin Yasin and A. Karam. 2006. A comparison of some population density sampling techniques for biodiversity, conservation, and environmental impact studies. 28 pp. Manuscript being reviewed.

Bakus, G.J., Schulte, B., Jhu, S., Wright, M., Green, G., and P. Gomez. 1990. Antibiosis and antifouling in marine sponges: laboratory vs. field studies. In: Proceedings of 3rd International Conference on the Biology of Sponges, 17–23 November 1986. Woods Hole, MA.

Bakus, G.J., Stebbins, T.D., Lloyd, K., Chaney, H., Wright, M.T., Bakus, G.E., Ponzini, C., Eid, C., Hunter, L., Adamczeski, M., and P. Crews. 1989/1990. Unpalatability and dominance on coral reefs in the Fiji and Maldive Islands. Ind. J. Zoolog. Aquat. Biol. 1:34–53 and 2 (No. 1) (end).

Bakus, G.J., Stillwell, W.G., Latter, S.M., and M.G. Wallerstein. 1982. Decision making: with applications for environmental management. Environ. Manag. 6(6):493–504.

Bakus, G.J., Targett, N.M., and B. Schulte. 1986. Chemical ecology of marine organisms: An overview. J. Chem. Ecol. 12(5):951–987.

Bale, A.J. and A.J. Kenny. 2006. Sediment analysis and seabed characteristics. pp. 43–86 in: A. Eleftheriou. and A. McIntyre (eds.). 2005. Methods for Study of Marine Benthos. Blackwell Science Ltd., Oxford, UK.

Banks, M., Rashbrook, V.K., Calavetta, M.J., Dean, C.A., and D. Hedgecock. 2000. Analysis of microsatellite DNA resolves genetic structure and diversity of Chinook salmon (*Onchorhynchus tshawytscha*) in California's Central Valley. Canadian J. Fish. Aquat. Sci. 57:915–927.

Banse, K. 1975. Pleuston and neuston: On the categories of organisms in the uppermost pelagial. Int. Rev. Ges. Hydrobiol. 60:439–447.

Barbour, M.G., Burk, J.H., Pitts, W.D., Gilliam, F.S., and M.W. Schwartz. 1999.

Terrestrial Plant Ecology. Addison-Wesley-Longman, Menlo Park.

Barker, H.R. and B.M. Barker. 1984. Multivariate Analysis of Variance (MANOVA): A Practical Guide to its Use in Scientific Decision Making. University of Alabama Press, Tuscaloosa.

Barnes, H. 1952. The use of transformations in marine biological statistics. J. du Conseil Int. Explor. de Mer 18:61–71.

Bartell, S.M., Gardner, R.H., and R.V. O'Neil. 1992. Ecological Risk Estimation. Lewis Publishers, Boca Raton, FL.

Bartholomew, G.A. 1986. The role of natural history in contemporary biology. BioScience 36: 324–329.

Batschelet, E. 1981. Circular Statistics in Biology. Academic Press, New York.

Bausell, R.B. and Y-F Li. 2002. Power Analysis for Experimental Research. Cambridge University Press, New York.

Baverstock, P.R. and C. Moritz. 1996. Project Design. pp. 17–27 in: Hillis D., Moritz C., and B. Mable (eds.). Molecular Systematics. Sinauer Associates, Inc. Sunderland, MA.

Beals, E.W. 1973. Ordination: Mathematical elegance and ecological naivete. J. Ecol. 61:23–35.

Beatley, T., Brower, D.J., and A.K. Schwab. 2002. An Introduction to Coastal Zone Management. Island Press, Covelo, CA.

Begon, M. 1979. Investigating Animal Abundance: Capture–Recapture for Biologists. University Park Press. Baltimore.

Berger, T. (ed.). 1993 or later. Artificial Reef Bibliography: A reference guide. Sport Fishing Institute, Washington, D.C.

Berkeley, S. 1989. Trends in Atlantic swordfish fisheries. Mar. Recreat. Fish. 13:47–60.

Bernhard, M., Ghibaudo, M., Lavarello, O., Peroni, C., and A. Zattera. 1974. A sampler for the aseptic collection of water samples in the sea. Mar. Bio!. (Berlin) 25:339–343.

Bernstein. A.L. 1966. A Handbook of Statistical Solutions for the Behavioral Sciences. Holt, Rinehart and Winston, Inc., New York.

Bernstein, B.B. and J. Zalinski. 1983. On optimal sampling design and power tests for environmental biologists. J. Environ. Manag. 16:35–43.

Best, J. 2001. Dammed Lies and Statistics. University of California Press, Berkeley.

Biondini, M.E., Bonham, C.D., and E.F. Redente. 1985. Secondary successional patterns in a sagebrush (*Artemisia tridentata*) community as they relate to soil disturbance and soil biological activity. Vegetatio 71:79–86.

Birch, D. 1981. Dominance in marine ecosystems. Amer. Nat. 118 (2):262–274.

Bissonette, J.A. and L. Storch (eds.). 2002. Landscape Ecology and Resource Management. Island Press, Covelo, CA.

Blackith, R.E. and R.A. Reyment (eds.) 1971. Multivariate Morphometrics. Academic Press, New York.

Blonquist, S. 1991. Quantitative sampling of soft-bottom sediments: problems and solutions. Mar. Ecol. Prog. Ser. 72:295–304.

Bloom, S.A., Santos, S.L., and J.G. Field. 1977. A package of computer programs for benthic community analysis. Bull. Mar. Sci. 27:577–580.

Bohnsack, J.A. and S.P. Bannerot. 1986. A stationary visual census technique for quantitatively assessing community structure of coral reef fishes. NOAA Tech. Rept. NMFS 41. July.

Boitani, L. and T.K. Fuller (eds.). 2000. Research Techniques in Animal Ecology. Columbia University Press, New York.

Borchers, D.L., Buckland, S.T., and W. Zucchini. 2002. Estimating Animal Abundance: Closed Populations. Springer, New York.

Bortone, S.A. and J.J. Kimmel. 1991. Environmental assessment and monitoring of artificial habitats. pp. 177–236 in: W. Seaman, Jr. and L.M. Sprague (eds.). Artificial Habitats for Marine and Freshwater Fisheries. Academic Press, San Diego.

Bortone, S.A., van Tasell, J., Brito, A., Falcon, J.M., and C.M. Bundrick. 1991. A visual assessment of the inshore fishes and fishery resources off El Hierro, Canary Islands: A baseline survey. Sci. Mar. 55(3):529–541.

Bouchon, C. 1981. Comparison of two quantitative sampling methods used in coral reef studies: The line transect and the quadrat methods. p. 375 in: Gomez, Birkeland, C.E., Buddemeier, R.W., Johannes, R.E., and R.T. Tsuda (eds.). The Reef and Man. Proceedings of the Fourth International Coral Reef Symposium. V. 2. Manila, Philippines. 18–22 May 1981.

Bour, W., Loubersac, L., and P. Rual. 1986. Thematic mapping of reefs by processing of simulated SPOT satellite data: application to the *Trochus niloticus* biotope on Tetembia Reef (New Caledonia). Mar. Ecol. Prog. Ser. 34: 243–249.

Bowen, B.W. and J.C. Avise. 1990. Genetic structure of Atlantic and Gulf of Mexico populations of sea bass, menhaden, and sturgeon: influences of zoogeographic factors and life-history patterns. Mar. Biol. 107:371–381.

Box, G.E.P. and G.M. Jenkins. 1970. Time Series Analysis: Forecasting and Control. Holden-Day, San Francisco.

Bradbury, R.H., Reichelt, R.E., and D.G. Green. 1984. Fractals in ecology: methods and interpretation. Mar. Ecol. Prog. Ser. 14:295–296.

Bradbury, R.H. and P.C. Young. 1983. Coral interactions and community structure: an analysis of spatial pattern. Mar. Ecol. 11:265–271.

Branden, K.L., Edgar, G.J. and S.A. Shepherd. 1986. Reef fish populations of the Investigator Group. South Australia: a comparison of two census methods. Trans. R. Soc. S. Aust. 110:69–76.

Bray, J.R. and J.T. Curtis. 1957. An ordination of the upland forest communities of southern Wisconsin. Ecol. Monogr. 27:325–349.

Brebbia, C.A. (ed.). 2002. Coastal Engineering V: Computer Modelling of Seas and Coastal Regions. WIT Press, Ashurst, Southhampton, U.K.

Brebbia, C.A. (ed.). 2004. Risk Analysis IV. WIT Press, Ashurst, Southhampton, U.K.

Breiman, L., et al. 1984. Classification and Regression Trees. Wadsworth, Belmont, Ca.

Brink, K.H. and T.J. Cowles. 1991. The Coastal Transition Zone Program. J. Geophy. Res. 96 (No. C8):14,637–14,647.

Brown, D.M. and L. Cheng. 1981. New net for sampling the ocean surface. Mar. Ecol. Prog. Ser. 5:225–228.

Brown, J.H. and G.B. West. 2000. Scaling in Biology. Oxford University Press, New York.

Buckland, S.T., Anderson, D.R., Burnham, K.P., Laake, J.L., Borchers, D.L., and L. Thomas. 2001. Distance Sampling: Estimating abundance of biological populations. Oxford University Press, New York.

Bunkin, A. and K. Voliak. 2001. Laser Remote Sensing of the Ocean: Methods and Applications. John Wiley and Sons, New York.

Burgess, R.F. 2000. Ships Beneath the Sea: A History of Subs and Submersibles. Spyglass Publications, Chattahoochee, FL.

Busch, D.E. and J.C. Trexler. 2002. Monitoring Ecosystems: Interdisciplinary Approaches for Evaluating Ecoregional Initiatives. Island Press, Covelo, CA.

Cade, B.S. and B.R. Noon. 2003. A gentle introduction to quantile regression for ecologists. Front. Ecol. Environ. 1(8):412–420.

Caddy, J.F. 1982. Provisional world list of computer programmes for fish stock assessment and their availability by country and fisheries institute. FAO Fish. Circ. No. 746. 51 pp.

Caldrin, S.X., Friedland, K.D., and J.R. Waldman (eds.). 2004. Stock Identification Methods: Applications in Fishery Science. Elsevier Academic Press, St. Louis.

Camerer, C.F. 2003. Behavioral Game Theory: Experiments in Strategic Interaction. Princeton University Press, Ewing, NJ.

Campbell, L. 2001. Flow cytometric analysis of autotrophic picoplankton. pp. 317–346, in Paul J.H. (ed.). Methods in Microbiology. Volume 30. Marine Microbiology. Academic Press, New York.

Campbell, R.C. 1969, 1979, 1989. Statistics for Biologists. Cambridge University Press, Cambridge.

Canham, C.D., Cole, J.J., and W.K. Lauenroth (eds.). 2003. Models in Ecosystem Science. Princeton University Press, Ewing, NJ.

Cao, Y., Bark, A.W., and W. P. Williams. 1996. Measuring the responses of macroinvertebrate communities to water pollution: a comparison of multivariate approaches, biotic and diversity indices. Hydrobiologia 341:1–19.

Cassie, R.M. 1963. Microdistnbution of plankton. In: Barnes, H. (ed.). Mar. Biol. & Ocean. Ann. Rev. 1:223–252.

Cassie. R.M. and A.D. Michael. 1968. Fauna and sediments of an intertidal mud flat: A muflivariate analysis. J. Exp. Mar. Biol. Ecol. 2:1–23.

Cates, J.R., Corbin, J.S., Crawford, J., and C.E. Helsley. 2001. Aquaculture: Beyond the Reef. Sea Technol. 42(10):10–13, 15. October.

Chapman M.G. and A.J. Underwood. 1999. Ecological patterns in multivariate assemblages: information and

interpretation of negative values in ANOSIM tests. Mar. Ecol. Prog. Ser. 180:257–265.

Chase, C. and Lt. R. Sanders II. 2005. POPEIE – Probe for Oil Pollution Evidence in the Environment. Sea Technol. 46(9):10–13.

Chatfield. C. 1996. The Analysis of Time Series: An Introduction. Chapman and Hall, London.

Chester, D. 1992. New trends in neural networks. Scientific Computing and Automation. May. pp. 43–44 and 46–49.

Christensen, C., Walters, C., and D. Pauly. 2004. EcoPath with EcoSim: A User's Guide. Fisheries Center, University of British Columbia, Vancouver.

Clapham, M.E. 2003. Multivariate analysis of variance (MANOVA) statistical testing of paleocommunity composition. Research Report, BISC 532 (Statistics), University of Southern California. December 1, 2003.

Clark, J.S. and M. Lavine. 2001. Bayesian statistics: Estimating plant demographic parameters. pp. 327–346 in: Scheiner S.M. and J. Gurevitch (eds.). Design and Analysis of Ecological Experiments. Oxford University Press, Oxford.

Clarke, K.C. 1993. Non-parametric multivariate analyses of change in community structure. Aust. J. Ecol. 18:117–143.

Clarke, K.C. 1999. Getting Started with Geographic Information Systems. 2nd edition. Prentice-Hall, Upper Saddle River, NJ.

Clarke, K.R. 1990. Comparisons of dominance curves. J. Exp. Mar. Biol. Ecol. 138:143–157.

Clarke, K.R. 1993. Non-parametric multivariate analyses of changes in community structure. Aust. J. Ecol. 18:117–143.

Clarke, K.R. and R.N. Gorley. 2001. PRIMER v5: User Manual/Tutorial. Plymouth Routines in Multivariate

Ecological Research. PRIMER-E, Plymouth Marine Laboratory, Plymouth, U.K.

Clarke, K.R. and R.M. Warwick. 2001. Change in Marine Communities: An Approach to Statistical Analysis and Interpretation. 2nd edition. PRIMER- E, Plymouth Marine Laboratory, Plymouth, U.K.

Clifford, H.T. and W. Stephenson. 1975. An Introduction to Numerical C!assification. Academic Press, New York, pp. 229.

Cochran, W.G. 1977. Sampling Techniques. Wiley, New York.

Cody, M.L. 1974. Optimization in ecology. Science 183:1156–1164.

Colwell, R. 1979. Native Aquatic Bacteria, Enumeration, Activity and Ecology. Symposium on Water, American Society for Testing Materials, Minneapolis.

Colwell, R.R. 2004. Marine biotechnology: A confluence of promise. Sea Technol. 45:32–33.

Collyer, C.E. and J.T. Enns. 1986. Analysis of Variance: The Basic Designs. Nelson-Hall, Chicago, pp. 310.

Conners, M.E. and S.J. Schwager. 2002. The use of adaptive cluster sampling for hydroacoustic surveys. ICES J. Mar. Sci. 59:1314–1325.

Connor, E.F. and D. Simberloff. 1979. The assembly of species communities: chance or competition? Ecology 60:1132–1140.

Connor, E.F. and D. Simberloff. 1983. Interspecific competition and species co-occurrence patterns on islands: Null models and the evaluation of evidence. Oikos 41:455–465.

Connor, E.F. and D. Simberloff. 1986. Competition, scientific method, and null models in ecology. Am. Sci. 74:155–162.

Cooley, W.W. and P.R. Lohnes. 1971. Multivariate Data Analysis. Wiley, New York.

Cooper, H.M. and L.V. Hedges (eds.). 1994. Handbook of Research Synthesis. Russell Sage Foundation, New York.

Costanza, R. and A. Voinov. 2003. Landscape Simulation Modeling. Springer-Verlag, New York.

Costanza, R., Wainger, L., Folke, C., and K-G. Mäler. 1993. Modeling complex ecological economic systems. BioScience 43(8):545–555.

Cote, I.M., Mosqueira, I., and J.D. Reynolds. 2001. Effects of marine reserve characteristics on the protection of fish populations: a meta.analysis. J. Fish. Biol. 59 (Suppl. A):178–189.

Coulson, R.N., Lovelady, C.N., Flamm, R.O., Spradling, S.L., and M.C. Saunders. 1991. Intelligent geographic information systems for natural resource management. pp. 153–173 in: Turner M.G. and R.H. Gardner (eds.). Quantitative Methods in Landscape Ecology. Springer-Verlag, New York.

Cousins, S.H. 1991. Species diversity measurement: Choosing the right index. Tree 6(66):190–192.

Cowen, R.K., Paris, C.B., and A. Srinivasin. 2006. Scaling connectivity in marine populations. Science 311 (No. 5760):522–527.

Cox, D.R. 1958. Planning of Experiments. John Wiley and Sons, New York.

Coyer, J. and J. Witman. 1990. The Underwater Catalog: A guide to methods in underwater research. Shoals Marine Laboratory, Cornell University, New York.

Crawley, M.J. 1993. GLIM for Ecologists. Blackwell Scientific Publications, London.

Crutchfield, J.P., Farmer, J.D., Packard, N.H., and R.S. Shaw. 1986. Chaos. Scient. Amer. December. V. 255:46–57.

Cuff, W. and C. Coleman. 1979. Optimal survey design: lessons from a stratified random sample of macrobenthos. J. Exp. Mar. Biol. Ecol. 36:351–361.

Culverhouse, P.F, Herry, V., Ellis, R., Williams, R., Reguera, B., Gonzalez-Gil, S., Umani, S.F., Cabrini, M., and T. Parisini. 2002. Dinoflagellate categorization by artificial neural network. Sea Technol. 43(12):39–46. December.

Cureton, E.E. and R.B. D'Agostino. 1983. Factor Analysis. Erlbaum, Hillsdale.

Cushing, J.M., Costantino, R.F., Dennis, B., Desharnais, R.A., and S.M. Henson. 2003. Chaos in Ecology: Experimental Nonlinear Dynamics. Academic Press - Elsevier Science, New York.

Czaran, T. and S. Bartha. 1992. Spatiotemporal dynamic models of plant populations and communities. Tree 7:38–42.

Czekanowski, J. 1913. Zarys Metod Staystycnck. E. Wendego, Warsaw.

Dale, M.R.T. 1999. Spatial Pattern Analysis in Plant Ecology. Cambridge Studies in Ecology. Cambridge University Press, New York.

Dale, V.H. 2003. Ecological Modeling for Resource Management. Springer, New York.

Daniel, W.W. 1978. Applied Nonparametric Statistics. Houghton Mifflin Co., Boston.

Dapson, R.W. 1971. Quantitative comparison of populations with different age structures. Annales Zoologici Fennici 8:75–79.

Davenport, D. 1955. Specificity and behavior in symbiosis. Q. Rev. Biol. 30:29–46.

Davis, J.C. 1973. Statistics and Data Analysis in Geology. Wiley, New York.

Davis, J.C. 1986. Statistics and Data Analysis in Geology. Wiley, New York.

Davis, G.E. and T.W. Anderson. 1989. Population estimates of four kelp forest fishes and an evaluation of three *in situ* assessment techniques. Bull. Mar. Sci. 44:1138–1151.

Dean, R.L. and J.H. Connell. 1987. Marine invertebrates in an algal succession. 3. Mechanisms linking habitat complexity with diversity. J. Exp. Mar. Biol. & Ecol. 109:249–273.

Delisi, C., Ianelli, M., and G. Koch (eds.). 1983. Stochastic methods in the life sciences: general aspects and specific models. Bull. Math. Biol. 45:439–658.

De Martini, E.E. and D. Roberts. 1982. An empirical test of bioses in the rapid visual technique for species-time censuses of reef fish assemblages. Mar. Biol. 70: 129–134.

DeNoyer, L.K. and J.D. Dodd. 1990. Maximum likelihood smoothing of noisy data. Am. Lab. (Fairfield) 22:21–27.

DeSalle, R., Giribet, G., and W. Wheeler. 2002. Molecular Systematics and Evolution: Theory and Practice. Birkhauser, Boston.

Dethier, M.N., Graham, E.S., Cohen, S., and L.M. Tear. 1993. Visual versus random-point percent cover estimations: 'objective' is not always better. Mar. Ecol. Prog. Ser. 96:93–100.

Diamond, J.M. 1973. Distributional ecology of New Guinea birds. Science 239:759–769.

Digby. P.G.N. and R.A. Kempton. 1987. Multivariate Analysis of Ecological Communities. Chapman & Hall, London, 206 pp.

Diggle, P. 1990. Time Series, A Biostatistical Introduction. Clarendon Press, Oxford.

Diserud, O.H. and K. Aagaard. 2002. Testing for changes in community structure based on repeated sampling. Ecology 83(8):2271–2277.

Dixon, P.M. 2001. The bootstrap and the jackknife: Describing the precision of ecological indices. pp. 267–288 in: Scheiner S.M. and J. Gurevitch (eds.). Design and Analysis of Ecological Experiments. Oxford University Press, Oxford.

Doherty, P.J. and P. Sale. 1985. Predation on juvenile coral reef fishes:an exclusion experiment. Coral Reefs 4:225–234.

Dopson, R.W. 1971. Quantitative comparison of populations with different age structures. Ann. Zoolog. Fennici. 8: 75–79.

Downes, B.J., Barmuta, L.A., Fairweather, P.G., Faith, D.P., Keough, M.J., Lake, P.S., and B.D. Mapstone. 2002. Monitoring Ecological Impacts. Cambridge University Press, New York.

Dufrene, M. and P. Legendre. 1997. Species assemblages and indicator species: the need for a flexible asymmetrical approach. Ecol. Monogr. 67:345–366.

Dugatkin, L.A. (ed.) 2000. Game Theory and Animal Behavior. Oxford University Press, New York.

Dugdale, R.C. 1982. Organization of persistent upwelling structures. Sumnary program description to accompany proposals submitted to 0cean Sciences Division, National Science Foundation. Univ. So. California. June.

Dytham, C. 1999. Choosing and Using Statistics: A Biologist's Guide. Blackwell Publishing, Williston, VT.

Eberhardt, L.L. 1978. Appraising variability in population studies. J. Wild. Manag. 42:207–238.

Eberhardt, L.L. and J.M. Thomas. 1991. Designing environmental field studies. Ecol. Monogr. 61(1):53–73.

Edmands, S., Moberg, P.E., and R.S. Burton. 1996. Allozyme and mitochondrial DNA evidence of population subdivision in the purple sea urchin *Strongylocentrotus purpuratus*. Mar. Biol. 126:443–450.

Edwards, A.W.F. 1972. Likelihood. Cambridge University Press, Cambridge.

Efron, B. and R.J. Tibshirani. 1993. An Introduction to the Bootstrap. Chapman and Hall, London.

Eleftheriou, A. and A. McIntyre. 2005. Methods for Study of Marine Benthos. Blackwell Science Ltd., Oxford, UK.

Elliott, J.M. 1977. Some Methods for the Statistical Analysis of Samples of Benthic Invertebrates. Sci. Publ. No. 25, Freshwater Biological Assoc. Ferry House, U.K.

Ellis, D.V. 1968. A series of conputer programs for summarizing data from quantitative benthic investigations. J. Fish. Res. Bd. Canada 25:1737–1738.

Ellison, A.M. 2001. Exploratory data analysis and graphic display. pp. 37–62 in: S.M. Scheiner (eds.). Design and Analysis of Ecological Experiments. Oxford University Press, Oxford.

Elseth, G.D. and K.D. Baumgardner. 1981. Population Biology. Van Nostrand Co., New York, 623 pp.

Eltringham, S.K. 1971. Life in Mud and Sand. English Universities Press, London.

Elzinga, C., Salzer, D., Willoughby, J.W. and J. Gibbs. 2001. Monitoring Plant and Animal Populations: A Handbook for Field Biologists. Blackwell Publishing, Williston, VT.

Emery, W.J. and R.E. Thomson. 1998. Data Analysis Methods in Physical Oceanography. Pergamon-Elsevier, New York.

English, S., Wilkinson, C., and V. Baker. 1997. Survey Manual for Tropical Marine Resources. 2nd edition. Australian Institute of Marine Science, Townesville.

Erlich, P.R. and J. Roughgarden. 1987. The Science of Ecology. Macmillan Publishing Company, New York.

Eleftherious, A. and A. McIntyre. 2006. Methods for the Study of Marine Benthos. Blackwell Publishing, Oxford.

Eleftheriou, A. and D.C. Moore. 2005. Macrofauna Techniques. pp. 160–228 in: Eleftheriou, A., and McIntyre A. (eds.). 2005. Methods for Study of Marine Benthos. Blackwell Science Ltd., Oxford, UK.

Emery, W.J. and R.E. Thompson. 1998. Data Analysis in Physical Oceanography. Pergamon Press, Oxford

Everhart, W.H., Eipper, A.W. and W.D. Youngs. 1980. Principles of Fishery Science. Cornell Univ. Press, Ithaca.

Everitt, B. 1980. Cluster Analysis. Halsted-Wiley, New York.

Fager, E.W. 1957. Determination and analysis of recurrent groups. Ecology 38:586–593.

Fager, E.W., Flechsig, A.O., Ford, R.F., Clutter, R.I., and R.J. Ghelardi. 1966. Equipment for use in ecological studies using SCUBA. Limnol. Oceanogr. 11:503–509.

Fairweather, P.G. 1991. Statistical power and design requirements for environmental monitoring. Aust. J. Freshwater Res. 42:555–567.

FAO. 1994. The FAO-ICLARM Stock Assessment Tools (FiSAT) User's Guide. Edited by. F.C. Gayanilo, Jr., P. Sparre, and D. Pauly. ICLARM Contribution No. 1048. Food and Agriculture Organization of the United Nations, Rome.

Feinsinger, P. 2001. Designing Field Studies for Biodiversity Conservation. The Nature Conservancy. Island Press, Covelo, CA.

Fennel, W. and T. Neumann. 2005. Introduction to the Modeling of Marine Ecosystems. Elsevier Oceanography Series 72. Elsevier, New York.

Ferson, S. and M. Burgman. 1998. Quantitative Methods for Conservation Biology. Springer-Verlag, New York.

Ferson, S., Downey, P., Klerks, P., Weissburg, M., Kroot, I., Stewart, S., Jacquez, O., Ssemakula, J., Malenky, R., and K. Anderson. 1986. Competing reviews, or why do Connell and Schoener disagree? Am. Nat. 127: 571–576.

Feynman, R.P., Leighton, RB. and M. Sands. 1963. The Feynman Lectures on Physics: Mainly Mechanics, Radiation, and Heat. V. 1, Addison-Wesley, Reading, MA.

Field, J.G., Clarke, K.R. and R.M. Warwick. 1982. A practical strategy for analyzing multispecies distribution patterns. Mar. Ecol. Prog. Ser. 8:37–52.

Fingas, M.F. and C.E. Brown. 2000. Oil-Spill remote sensing. Sea Technol., October. pp. 21–26.

Fireman, J. 2000. Data Acquisition Processing. Scientific Computing And Instrumentation. November. pp. 15–16, 67. Cahners Business Information, Highlands Ranch, CO.

Firestone, M.A. 1976. Computer Programs in Marine Science. National Oceanographic Data Center, Washington, D.C.

Fisher, R.A. 1932. Statistical Methods for Research Workers. Oliver and Boyd, Edinburgh.

Fisher, N.I. 1996. Statistical Analysis of Circular Data. Cambridge University Press, Cambridge.

Fisher, R.A. 1932. Statistical Methods for Research Workers. Oliver and Boyd, Edinburgh.

Fisher, R.A. 1971. The Design of Experiments. Hafner, New York.

FitzGerald, S. 2004. Automatic Time Series Forecasting. Scientific Computing. p. 34. January.

Floyd, T. 2001. Logic modeling and logistic regression: Aphids, ants, and plants. pp. 197–216 in: Scheiner S.M. and J. Gurevitch (eds.). Design and Analysis of Ecological Experiments. Oxford University Press, Oxford.

Fogarty, M.J. 1989. Forecasting yield and abundance of exploited invertebrates. Chapter 31, in: Caddy J.F. (ed.). Marine Invertebrate Fisheries, Their Assessment and Management. John Wiley and Sons, New York.

Fortin, M-J and M.R.T. Dale. 2005. Spatial Analysis: A Guide for Ecologists. Cambridge University Press, Cambridge.

Fortin, M-J. and J. Gurevitch. 2001. Mantel tests: Spatial structure in field experiments. pp. 308–326 in: Scheiner S.M. and J. Gurevitch (eds.). Design and Analysis of Ecological Experiments. Oxford University Press, Oxford.

Fowler, J., Cohen, L., and P. Jarvis. 1998. Practical Statistics for Field Biology. John Wiley & Sons, New York.

Fox, G.A. 2001. Failure-time analysis: Studying times to events and rates at which events occur. pp. 235–266 in: Scheiner S.M. and J. Gurevitch (eds.). Design and Analysis of Ecological Experiments. Oxford University Press, Oxford.

Frontier, S. (ed.). 1983. Strategies d'echantillonnage en ecologie. Collection D'Ecologie No. 7. Masson. Paris, France.

Frontier, S. 1985. Diversity and structure in aquatic ecosystems. Oceanogr. Mar. Biol. Ann. Rev. 23:253–312.

Fryer, R.J. and M.D. Nicholson. 2002. Assessing covariate-dependent contaminant time-series in the marine environment. ICES J. Mar. Sci. 59:1–14.

Gage, J.D. and B.J. Bett. 2005. Deep-Sea Benthic Sampling. pp. 273–325 in: Eleftheriou A. and A. McIntyre (eds.). 2005. Methods for Study of Marine Benthos. Blackwell Science Ltd., Oxford, UK.

Gamito, S. and D. Raffaelli. 1992. The sensitivity of several ordination methods to sample replication in benthic surveys. J. Exp. Mar. Biol. Ecol. 164(2):221–232.

Gauch, H.G. 1977. ORDIFLIX—A flexible computer program for four ordination techniques: weighted averages, polar ordination, principal components analysis, and reciprocal averaging. In: Ecology and Systematics, Cornell University, Ithaca, N.Y.

Gauch, H.G., Jr. 1982. Multivariate Analysis in Community Ecology. Cambridge University Press, New York.

Gauch, H.G. and R.N. Whittaker. 1981. Hierarchical classification of community data. J. Ecol. 69:135–152.

Geiger, D.L. 1999. A Total Evidence Cladistic Analysis of the Haliotidae (Gastropoda: Vetigastropoda). Ph.D. Dissertation, Department of Biological Sciences, University of Southern California, Los Angeles. California. December.

Gelman, A., Carlin, J.B., Stern, H.S., and D.B. Rubin (eds.). 2003. Bayesian Data Analysis. CRC Press, Boca Raton, Florida.

Getz, W.M. (ed.) 1980. Mathematical Modelling in Biology and Ecology. Springer-Verlag, New York, 356 pp.

Giere, O. 1993. Meiobenthology: The Microscopic Fauna in Aquatic Sediments. Springer-Verlag, New York.

Gilbert, R.O. 1987. Statistical methods for environmental pollution monitoring. Van Nostrand Reinhold, New York.

Gittins, R. 1985. Canonical analysis: a review with applications in ecology. Springer-Verlag, Berlin.

Gleich, J. 1987. Chaos: Making a New Science. Penguin (Non-Classics), New York.

Glover, T. and K. Mitchell. 2002. An Introduction to Biostatistics. McGraw-Hill, Boston.

Goedickemeier, I., Wildi, O., and F. Kienast. 1997. Sampling for vegetation survey: Some properties of a GIS-based stratification compared to other statistical sampling methods. Coenoses 12(1):43–50.

Gold, J.R., Richardson, L. R., Furman, C., and F. Sun. 1994. Mitochondrial DNA diversity and population structure in marine fish species from the Gulf of Mexico. Canadian J. Fish. Aquat. Sci. 51 (Suppl. 1):205–214.

Goldberg, D.E. and S. M. Scheiner. 2001. ANOVA and ANCOVA: Field competition experiments. pp. 77–98 in: Scheiner S.M. (ed.). Design and Analysis of Ecological Experiments. Oxford University Press, Oxford.

Gonor, J.J. and P.F. Kemp. 1978. Procedures for quantitative ecological assessments in intertidal environments. EPA-600/3–78-087. Corvallis, OR. pp. 103.

Gore, A., Paranjpe, S., and S.A. Paranjpe. 2000. A Course in Mathematical and Statistical Ecology. Springer, New York.

Gotelli, N.J. 2001. A Primer of Ecology. Sinauer Associates, Sunderland.

Gotelli, N.J. and A.M. Ellison. 2004. A Primer of Ecological Statistics. Sinauer Associates, Sunderland, Maine.

Gotelli, N.J. and G.R. Graves. 1996. Null Models in Ecology. Smithsonian Institution Press, Washington, D.C.

Gould, J.L. and G.F. Gould. 2002. Biostats Basics: A Student Handbook. W.H. Freeman and Company, New York.

Grace, J. 2005. A Multivariate Perspective on Ecological Systems. Cambridge University Press, Cambridge.

Grafen, A. and R. Hails. 2002. Modern Statistics for the Life Sciences: Learn How to Analyze Your Experiments. Oxford University Press, New York.

Graham, M.H. and M.S. Edwards. 2001. Statistical significance versus fit: Estimating the importance of individual factors in ecological analysis of variance. Oikos 93(3):505–513.

Grassle, J.F., Sanders, H.L., Hessler, R.R., Rowe, G.T., and T. McLellan. 1975. Pattern and zonation: A study of the bathyal megafauna using the research submersible Alvin. Deep Sea Res. 22: 457–481.

Green, R.H. 1979. Sampling Design and Statistical Methods for Environmental Biologists. John Wiley and Sons, New York.

Green, R.H. 1989. Power analysis and practical strategies for environmental monitoring. Environ. Res. 50:195–205.

Grieg-Smith, P. 1964. Quantitative Plant Ecology. Butterworth, London. Second edition.

Grimm, V. and S.F. Railsback. 2005. Individual-based Modeling and Ecology. Princeton University Press, Ewing NJ.

Grossman, G.D. 1982. Dynamics and organization of a rocky intertidal fish assemblage: The persistence and resilience of taxocene structure. Amer. Natur. 119(5):611–637.

Grussendorf, Mi. 1981. A flushing coring device for collecting deep burrowing infaunal bivalves in intertidal sand. Fish. Bull. 79:383–385.

Gudbjornsson, S., Goda, O.R., and O.K. Palsson. 2004. Mini GPS fish tags contributing to fisheries management. Sea Technol. 45(6): 23–27.

Guest, P.O. 1961. Numerical Methods of Curve Fitting. Cambridge University Press.

Guisan, A., Edwards, T.C., and T. Hastie. 2002. Generalized Linear and Generalized Additive Models in Studies of Species Distributions: Setting the Scene. Ecol. Model. 157:89–100.

Gurevitch, J. and L.V. Hedges. 2001. Meta-analysis: Combining the results of independent experiments. pp. 347–369 in: Scheiner S.M. and J. Gurevitch (eds.). Design and Analysis of Ecological Experiments. Oxford University Press, Oxford.

Gurevitch, J., Morrow, L.L., Wallace, A., and J.S. Walsh. 1992. A meta-analysis of field experiments on competition. Amer. Natur. 155:435–453.

Gutowitz, H. (ed.). 1991. Cellular Automata: Theory and Experiment. Labringth Books, Millwood, New York.

Gutzwiller, K.J. and R.T.T. Forman. 2002. Applying Landscape Ecology in Biological Conservation. Springer-Verlag, New York.

Hadamard, J. 1898. Les surfaces a curbures opposes et leurs lignas geodesiques. J. Math. Anal. Appl. 4:27–73.

Haddon, M. 2001. Modelling and Quantitative Methods in Fisheries. CRC Press, Boca Raton, Florida.

Hall, C.A.S. 1985. Models in ecology: paradigms found or paradigms lost? Bull. Ecol. Soc. Am. 66(3):339–346.

Hall, C.A.S. 1988. An assessment of several of the historically most influential theoretical models used in ecology and of the data provided in their support. Ecol. Model. 43:5–31.

Hall, C.A.S. 1991. An idiosyncratic assessment of the role of mathematical models in environmental sciences. Environ. Int.: 507–517.

Hall, C.A.S. 2003. Making ecology more relevant and powerful for millennia III. Manuscript. 23 pp.

Hall, C.A.S. and J.W. Day, Jr. (eds.) 1977. Ecosystem Modeling in Theory and Practice: An Introduction with Case Histories. John Wiley and Sons, New York.

Hallacher, L.E. 2004. Underwater Sampling Techniques. Quest. March. Internet address: www.kmec.uhh.hawaii.edu/QUESTInfo

Halpern, D. (ed.). 2001. Satellites, Oceanography and Society. Elsevier Science Inc., New York.

Hamner, W.M., Madin, L.P., Aldredge, A.L., Gilmer, R.W., and P.P. Hamner. 1975. Underwater observations of gelatinous zooplankton: sampling problems, feeding biology and behavior. Limnol. Oceanogr. 20:907–917.

Hamner, R.M. 1980. Ecology of kelp forest demersal zooplankton. Ph.D. Dissertation, Department of Biological Sciences, University of Southern California, Los Angeles. September.

Hamner, R.M. 1981. Day-night differences in the emergence of demersal zooplankton from a sand substrate in a kelp forest. Mar. Biol. 62: 275–280.

Hammond, L.S. 1982. Patterns of feeding and activity in deposit-feeding holothurians and echinoids (Echinodermata) from a shallow back-reef lagoon, Discovery Bay, Jamaica. Bull. Mar. Sci. 32: 549–571.

Hampton, R. 1994. Introductory Biological Statistics. W.C. Brown, Dubuque.

Hannon, B.M. and M. Ruth. 1997. Modeling Dynamic Biological Systems. Springer Verlag, New York.

Hannon, B.M. and M. Ruth. 2001. Dynamic Modeling. Springer-Verlag, New York.

Hardy, A.C. 1958. The Open Sea: The World of Plankton. Collins, London.

Harding, J.P. 1949. The use of probability paper for the graphical analysis of polymodal frequency distributions. J. Mar. B. Assoc. U. K. 28:141–153.

Harris, R.P., Wiebe, P.A., Lenz, J., Skjoldal, H.R., and M. Huntley. 2000. ICES Zooplankton Methodology Manual. Academic Press, New York.

Hastings, H.M. and G. Sugihara. 1993. Fractals: A User's Guide for the Natural Sciences. Oxford Univ. Press, Oxford & New York.

Hatcher, B.G., Imberger, J. and S.V. Smith. 1987. Scaling analysis of coral reef systems: an approach to problems of scale. Coral Reefs 5:171–181.

Hauer, F.R. and G.A. Lamberti (eds.). 1998. Methods in Stream Ecology. Elsevier Academic Press, St. Louis.

Hayek, L-A.C. and M.A. Buzas. 1996. Surveying Natural Populations. Columbia University Press, New York.

Hedgpeth, J.W. 1977. Models and muddles: some philosophical obervations. Helgol. Wiss. Meeres. 30:92–104.

Heip, C., Herman, P.M.J., and K. Soeteart. 1988. Data processing, evaluation, and analysis. pp. 197–231 in: Higgins R.P., and H. Thiel (eds.). Introduction to the Study of Meiofauna. Smithsonian Institution, Washington, D.C.

Helvarg, D. (ed.). 2005. The Ocean and Coastal Conservation Guide 2005–2006: The Blue Movement Directory. Island Press, Washington, D.C.

Hempel, G. and K. Sherman (eds.). 2003. Large Marine Ecosystems of the World: Trends in Exploitation, Protection and Research. V. 12. Elsevier, New York.

Henderson, F.M. and A.J. Lewis (eds.). 1998. Manual of Remote Sensing: Principles and Applications of Imaging Radar. John Wiley and Sons, New York.

Hennessey T.M. and J.G. Sutinen (eds.). 2005. Sustaining Large Marine Ecosystems: The Human Dimension. V. 13. Elsevier, New York.

Hessler, R.R. and P.A. Jumars. 1974. Abyssal community analysis from replicate box cores in the central North Pacific. Deep-Sea Res. 21: 185–209.

Heyer, W.R., Donnelly, M., and R.W. Heyer. 1994. Measuring and Monitoring Biodiversity. Smithsonian Books, Washington, D.C.

Hillis, D.M., Moritz, C., and B.K. Maple (eds.). 1996. Molecular Systematics. Sinauer Associates, Sunderland.

Hintze, J.L. 1996. PASS User's Guide. NCSS, Kaysville.

Hobson, E.S. 1974. Feeding relationships of teleostean fishes on coral reefs in Kona, Hawaii. Fish. Bull. 72:915–1031.

Høisaeter, T. (ed). 2001. Special Issue: Marine Spatial Modelling. V. 86(6): 405–560.

Holden, H. and E. LeDrew. 2001. Coral reef features: multi- vs. hyper-spectral characteristics. Sea Technol. February. pp. 63–70.

Hollander, M. and D.A. Wolfe. 1998. Nonparametric Statistical Methods. Wiley, New York.

Holm-Hansen, O. and B. Riemann. 1978. Chlorophyll a determination: improvements in methodology. Oikos 30: 438–447.

Holme, N.A. and R.L. Barret. 1977. A sledge with television and photographic cameras for quantitative investigation of the epifauna in the continental shelf. J. Mar. Biol. Ass. U.K. 57:391–403.

Holme, M.A. and A.D. Mcintyre (eds.) 1984. Methods for the Study of Marine Benthos. Blackwell Scientific, Palo Alto.

Holyoak, M., Leibold, M.A., and R.D. Holt. 2005. Metacommunities: Spatial Dynamics and Ecological Communities. University of Chicago Press, Chicago, IL.

Hooke, R. 1983. How to Tell the Liars from the Statisticians. Dekker, New York.

Houston, T.R. 1985. Why models go wrong. Byte Mag. 10 (10):151–161. October.

Hoyt, E. 2005. Marine Protected Areas for Whales, Dolphins and Porpoises. James and James/Earthscan, London, U.K.

Hudson, P.F. and R.R. Colditz. 2003. Relations between floodplain environments and land use – Land cover in a large humid tropical river valley: Panuco Basin, Mexico. J. Hydrol. 280:229–245.

Huff, D. 1954. How to Lie with Statistics. WW Norton, New York.

Hughes, J.B., Hellmann, J.J., Ricketts, T.H., and B.J.M. Bohannan. 2001. Counting the uncountable: Statistical approaches to estimating microbial density. App. Environ. Microbiol. 67(10):4399–4406.

Hulings, N.C. and J.S. Gray. 1971. A Manual for the Study of the Meiofauna. Smithson. Contrib. Zoolog. 78.

Hunsaker, C.T., Friedl, M.A., Goodchild, M.F., and T.J. Case. 2001. Spatial Uncertainty in Ecology: Implications for Remote Sensing and GIS Applications. Springer, New York.

Hurlbert, S.H. 1971. The non-concept of species diversity: A critique and alternative parameters. Ecology 52: 557–586.

Hurlbert. S.H. 1984. Pseudoreplication and the design of ecological field experiments. Ecol. Monogr. 54:187–211.

Hurlbert, S.H. 1990a. Spatial distribution of the momtane unicorn. Oikos 58:257–271.

Hurlbert, S.H. 1990b. Pastor binocularis: Now we have no excuse. Ecology 71(3):1222–1223.

Hurlbert, S.H. and M.D. White. 1993. Experiments with freshwater invertebrate zooplanktivores: quality of statistical analyses. Bull. Mar. Sci. 53:128–153.

Hurlbert, S.H. 1994. Old shibboleths and new syntheses. TREE 9:495–496.

Hurlbert, S.H. 1997. Functional importance *vs* keystoneness: Reformulating some questions in theoretical biocenology. Aust. J. Ecol. 22:369–382.

Hunsaker, C.T., Goodchild, M.F., Friedl, M.A., and T.J. Case. 2001. Spatial Uncertainty in Ecology. Springer-Verlag, New York.

Huston, M., DeAngelis, D., and W. Post. 1988. New computer models unify ecological theory. Bioscience 38: 682–691.

Hutchings, P.A. 1986. Biological destruction of coral reefs. Coral reefs 4:239–252.

ICLARM. 2000. ReefBase 2000: C-NAC: Coral Navigator. Australian Institute of Marine Sciences, Townesville, Australia. ICLARM has moved from Manila, Philippines to Penang, Malaysia.

Interoceanic Canal Studies. 1970. Appendix 16–Marine Ecology. Inclosure A— Possible Effects on Marine Ecology. Battelle Memorial Institute, Columbus, Ohio.

Isaacs, A. 1981. The Multilingual Computer Dictionary. Facts on File, New York.

Isagi, Y. and N. Nakagashi. 1990. A Markov approach for describing post-fire succession of vegetation. Ecol. Res. 5(2):163–172.

Ivlev, V.S. 1961. Experimental Ecology of the Feeding of Fishes. Yale University Press, New Haven.

Jaccard, P. 1908. Nouvelles recherches sur la distribution florale. Bull. Soc. Vaud. Sci. Natur. 44:223–270.

Jacobs, J. 1974. Quantitative measurement of food selection. A modification of the forage ratio and Ivlev's electivity index. Oecologia 14(4):413–417.

Jaffe, A.J. and H.F. Spirer. 1987. Misused Statistics: Straight Talk for Twisted Numbers. Dekker, New York.

Jähne, B. 1997. Practical Handbook on Image Processing for Scientific Applications. CRC Press, Boca Raton.

Jamieson, A.J. and P.M. Bagley. 2005. Biodiversity Survey Techniques: ROBIO and DOBO Landers. Sea Technol. 46(1):52–54. January.

Janson, S. and J. Vegelius. 1981. Measures of ecological association. Oecologia 49:371–376.

Jeffers, J.N.R. 1978. An Introduction to Systems Analysis: with Ecological Applications. University Park Press, London.

Jeffers, J.N.R. 1982. Modelling. Chapman & Hall, London.

Jensen, A.L. 1985. Time series analysis and the forecasting of menhaden catch and CPUE. N. Amer. J. Fish, Mgmt. 5: 78–85.

Jensen, J.R. 2000. Remote Sensing of the Environment: An Earth Resource Perspective. Prentice Hall, Indianapolis.

Jensen, M. and P.S. Bourgeron. 2001. A Guidebook for Integrated Ecological Assessments. Springer-Verlag, New York.

Johnson, D.H. 1980. The comparison of usage and availability measurements for evaluating resource preference. Ecology 61:65–71.

Johnson, D.H. 1999. The insignificance of statistical significance testing. J. Wildl. Manag. 63:763–772.

Johnson, M.S. and R. Black. 1982. Chaotic genetic patchiness in an intertidal limpet, *Siphonia* sp. Mar. Biol. 70:157–164.

Jones, A.R. and C. Watson-Russell. 1984. A multiple coring system for use with scuba. Hydrobiol. 109:211–214.

Jongman, R. and G. Pungetti. 2003. Ecological Networks and Greenways. Cambridge University Press, New York.

Jongman, R.H.G., Ter Braak, C.J.F., and O.F.R. van Tongeren (eds.). 1995. Data Analysis in Community and Landscape

Ecology. Cambridge University Press, Cambridge.

Jorgensen, S.E., Halling-Sorenson, B., and S.N. Nielsen. 1996. Handbook of Environmental and Ecological Modeling. Lewis Publishers, Boca Raton.

Juliano, S.A. 2001. Nonlinear curve fitting: Predation and functional response curves. pp. 178–196 in: Scheiner S.M., and J. Gurevitch (eds.). Design and Analysis of Ecological Experiments. Oxford University Press, Oxford.

Kareiva, P.M. and N. Shigesada. 1983. Analyzing insect movement as a correlated random walk. Oecologia 56: 234–238.

Kauwling, T.J. and G.J. Bakus. 1979. Effects of Hydraulic Clam Harvesting in the Bering Sea. North Pac. Fish. Manag. Counc. Doc. 5.

Keeney, R.L. 1992. Value-Focused Thinking: A Path to Creative Decisionmaking. Harvard University Press, Cambridge.

Keith, L.H. 1991. Environmental Sampling and Analysis: a practical guide. Lewis Publications, Boca Raton.

Keltunen, J. 1992. Estimating the number of samples and replicates on the basis of statistical reasoning. Vesitalous 33(3):23–33.

Kendall, M.G. 1955. Rank Correlation Models. Griffin, Ltd.

Kendall, M.G. and A. Stuart. 1961. The Advanced Theory of Statistics. v. 2, Hafner, New York.

Kendall, M. 1980. Multivariate Analysis. Macmillan, New York.

Kenny, B.C. 1982. Beware of spurious self-correlations. Water Resources Res. 18:1041–1048.

Kershaw, K.A. 1973. Quantitative and Dynamic Plant Ecology. American Elsevier Publishing Company, Inc., New York.

Kingsford, M.J. and C. N. Battershill. 1998. Studying Temperate Marine Environments. CRC Press, Boca Raton.

Kirk, J.T.O. 1994. Light and Photosynthesis in Aquatic Ecosystems. Cambridge University Press, Cambridge.

Knight, D.B., Knutson, P.L., and E.J. Pullen. 1980. An Annotated Bibliography of Seagrasses with Emphasis on Planting and Propagation Techniques. CERC Misc. Rept. No. 80-7:1–46.

Knox, G.A. 2001. The Ecology of Seashores. CRC Press, Boca Raton, FL.

Kolata, G. 1986. What does it mean to be random? Science 231:1068–1070.

Kot, M. 2001. Elements of Mathematical Ecology. Cambridge University Press, New York.

Kozlov, M.V. and S.H. Hurlbert. 2006. Pseudoreplication, chatter, and the international nature of science: A response to D.V. Tatarnikov. J. Fundam. Biol. (Moscow) 67(2):128–135. Available in English from Stuart Hurlbert (shurlbert@sunstroke.sdsu.edu).

Krebs, C.J. 1989. Ecol. Method. Harper Collins, New York.

Krebs, C.J. 1999. Ecological Methodology. Harper & Row, New York.

Krebs, C.J. 2000a. Programs for Ecological Methodology. Department of Zoology, University of British Colombia, Vancouver, British Colombia, Canada.

Krebs, C.J. 2000b. Hypothesis testing in ecology. pp. 1–14 in: Boitani, L. and T.K. Fuller (eds.). 2000. Research Techniques in Animal Ecology. Columbia University Press, New York.

Krebs, C.J. 2001. Ecology: The Experimental Analysis of Distribution and Abundance. Addison, Wesley and Longman, Inc., San Francisco.

Kremer, J.N. 1983. Modelling of marine ecosystems: formulation of processes. pp. 40–45. In: Quantitative Analysis and Simulation of Mediterranean Coastal Systems: The Gulf of Naples, a Case

Study. UNESCO, UNESCO Repts. Mar. Sci. 20.

Kremer, J.N. and S.W. Nixon. 1978. A Coastal Marine Ecosystem: Simulation and Analysis. Ecol. Stud. v. 24, Springer-Verlag, Heidelberg, 217 pp.

Krohne, D.T. 1998. General Ecology. Wadsworth Publishing Company, Belmont, CA.

Kuehe, R.O. 2000. Design of experiments: statistical principles of research design and analysis. Duxbury, Pacific Grove, CA.

Kutner, M.H., Nachtscheim, C.J., Wasserman, W., and J. Neter. 1996. Applied Linear Statistical Models. WCB/McGraw-Hill, New York.

Lambshead, P.J.D., Platt, H.M. and K.M. Shaw. 1983. The detection of differences among assemblages of marine benthic species based on an assessment of dominance and diversity. J. Nat. His. 17:859–874.

Lammons, G. 2005. Naval community pursues new autonomous surface vehicles. Sea Technol. 46(12):27–31.

Lance, G.N. and W.T. Williams. 1966a. Computer programs for hierarchical polythetic classifications ("similarity analyses"). Comput. J. 9:60–64.

Lance, G.N. and W.T. Williams. l966b. A generalised sorting strategy for computer classifications. Nature (London) 212: 218.

Lance, G.N. and W.T. Williams. l967a. A general theory of classificatory sorting strategies. 1. Hierarchical systems. Comput. J. 9:373–380.

Lance, G.N. and W.T. Williams. 1967b. Mixed-data classificatory programs. I. Agglomerative systems. Aust. Comput. J. 1:15–20.

Lance, G.N. and W.T. Williams. 1968a. Mixed-data classificatory programs. II. Divisive systems. Aust. Comput. J. 1: 82–85.

Lance, G.N. and W.T. Williams. 1968b. Note on a new information-statistic classificatory program. Comput. J. 11: 195.

Lance, G.N. and W.T. Williams. 1970. Computer programs for hierarchical polythetic classification ("similarity analysis"). Aust. Comput. J. 2:60–64.

Lance, G.N. and W.T. Williams. 1971. A note on a new divisive classificatory program for mixed data. Comput. J. 14:154–155.

Lang, J.C. (ed.). 2003. Status of coral reefs in the Western Atlantic: Results of initial surveys, Atlantic and Gulf Rapid Reef Assessment (AGRRA) program. Atoll Research Bulletin No. 496:1–630.

LaPointe, B.E., P.J. Barile, C.S. Yentsch, M.M. Littler, D.S. Littler, and B. Kakuk. 2003. The relative importance of nutrient enrichment and herbivory on macroalgal communities near Norman's Pond Cay, Exumas Cays, Bahamas: a "natural" enrichment experiment. J. Exper. Mar. Biol. Ecol. 298(2004):275–301.

LaPointe, B.E., Littler, M.M. and D.S. Littler. 1997. Macroalgal overgrowth of fringing coral reefs at Discovery Bay, Jamaica: bottom-up versus top-down control. Proceedings of 8th International Coral Reef Symposium 1:927–932.

Lawrence, J. 1992. Introduction to Neural Networks and Expert Systems. California Scientific Software, Nevada City, CA.

Lee, Y-W and D.B. Sampson. 2000. Spatial and temporal stability of commercial groundfish assemblages off Oregon and Washington as inferred from Oregon travel logbooks. Canadian J. Fish. Aqua. Sci. 57:2443–2454.

Legendre, P. and L. Legendre. 1998. Numerical Ecology. 2nd Edition. Elsevier, Amsterdam.

Lenfant, P. and S. Planes. 2002. Temporal genetic changes between cohorts in a natural population of a marine fish, *Diplocus sargus*. Biol. J. Linnean Soc. 76:9–20.

Leps, J. and P. Smilauer. 2003. Multivariate Analysis of Ecological Data using CANOCO. Cambridge University Press, New York.

Levashov, D.E., Mikheychic, P.A., Sedov, A.Y., Kantakov, G.A. and A.P. Voronkov. 2004. Laser plankton meter TRAP-7A, a new sensor for CTD probing. Sea Technol. 45(2):61–65. February.

Li, G. and D. Hedgecock. 1998. Genetic heterogeneity, detected by PCR-SSCP, among samples of larval Pacific oysters (*Crassostrea gigas*) supports the hypothesis of large variance in reproductive success. Canadian J. Fish. Aquat. Sci. 55(4):1025–1033.

Licuanan, W.R.Y. 1995. Analysis and simulation of spatial distributions of corals on unconsolidated reef substrates. Ph.D. dissertation. Department of Biological Sciences, University of Southern California, Los Angeles. May.

Licuanan, W.Y. and G.J. Bakus. 1994. Coral spatial distributions: the ghost of competition past roused? Proceedings of 7th International Coral Reef Symposium, June 22–27, 1992. Agana, Guam. V. 1, pp. 545–549.

Lindsey, J.K. 2003. Introduction to Applied Statistics: A Modelling Approach. Oxford University Press, New York.

Little, C. 2000. The Biology of Soft Shores and Estuaries. Oxford University Press, New York, NY.

Littler, D.S. and M.M. Littler. 1987. Rocky intertidal aerial survey methods utilizing helicopters. pp. 31–34 in: Dubois J.M. (ed.). Remote Sensing of Pluricellular Marine Algae: Photo-interpretation. Editions Technip, Paris.

Littler, M.M. and D.S. Littler. 1985. Nondestructive sampling. pp. 161–175 in: Littler M.M. and D.S. Littler (eds.). Handbook of Phycological Methods. Ecological Field Methods: macroalgae. Cambridge University Press, Cambridge.

Littler, M.M., Littler, D.S., Brooks, B.L. and J.F. Koven. 1996. Field expedient methods: A comparison between limited vs. optimal resources for the study of a unique coral reef. pp. 159–165 in: Lang M.M. and C.C. Baldwin (eds.). Proceedings of the American Association of Underwater Scientists. 1966 Scientific Diving Symposium. Smithsonian Institution, Washington, D.C.

Litvaitis, J.A. 2000. Investigating food habits of terrestrial vertebrates. pp. 165–190 in: Boitani, L. and T.K. Fuller (eds.). 2000. Research Techniques in Animal Ecology. Columbia University Press, New York.

Lombardi, C.M. and S.H. Hurlbert. 1996. Sunfish cognition and pseudoreplication. Anim. Behav. 52:419–422.

Lombardi, M. 2004. Rebreather Technology: There is a Market, but can anyone Deliver? Sea Technol. 45(12):15–19. December.

Longhurst, A.R. and D. Pauly. 1987. Ecology of Tropical Oceans. Academic Press, New York.

Longhurst, A.R. and D.L.R. Siebert. 1967. Skill in the use of Folson's plankton sample splitter. Limnol. Oceanogr. 12: 334.

Looney, S.W. (ed.). 2002. Biostatistical Methods. Humana Press, Totowa, NJ.

Love, M. (1996). Probably More than You Want to Know About the Fishes of the Pacific Coast. Really Big Press, Santa Barbara, CA.

Loya, Y. 1978. Plotless and transect methods. pp. 197–218 in: Stoddart D.R. and R.E. Johannes (eds.). Coral Reefs: research methods. Unesco, Paris.

Ludwig, J.A. and J.F. Reynolds. 1988. Statistical Ecology: A primer on methods and computing. John Wiley, New York.

Lunetta, R.S. and C.D. Elvidge (eds.). 2002. Remote Sensing Change Detection: Environmental Monitoring Methods and

Applications. Ann Arbor Press, Ann Arbor.

MacArthur, R.M. 1955. Fluctuations of animal populations and a measure of community stability. Ecology 36:533–536.

MacKenzie, D.I., Nichols, J.D., Royle, A.J., Pollock, K.H., Bailey, L.L. and J.E. Hines. 2005. Occupancy Estimation and Modeling: Inferring Patterns and Dynamics of Species Occurrence. Academic Press, New York.

Magnell, B., Vu, M.C. and V. Nordahl Jr. 2005. NOAA Fisheries Service Survey Sensor Package. Sea Technol. 46(9):42–45. September.

Magnusson, W.E. and G. Mourão. 2004. Statistics without Math. Sinauer Associates, Sunderland, MA.

Magurran, A.E. 1988. Ecological Diversity and its Measurement. Princeton University Press, Princeton.

Magurran, A.E. 2003. Measuring Biological Diversity. Blackwell Publishing, Oxford U K.

Magurran, A.E. and R.M. May. 1999. Evolution of Biological Diversity. Oxford University Press, Oxford.

Mahloch, J.L. 1974. Multivariate techniques for water quality analysis. J. Envir. Engin. Div. EE5, Oct. 1974. pp. 1119–1132.

Mandelbrot, B.B. 1982. The Fractal Geometry of Nature. W.H. Freeman, New York.

Manly, B.F.G. and M.J. Parr. 1968. A new method of estimating population size, survivorship, and birth-rate from capture-recapture data. Trans. Soc. Brit. Ent.18: 81–89.

Manly, B.F.J. 1986. Multivariate Statistical Methods—A Primer. Chapman & Hall, London.

Manly, B.F.J., McDonald, L.L., Thomas, D.L., McDonald, T.L. and W.P. Erickson. 2002. Resource Selection by Animals: Statistical Design and Analysis for Field Studies. Kluwer Academic Publishers, Dordrecht, Netherlands.

Manly, B.F.G. 1992. Bootstrapping for determining sample sizes in biological studies. J. Exp. Mar. Biol. Ecol. 158(2):189–196.

Manly, B.F.G. 1997. Randomization, Bootstrap and Monte Carlo Methods in Biology. Chapman and Hall, London.

Mann, K.H. 2000. Ecology of Coastal Waters: With Implications for Management. Blackwell Publishing, Malden, MA.

Mann, K.H. and J.R. Lazier 2005. Dynamics of Marine Ecosystems: Biological-Physical Interactions in the Oceans. Blackwell Publishing, Malden, MA.

Mantel, N. 1967. The detection of disease clustering and generalized regression approach. Cancer Res. 27:209–220.

Mantel, N. and R.S. Valand. 1970. A technique of nonparametric multivariate analysis. Biometrics 26:547–558.

Margalef, R. 1958. Temporal succession and spatial heterogeneity in phytoplankton. pp. 323–349. In: Buzzati-Traverso, U., and U. Calif (eds.). Perspectives in Marine Biology. Press, Berkeley.

Margalef, R. 1972. Homage to Evelyn Hutchinson, or why there is an upper limit to diversity. Trans. Conn. Acad. Arts & Sci. 44:213–235.

Marsh, L.M., Bradbury, R.H. and R.E. Reichelt. 1984. Determination of the physical parameters of coral distributions using line transect data. Coral Reefs 2:175–180.

May, R.M. 1974. Biological populations with nonoverlapping generations: stable points, stable cycles and chaos. Science 186:645–647.

May, R.M. 1986. Species interactions in ecology. Science 231:1451–1452.

May, R.M. 2001. Stability and Complexity in Model Ecosystems. Princeton University Press, Ewing, NJ.

Mayer, L., Li, Y. and G. Melvin. 2002. 3D visualization for pelgic fisheries research and assessment. ICES J. Mar. Sci. 59:216–225.

McConnell, A. 1982. No Sea Too Deep: The History of Oceanographic instruments. Hilger, Bristol.

McCormick, M.I. and J.H. Choat. 1987. Estimating total abundance of a large temperate-reef fish using visual strip-transects. Mar. Biol. 96:469–478.

McCune, B. and M.J. Mefford. 1999. PC-ORD. Multivariate Analysis of Ecological Data, Version 4. MjM Software Design, Gleneden Beach, Oregon, USA.

McCune, B., Grace J.B. and D.L. Urban. 2002. Analysis of Ecological Communities. MjM Software Design, Gleneden Beach, Oregon.

McDonald, L.L. 2004. Sampling Rare Populations. pp. 11–42 in: Thompson W.L. (ed.). Sampling Rare or Elusive Species. Island Press, Washington, D.C.

McGarigal, K., Cushman, S. and S. Stafford. 2000. Multivariate Statistics for Wildlife and Ecology Research. Springer-Verlag, New York.

McKellar, H.N., Jr. 1977. Metabolism and model of an estuarine bay ecosystem affected by a coastal power plant. Ecol. Model. 3:85–118.

Mead, R. 1988. The Design of Experiments: statistical principles for practical applications. Cambridge University Press, New York.

Mendelssohn, R. 1981. Using Box-Jenkins models to forecast fishery dynamics: identification, estimation, and checking. Fish Bull. 78(4):887–896.

Mertes, L.A.K., Hickman, M., Waltenberger, B., Bortman, A.L., Inlander, E., McKenzie, C. and J. Dvorsky. 1998. Synoptic views of sediment plumes and coastal geography of the Santa Barbara Channel, California. Hydrol. Process. 12:967–979.

Mertes, L.A.K. and J.A. Warrick. 2001. Measuring flood output from 110 coastal watersheds in California with field measurements and SeaWiFS. Geology 29(7):659–662.

Mielke, P.W. Jr., Berry, K.J. and E.S. Johnson. 1976. Multi-response permutation procedures for a priori classifications. Commun. Stat. – Theory and Methods 5(14):1409–1424.

Miller, J.E. 2005. The Chicago Guide to Writing about Multivariate Analysis. University of Chicago Press, Chicago, IL.

Millero, F.J. 2006. Chemical Oceanography. CRC Press, Boca Raton, FL.

Milligan, B.G. 2003. Maximum-likelihood estimation of relatedness. Genetics 163(3):1153–1167.

Millington, A.C., Walsh, S.J. and P.E. Osborne. 2001. GIS and Remote Sensing Applications in Biogeography and Ecology. Kluwer, Accord Station, Hingham, MA.

Milne, B.T. 1991. Lessons from applying fractal models to landscape patterns. pp. 199–235 in: Turner M.G. and R.H. Gardner (eds.). Quantitative Methods in Landscape Ecology. Springer-Verlag, New York.

Minchin, P.R. 1987. Simulation of multidimensional community patterns: towards a comprehensive model. Vegetatio 71:145–156.

Mitchell, K. 2001. Quantitative analysis by the Point-centered Quarter method. Internet. Contact: Mitchell@hws.edu

Mitchell, R.J. 2001. Path analysis: Pollination. pp. 217–234 in: Scheiner S.M. and J. Gurevitch (eds.). Design and Analysis of Ecological Experiments. Oxford University Press, Oxford.

Moriarty, F. 1999. Ecotoxicology: The Study of Pollutants in Ecosystems. Academic Press, New York.

Morris, T.R. 1999. Experimental Design and Analysis in Animal Sciences. Oxford University Press, New York.

Morrison, D.F. 1976. Multivariate Statistical Methods. McGraw-Hill, New York.

Mosteller, F., W.H. Kruskal, R.F. Link, R.S. Pieters, and G.R. Rising (eds.) 1973. Statistics by Example: Weighing Chance. Addison-Wesley, Reading.

Motyka, J., Dobrzanski, B. and S. Zawazki. 1950. Wstspne badania had lagami poludnio-wowschodniez Lubelszczyzny. (Preliminary studies on meadows in the southeast of the province Lublin.) Univ. Mariae Curie-Sklodowska Ann. Sect. E. 5:367–447.

Mudroch, A. and S.D. MacKnight. 1994. CRC Handbook of Techniques for Aquatic Sediments Sampling. Boca Raton, FL.

Mueller, W., Smith, D.L. and L.H. Keith. 1991. Compilation of EPA's Sampling and Analysis Methods. Lewis Publications, Boca Raton.

Mueksch, M.C. 2001. Full-frame video multispectral airborne remote sensing. Sea Technol. February. pp. 21–27.

Mumby, P.J. 2002. Statistical power of non-parametric tests: A quick guide for designing sampling strategies. Mar. Pollut. Bull. 44:85–87.

Mumby, P.J., Green, E.P., Edwards, A.J., and C.D. Clark. 1999. The cost-effectiveness of remote sensing for tropical coastal resources assessment and management. J. Environ. Manag. 55:157–166.

Munro, C. 2005. Diving Systems. pp. 112–159 in: Eleftheriou. A. and A. McIntyre (eds.). 2005. Methods for Study of Marine Benthos. Blackwell Science Ltd., Oxford, UK.

Murray, D.L. and M.R. Fuller. 2000. A critical review of the effects of marking on the biology of vertebrates. pp. 15–64 in: Boitani, L. and T.K. Fuller (eds.). 2000. Research Techniques in Animal Ecology. Columbia University Press, New York.

Murray, S.N., Ambrose, R. F., and M.N. Dethier. 2006. Monitoring Rocky Shores. University of California Press, Berkeley.

Neal, D. 2004. Introduction to Population Biology. Cambridge University Press, New York.

Newell, S.Y. 2001. Fungal biomass and productivity. pp. 357–374, in: Paul J.H. (ed.). Methods in Microbiology. V. 30. Marine Microbiology. Academic Press, New York.

Newman, M.C. 2002. Fundamentals of Ecotoxicology. CRC Press, Boca Raton, FL.

Newman, M.C. and D.C. Adriano. 1995. Quantitative Methods in Aquatic Ecotoxicology. CRC Press, Boca Raton.

Nie, U.N., Hull, C.U., Jenkins, J.G., Steinbrenner, K. and D.H. Bent. 1975. SPSS-Statistical Package for the Social Sciences. McGraw-Hill, New York.

Nielsen, A. and P. Lewy. 2002. Comparison of the frequentist properties of Bayes and the maximum likelihood estimators in an age-structured fish stock. Canadian J. Fish. Aquat. Sci. 59(1):136–143.

Nihoul,. J.C.J. (ed.) 1975. Modelling of Marine Systems. Elsevier Sci. Pubi. Co., New York, p. 272.

Nimis, P.L. and T.J. Crovello. 1991. Quantitative Approaches to Phytogeography (Tasks for Vegetation Science No. 24). Kluwer Academic Publishers, Boston.

Nisbet, R.M. and W.S.C. Gurney. 1982. Modelling Fluctuating Populations. Wiley, p. 380.

Nishiyama, G.K. 1999. Determination of Densities, Species Richness, and Dispersions of Macroinvertebrates in the Rocky Intertidal Habitat Using Line Transect Methods. Master's Thesis, Department of Biological Sciences, University of Southern California, Los Angeles, California. December.

Nishiyama, G.K., Bakus G.J. and C. Kay. 2004. The effect of sponge allelochemicals on the settlement, growth and associations of substratum competitors. Proceedings Fifth

International Boll. Mus. Ist. Biol. Univ. Genova 68: 499–508.

Noble, R.T. 2001. Enumeration of viruses. pp. 43–52 in: Paul J.H. (ed.). Methods in Microbiology. V. 30. Marine Microbiology. Academic Press, New York.

Noble, R.T. and J.A. Fuhrman. 1998. Use of SYBR Green I for rapid epifluorescence counts of marine viruses and bacteria. Aquat. Microbiol. Ecol. 14:113–118.

Normile, D. 2000. Reef migrations, bleaching effects stir the air in Bali. Science 290:1281–1283.

Norse, E.A. and L.B. Crowder (eds.). 2005. Marine Conservation Biology. Island Press, Washington, D.C.

Noy-Meir, I. 1970. Component analysis of semi-arid vegetation in southeastern Australia. Ph.D. Thesis, Australian National University, Canberra, Australia.

Noy-Meir, I. 1971. Multivariate analysis of semi-arid vegetation in southeastern Australia. Nodal ordination by component analysis. Quantifying Ecol. Proc. Ecol. Soc. Aust. 6:159–193.

Noy-Meir, I. 1973. Data transformations in ecological ordination. I. Some advantages of non-centering. J. Ecol. 61:329–341.

Ochavillo, D.G. 2002. The Dispersal of Tropical Coral Reef Fish Larvae: Genetic Markers, Planktonic Duration and Behavior. Ph.D. Dissertation. Department of Biological Sciences, University of Southern California, Los Angeles. June.

Odeh, R.E. 1991. Sample Size Choice: charts for experiments with linear models. CRC Press, Boca Raton, FL.

Odum, H.T. 1982. Systems Ecology. Wiley, New York, p. 592.

Odum, H.T. 1994. Ecological and General Systems: An Introduction to Systems Ecology. University Press of Colorado, Niwot, CO.

Odum, E.P. and G.W. Barrett. 2005. Fundamentals of Ecology. Thomas Brooks/Cole, Belmont, CA.

Odum, H.T. and E.C. Odum. 2000. Modeling for all Sciences: An Introduction to System Simulation. Elsevier Academic Press, St. Louis.

Okamoto, A., Edwards, D.B. and M.J. Anderson. 2005. Robust control of a platoon of underwater autonomous vehicles. Sea Technol. 46(12):10–13.

Oksanen, J. 1997. Plant neighbour diversity. J. Veg. Sci. 8:255–258.

Oksanen, L. 2001. Logic of experiments in ecology: Is pseudoreplication a pseudoissue? Oikos 94(1):27–38.

Omori, M and T. Ikeda. 1984. Methods in Marine Zooplankton Ecology. John Wiley and Sons, New York.

OPL Staff (Oilfield Publications Ltd. Staff, eds). 2002. Remotely Operated Vehicles of the World – Fifth Edition. Oilfield Publications, Inc., Ledburg, Hereford and Worcester, U.K.

Oquist, G., Hagstrom, A., Alm, P., Samuelsson, G. and K. Richardson. 1982. Chlorophyll a fluorescence: An alternative method for estimating primary production. Mar. Biol. (Berlin) 68:7 1–76.

Orloci, L. 1967. Data centering: A review and evaluation with reference to component analysis. Syst. Zoolog. 16:208–212.

Orloci, L. 1975. Multivariate Analysis in Vegetation Research. Junk, The Hague.

Ornduff, R., Faber, P.M. and T.Keeler-Wolf. 2003. Introduction to California Plant Life. University of California Press, Berkeley.

Otnes, R. and L. Enochson. 1978. Applied Time Series Analysis. Vol. 1. Basic Techniques. Wiley, New York.

Ott, l.A. and A. Losert. 1979. A new quantitative sampler for submerged macrophytes, especially seagrass. Senckenbergiana Maritima 11:39–45.

Page, R.D.M. and E.C. Holmes. 1998. Molecular Evolution: A phylogenetic approach. Blackwell Science, Oxford.

Palmer, M.W. 1988. Fractal geometry: a tool for describing spatial patterns of plant communities. Vegetatio 75:91–102.

Parkratz, A. 1983. Forecasting with Univariate Box-Jenkins Models. Wiley, New York, p. 562.

Parsons, D.G. and E.B Colbourne. 2000. Forecasting fishery performance for northern shrimp (*Pandalus borealis*) on the Labrador Shelf (NAFO Divisions 2HJ). J. Northw. Atl. Fish. Sci. 27:11–20.

Paul, J.H. (ed.). 2001. Methods in Microbiology. V. 30. Marine Microbiology. Academic Press, New York.

Paukert, C.P. and T.A. Wittig. 2002. Applications of multivariate statistical methods in fisheries. Fisheries 27(9): 16–22.

Peet, R.K. 1974. The measurement of species diversity. Ann. Rev. Ecol. Syst. 5:285–307.

Peet, R.K., Knox, R.G., Case J.S. and R.B. Allen. 1988. Putting things in order: the advantages of detrended correspondence analysis. Amer. Nat. 131:924–934.

Peters, R.H. 1986. The role of prediction in limnology. Limnol. Oceanog. 31(5):1143–1159.

Peterson, C.H., MacDonald, L.L., Green, R.H., and W.P. Erickson. 2001. Sampling design begets conclusions: The statistical basis for detection of injury to and recovery of shoreline communities after the 'Exxon Valdez' oil spill. Mar. Ecol. Prog. Ser. 210:255–283.

Peterson, C.H., MacDonald, L.L., Green, R.H., and W.P. Erickson. 2002. The joint consequences of multiple components of statistical sampling designs. Mar. Ecol. Prog. Ser. 231:309–314.

Peterson, D.L. and V.T. Parker (eds.). 1998. Ecological Scale: Theory and Applications. Columbia University Press, New York.

Petratis, P.S., Beaupre, S.J., and A.E. Dunham. 2001. ANCOVA: Nonparametric and Randomization Approaches. pp. 116–133 in: Scheiner S.M. and Gurevitch (eds.). Design and Analysis of Ecological Experiments. Oxford University Press, Oxford.

Phalen, C. M. (1999). Genetic variation among kelp bass (*Paralabrax clathratus*) from seven locations throughout their natural range. Masters Thesis, Department of Biology. California State University, Northridge, CA.

Phillips, R.C. and C.P. McRay (eds.) 1980. Handbook of Seagrass Biology: An Ecosystem Perspective. Garland STPM Press, New York, p. 353.

Pielou, E.C. 1984. The interpretation of Ecological Data. John Wiley & Sons, New York, p. 263.

Platt, T. and K.L. Denman. 1975. Spectral analysis in ecology. Ann. Rev. Ecol. Sys. 6:189–210.

Podani, J. 1994. Multivariate Analysis in Ecology and Systematics. SPB Publishing, The Hague, Netherlands.

Podani, J., Czaran, T. and S. Bartha. 1993. Pattern, area, and diversity: the importance of spatial scale in species assemblages. Abst. Bot. 17:37–52.

Pollard, J.H. 1971. On distance estimators of density in randomly distributed forests. Biometrics 27:991–1002.

Poole, R.W. 1974. An Introduction to Quantitative Ecology. McGraw-Hill, New York.

Potvin, C. 2001. ANOVA: Experimental layout and analysis. pp. 63–76 in: Scheiner S.M. and J. Gurevitch (eds.). Design and Analysis of Ecological Experiments. Oxford University Press, Oxford.

Powell, R.A. 2000. Animal home ranges and territories and home range estimators. pp. 65–110 in: Boitani, L. and T.K. Fuller (eds.). 2000. Research

Techniques in Animal Ecology. Columbia University Press, New York.

Powell, T.M. and J.H. Steele (eds.). 1994. Ecological Time Series. Springer-Verlag, New York.

Power, M.E., Tilman, D., Estes, J.A., Menge, B.A., Bond, W.J., Mills, L.S., Daiily, G., Castilla, J.C., Lubchenko, J. and R.T. Paine. 1996. Challenges in the quest for keystones. BioScience 46:609–620.

Pringle, J.D. 1984. Efficiency estimates for various quadrat sizes used in benthic sampling. Can. J. Fish. Aquat. Sci. 41: 1485–1489.

Pugesek, B, Tomer, A. and A. von Eye (eds.). 2002. Structural Equations Modeling: Applications in Ecological and Evolutionary Biology Research. Cambridge University Press, Cambridge, U.K.

Quinn, G.P. and M.J. Keough. 2002. Experimental Design and Data Analysis for Biologists. Cambridge University Press, New York.

Quinn, T.J. and R.B. Deriso. 1999. Quantitative Fish Dynamics. Oxford University Press, New York.

Raftery, A.E. and J.E. Zeh. 1993. Estimation of Bowhead Whale, *Balaena mysticetus*, population size. pp. 163–240 in: Gatsonis, C., Hodges, J., Kass, R. and N. Singpurwalla (eds.). V. I Case Studies in Bayesian Statistics. Springer-Verlag, New York.

Rasmussen, P.W., Heisey, D.H., Nordheim, E.V., and T.M. Frost. 2001. Time series intervention analysis: Unreplicated large-scale experiments. pp. 158–177 in: Scheiner S.M. and J. Gurevitch (eds.). Design and Analysis of Ecological Experiments. Oxford University Press, Oxford.

Reeb, C.A., Arcangeli, L. and B.A. Block. 2000. Structure and migration corridors in Pacific populations of the swordfish *Xiphius gladius*, as inferred through analyses of mitochondrial DNA. Mar. Biol. 136:1123–1131.

Reiners, W.A. and K.L. Driese. 2004. Propagation of Ecological Influences through Environmental Space. Cambridge University Press, New York.

Resetarits, Jr., W.J. (ed.). 2001. Experimental Ecology: Issues and Perspectives. Oxford University Press, New York.

Ricker, W.E. 1958. Handbook of computation for biological statistics of fish populations. Publ. Fish. Res. Bd. Canada, Bull. No. 119.

Rickleffs, R.E. and G. Miller. 2000. Ecology. Freeman, New York.

Riley, G.A. 1946. Factors controlling phytoplankton populations on Georges Bank. J. Mar. Res. 6:54–73.

Risk, M. 1971. Intertidal Substrate Rugosity and Species Diversity. Ph.D. Dissertation, Department of Biological Sciences, University of Southern California, Los Angeles. 70 p. October.

Rohlf, F.J. 1995. BIOM. A Package of Statistical Programs to Accompany the Text Biometry. Exeter Software, Setauket, New York.

Rosenzweig, M.L., Turner, W.R., Cox, J.G. and T.H. Ricketts. 2003. Estimating diversity in unsampled habitats of a biogeographical province. Conservation Biol. 17(3):864–874.

Roughgarden, J. 1998. Primer of Ecological Theory. Prentice-Hall, Upper Saddle River, NJ.

Russ, J.C. 1998. The Image Processing Handbook. CRC Press, Boca Raton.

Ruth, M. and J. Lindholm (eds.). 2002. Dynamic Modeling for Marine Conservation. Springer, New York.

Ruxton, G.D. and N. Colegrave. 2003. Experimental Design for the Life Sciences. Oxford University Press, New York.

Safer, A. 2001. Spatial visualization of the marine environment. Sea Technol. 42(2):48–54.

Saila, S.B., Pikanowski, R.A. and D.S. Vaughan. 1976. Optimum allocation strategies for sampling benthos in New York Bight. Estuarine Coastal Mar. Sci. 4:119–128.

Saila, S.B., Wigbout, M. and RJ. Lermit. 1980. Comparison of some time series models for the analysis of fisheries data. J. Cons. int. explor. Mer. 39(1):44–52.

Sala, O.E., Jackson, R.B., Mooney, H.A. and R.W. Howarth (eds.). 2000. Methods in Ecosystem Science. Springer, New York.

Sanders, H.L. 1968. Marine benthic diversity: a comparative study. Am. Nat. 102:243–282.

Schaffer, W.M. and M. Kot. 1986. Chaos in ecological systems–the coals that Newcastle forgot. Trends Ecol. Evol. 1: 58–63.

Schaffer, W.M. and G.L. Truty. 1988. Chaos in the classroom: nonlinear dynamics and dynamical software. Academic Computing. March-April. pp. 59–63.

Scheiner, S.M. 2001. Theories, hypotheses, and statistics. pp. 3–13 in: Scheiner S.M. and J. Gurevitch (eds.). Design and Analysis of Ecological Experiments. Oxford University Press, Oxford.

Scheiner, S.M. and J. Gurevitch. (eds). 2001. Design and Analysis of Ecological Experiments. Oxford University Press, New York.

Schefler, W.C. 1969. Statistics for the Biological Sciences. Addison-Wesley, Menlo Park.

Schmitt, R.J. and C.W. Osenberg. 1996. Detecting Ecological Impacts: Concepts and Applications in Coastal Habitat. Academic Press, New York.

Schneider, D.C. 1994. Quantitative Ecology: Spatial and Temporal Scaling. Academic Press, New York.

Schram, T.A., Svelle, M. and M. Opsahl. 1981. A new divided neuston sampler in 2 modifications: description, tests, and biological results. Sarsia 66:273–282.

Schulte, B.A., de Nys, R., Bakus, G.J., Crews, P., Eid, C., Naylor, S. and L.V. Manes. 1991. A modified allomone collecting apparatus. J. Chem. Ecol. 17:1327–1332.

Scott, J.M., Heglund, P.J., Morrison, M.I., Haufler, J.B., Raphael, M.G., Wall, W.A. and F. B. Samson (eds.). 2002. Predicting Species Occurrences: Issues of Accuracy and Scale. Island Press, St. Louis.

Sea Technology. 2001. Compass Publications Inc. Arlington, VA. February issue. Vol. 42 (2).

Sea Technology. 2006. High-Tech Tags on Marine Animals Yield Valuable Data. Oceanresearch. Vol. 47(3):56. March.

Seal, H. 1964. Multivariate Statistical Analysis for Biologists. Methuen, London.

Seaman Jr., W. 2000. Artificial Reef Evaluation with Application to Natural Marine Habitats. CRC Press, Boca Raton, FL.

Seaman Jr., W. and L.M. Sprague (eds.). 1991. Artificial Habitats for Marine and Freshwater Fisheries. Academic Press, San Diego.

Seber, G.A.F. 1982. The Estimation of Animal Abundance. Charles Griffin and Company, London.

Segal, S. and N.J. Castellan, Jr. 1988. Nonparametric Statistics for the Behavioral Sciences. McGraw-Hill, New York.

Seuront, L. and P.G. Strutton (eds.). 2003. Handbook of Scaling Methods in Aquatic Ecology. CRC Press, Boca Raton, FL.

Shenk, T.M. and A.B. Franklin. 2001. Modeling in Natural Resource Management. Island Press, Covelo, CA.

Sheppard, C. 2000. Seas at the Millenium: An Environmental Evaluation. Elsevier Academic Press, St. Louis.

Sheppard, C.R.C. 1999. How large should my sample be? Some quick guides to sample size and the power of tests. Mar. Poll. Bull. 38:439–447.

Shin, P.K.S. 1982. Multiple discriminant analysis of macrobenthic infaunal assemblages. J. Exp. Mar. Biol. Ecol. 59: 39–50.

Shipley, B. 2000. Cause and Correlation in Biology. Cambridge University Press, Cambridge, U.K.

Short, F.T. and R.G. Coles (eds.). 2001. Global Seagrass Research Methods. Elsevier Academic Press, St. Louis.

Shugart, H.H. (ed.). 1978. Time Series and Ecological Processes. Soc. for Indust. & Appl. Math., Philadelphia.

Shugart, H.H. 2000. Ecosystem modeling. pp. 373–386 in: Sala., O.E, Jackson, R.B., Mooney, H.A. and R.W. Howarth (eds.). 2000. Methods in Ecosystem Science. Springer, New York.

Sibly, R., Hone, J. and T. Clutton-Brock. 2003. Wildlife Population Growth Rates. Cambridge University Press, New York.

Simonson, M.S. 1986. The software void: Creating custom software. Bioscience 36(5):330–335.

Skalski, J.R. 1987. Selecting a random sample of points in circular field plots. Ecology 68(3):749.

Skalski, J.R., Ryding, K.E. and J.J. Millspaugh. 2005. Wildlife Demography: Analysis of Sex, Age, and Count Data. Academic Press, New York.

Smith, B. and J.B. Wilson. 1996. A consumer's guide to evenness indices. Oikos 76:70–82.

Smith, C.J. and H. Rumohr. 2005. Imaging Techniques. pp. 87–111 in: Eleftheriou. A. and A. McIntyre (eds.). 2005. Methods for Study of Marine Benthos. Blackwell Science Ltd., Oxford, UK.

Smith, E.P. 1982. Niche breadth, resource availability, and inference. Ecology 63(6): 1675–1681.

Smith, G. 1988. Statistical Reasoning. Allyn and Bacon, Inc., Boston.

Smith, M. 1982. Evolution and the Theory of Games. Cambridge University Press, Cambridge.

Smith, R.W. 1976. Numerical Analysis of Ecological Survey Data. Ph.D. Dissertation, Dept. of Biological Sciences, University of Southern California, Los Angeles.

Smith, R.W. 1981. The Re-estimation of Ecological Distance Values Using the Step-across Procedure. EAP Ecological Analysis Package. Tech. Rept. No. 2 Ecological Data Analysis, Ojai, CA.

Smith, R.W. and B.B. Bernstein. 1985. A Multivariate Index of Benthic Degradation. Prepared under contract to Brookhaven National Laboratory for NOAA, Ocean Assessments Division, Stony Brook, New York.

Snow, D. 2005. Oceans: An Illustrated Reference. University of Chicago Press, Chicago, IL.

Sobel, J. and C. Dahlgren. 2004. Marine Reserves: A Guide to Science, Design, and Use. Island Press, Covelo, CA.

Soetaert, K. and C. Heip. 2003. Sample-size dependence of diversity studies and the determination of sufficient sample size in a high-diversity deep-sea environment. Mar. Ecol. Prog. Ser. 59(3):305–307.

Sokal, R.P. and C.D. Michener. 1958. A statistical method for evaluating systematic relationships. Univ. Kans. Sci. Bull. 38:1409–1438.

Sokal, R.R. and F.J. Rohlf. 1981. Biometry. W.H. Freeman, San Francisco.

Sokal, R.R. and F.J. Rohlf. 1995. Biometry. W.H. Freeman, San Francisco.

Somerfield, P.J., Warwick, R.M. and T. Moens. 2005. Meiofauna Techniques. pp. 160–228 in: Eleftheriou. A. and A. McIntyre (eds.). Methods for Study of Marine Benthos. Blackwell Science Ltd., Oxford, UK.

Song, G-S. 2005. Water dams scanned by side scan sonar. Sea Technol. 46(12): 57–60.

Sorensen, I. 1948. A method of establishing groups of equal amplitude in plant sociology based on similarity of species content and its application to analyses of the vegetation on Danish commons. Biol. Skr. 5:1–34.

Sornette, D. 2004. Critical Phenomena in Natural Sciences: Chaos, Fractals, Selforganization and Disorder: Concepts and Tools. Springer, New York.

Southwood, R. and P.A. Henderson. 2000. Ecological Methods. Blackwell Publishing, Williston, VT.

Southwood, T.R.E. 1978. Ecological Methods. Chapman and Hall, New York.

Spain. J.D. 1982. Basic Microcomputer Models in Biology. Addison-Wesley, Reading, MA.

Spalding, M.D., Ravilious, C. and E.P. Green. 2001. World Atlas of Coral Reefs. University of California Press, Berkeley.

Späth, H. 1985. Cluster Analysis Algorithms for Data Reduction and Classification of Objects. Halsted-Wiley, New York.

Spellerberg, I.F. 2005. Monitoring Ecological Change. Cambridge University Press, New York.

Stanciulescu, F. 2005. Modelling of High Complexity Systems with Applications. WIT Press, Billerica, ME.

Statsoft, Inc. 1995. STATISTICA for Windows. 2nd edition. Tulsa, OK.

Stebbins, T.D. 1988. Ecology of the commensal isopod Colidotea rostrata (Benedict, 1898) in southern California. Ph.D. Dissertation, Department of Biological Sciences, University of Southern California, May.

Steele, J.H., Thorpe, S.A. and K.A. Turekian. 2001. Encyclopedia of Ocean Sciences. Academic Press, New York.

Steele, M.A. 1996. Effects of predators on reef fishes: separating cage artifacts from effects of predation. J. Exp. Mar. Biol. Ecol. 198:249–267.

Steidl, R.J. and L. Thomas. 2001. Power analysis and experimental design. pp. 14–36 in: Scheiner S.M. and J. Gurevitch (eds.). Design and Analysis of Ecological Experiments. Oxford University Press, Oxford.

Stein, A. and C. Ettema. 2003. An overview of spatial sampling procedures and experimental design of spatial studies for ecosystem comparisons. Agricultural Ecosystems Environ. 94(1):31–47.

Stephenson, T.A. and A. Stephenson. 1977. Life between Tidemarks on Rocky Shores. W.H. Freeman & Co., San Francisco.

Stephenson, W. 1973. The validity of the community concept in marine biology. Proc. Roy. Soc. Qsld. 84:73–86.

Stephenson, W. 1978. Analysis of periodicity in macrobenthos using constructed and real data. Aust. J. Ecol. 3:321–336.

Stepien, C. A. and T. D. Kocher (1997). Molecules and morphology, in Studies of Fish Evolution. Molecular Systematics of Fishes. Kocher, T.D. and C.A. Stepien (eds.). pp. 1–11. Academic Press, Inc., San Diego, CA.

Stewart-Oaten, A. and J.R. Bence. 2001. Temporal and spatial variation in environmental impact assessments. Ecol. Monogr. 71(2):305–339.

Stiling, P. 2002. Ecology: Theories and Applications. Prentice Hall, Upper Saddle River, N.J.

Stoddart, D. and R.E. Johannes. 1978. Coral Reefs: Research Methods. UNESCO, Paris.

Stow, D. 2006. Oceans: An Illustrated Reference. Univ. Chicago Press, Chicago.

Strauss, R.E. 1979. Reliability estimates for Ivlev's electivity index, the forage ratio, and a proposed linear index of food selection. Trans. Amer. Fish. Soc.108: 344–352.

Strickland, J.D.H. 1966. Measuring the Production of Marine Phytoplankton. Bulletin No. 122. Fisheries Research Board of Canada, Ottawa.

Strong, C.W. 1966. An improved method of obtaining density from line transect data. Ecology 47:311–313.

Sugihara, G. & R.M. May. 1990. Applications of fractals in ecology. Tree 5(3):79–86.

Sund, P.N., Blackburn, M., and F. Williams. 1981. Tunas and their environment in the Pacific Ocean: a review. Ocean. Mar. Biol. Ann. Rev. 19:443–512.

Suter, G.W. II (ed.). 1993. Ecological Risk Assessment. Lewis Publishers, Boca Raton, FL.

Sutherland, W.J. (ed.) 1996. Ecological Census Techniques: A Handbook. Cambridge University Press, New York.

Sverdrup, H.U., Johnson, M.W. and R.H. Fleming. 1942. The Oceans: Their Physics, Chemistry, and General Biology. Prentice-Hall Inc., Englewood Cliffs, N.J.

Swan, J.M.A. 1970. An examination of some ordination problems by use of simulated vegetational data. Ecology 51: 89–102.

Tabor, P.S., Deming, J.W., Ohwada, K., Davis, H., Waxman, M. and R.R. Colwell. 1981. A pressure-retaining deep ocean sampler and transfer system for measurement of microbial activity in the deep sea. Microb. Ecol. 7:51–66.

Tate, M.W. and R.C. Clelland. 1959. Non-parametric and Shortcut Statistics. Interstate, Danville.

Taylor, L.R. 1961. Aggregation, variance and the mean. Nature 189(4766): 732–735.

Taylor, L.R. 1971. Aggregation as a species characteristic. In: Statistical Ecology, Patel, G.P., Pielou, E.C. and W.E. Waters (eds.) Vol. I. Spatial Patterns and Statistical Distributors, Penn. State Univ., pp. 356–377.

Tengberg, A., Hovdenes, J., Diaz, R., Sarkkula, J., Huber, C., and A. Stangelmayer. 2003. Optodes to measure oxygen in the aquatic environment. Sea Technol. 44(2):10–15.

ter Braak, C.J.F. 1986. Canonical correspondence analysis: A new eigenvector technique for multivariate direct gradient analysis. Ecology 67:1167–1179.

ter Braak, C.J.F. 1987. The analysis of vegetation-environment relationships by canonical correspondence analysis. Vegetatio 69:69–77.

Thieme, H.R. 2003. Mathematics in Population Biology. Princeton University Press, Ewing NJ.

Thompson, S.K. 1992. Sampling. John Wiley and Sons, New York. First Edition.

Thompson, S.K. 2002. Sampling. John Wiley & Sons, New York. Second Edition.

Thompson, W.L. (ed.). 2004. Sampling Rare or Elusive Species: Concepts, Designs, and Techniques for Estimating Population Parameters. Island Press, Covelo, CA.

Thomson, J.M. 1962. The tagging and marking of marine animals in Australia. Australian Commission of Science and Industry, Research Organization, Division of Fish and Oceanography Technical Paper 13:3–39.

Tilman, D. and P. Kareiva. 1997. Spatial Ecology: The Role of Space in Population Dynamics and Interspecific Interactions. Princeton University Press, Princeton, NJ.

Tkachenko, K.S. 2005. An evaluation of the analysis system of video transects used to sample subtidal biota. J. Exp. Mar. Biol. Ecol. 318:1–9.

Tomaszeski, S.J. 2004. Naval transformation and the oceanography community. Sea Technol. 45(1):10–13.

Townend, J. 2002. Practical Statistics for Environmental and Biological Sciences. Wiley, New York.

Trochim, W.M.K. 2002. General Linear Model. Research Methods Knowledge Base. Internet.

Tukey, J.W. 1977. Exploratory Data Analysis. Addison-Wesley, Reading, Massachusetts.

Turchin, P. 2003. Complex Population Dynamics: A theoretical/empirical synthesis. Princeton University Press, Ewing, NJ.

Turner, M.G. and R.H. Gardner. 1991. Quantitative Methods in Landscape Ecology: The Analysis and Interpretation of Landscape Heterogeneity. Springer-Verlag, New York.

Turner, M.G., Gardner, R.H. and R.V. O'Neill. 2001. Landscape Ecology in Theory and Practice: Pattern and Process. Springer-Verlag, New York.

Ulanowiez, R.E. 2005. The Complex Nature of Ecodynamics. pp. 303–329 in: Bonchev, D. and D.H. Rouvray (eds.). Complexity in Chemistry, Biology, and Ecology. Springer, New York.

Underwood, A.J. 1976. Nearest neighbor analysis of spatial dispersion of intertidal prosobranch gastropods within two substrata. Oecologia (Ben.) 26: 257–266.

Underwood, A.J. 1986. What is a community? pp. 351–367 in: Raup, D.B. and D. Jablonski (eds.). Patterns and Processes in the History of Life. Springer-Verlag, Berlin.

Underwood, A.J. 1997. Experiments in Ecology, their Logical Design and Interpretation using Analysis of Variance. Cambridge University Press, Cambridge.

Underwood, A.J. and M.G. Chapman. 1985. Multifactorial analysis of directions of movement of animals. J. Exp. Mar. Biol. Ecol. 91:17–44.

Underwood, A.J. and M.G. Chapman. 2003. Power, precaution, Type II error and sampling design in assessment of environmental impacts. J. Exper. Biol. Ecol. 296(1):49–90.

Underwood, A.J. and M.G. Chapman. 2005. Design Analysis in Benthic Surveys. pp. 1–32 in: Eleftheriou. A. and A. McIntyre (eds.). Methods for Study of Marine Benthos. Blackwell Science Ltd., Oxford, UK.

UNESCO. 1984. Comparing Coral Reef Survey Methods. Unesco Repts. Mar. Sci. No. 21.

Unwin, D.J. (ed.). 2005. Re-Presenting GIS. Wiley, New York.

Van den Belt, M. 2004. Mediated Modeling: A System Dynamic Approach to Environmental Consensus Building. Island Press, Washington, D.C.

Van der Maarel (ed.) 1980. C!assification and Ordination. Junk, The Hague (Kluwer, Boston).

van der Meer, J., Heip, C.H., Herman, P.J.M., Moens T. and D. van Oevelen. 2005. Measuring the Flow of Energy and Matter in Marine Benthic Animal Populations. pp. 326–408 in: Eleftheriou. A. and A. McIntyre (eds.). 2005. Methods for Study of Marine Benthos. Blackwell Science Ltd., Oxford, UK.

van Vleet, E.S. and P.M. Williams. 1980. Sampling sea surface films: a laboratory evaluation of techniques and collecting materials. Limnol. Oceanogr. 25:764–770.

Vandermeer, J.H. and D.E. Goldberg. 2003. Population Ecology: First Principles. Princeton University Press, Ewing, NJ.

Venables, W.N. and B.D. Ripley. 1997. S-PLUS. Springer-Verlag, New York.

Venrick, E.L. 1971. The statistics of subsampling. Limnol. Oceanogr. 16: 811–818.

Ver Hoef, J.M. and N. Cressie. 2001. Spatial statistics: Analysis of field experiments. pp. 289–307 in: Scheiner, S.M. and J. Gurevitch (eds.). Design and Analysis of Ecological Experiments. Oxford University Press, Oxford.

Vinogradov. 1953. The Elementary Chemical Composition of Marine Organisms. Sear. Found. Mar. Mem. 2, Yale Univ., New Haven.

Vodopich, D.S. and J.J. Hoover. 1981. A computer program for integrated feeding ecology analyses. Bull. Mar. Sci. 31(4):922–924.

Von Brandt, A. 1985. Fish Catching Methods of the World. Fishing News Books, Ltd., Farnham, Surrey, England.

Von Ende, C.N. 2001. Repeated-measures Analysis: Growth and Other Time-dependent Measures. pp. 134–157 in: Scheiner, S.M. and J. Gurevitch (eds.). Design and Analysis of Ecological Experiments. Oxford University Press, Oxford.

Von Koch, H. 1904. Su rune courbe continue sans tangente, obtenue par une construction geometrique elementaire, Archiv. Matemat., Astron. Och Fys. 1:681–702.

Von Winterfeldt, D. 1999. On the relevance of behavioral decision research for decision analysis. pp. 133–154 in: Shanteau, J., Mellers, B.A. and D. A. Schum (eds.). Decision Science and Technology: Reflections on the Contributions of Ward Edwards. Kluwer Academic Publishers, Boston.

Von Winterfeldt, D. and W. Edwards. 1986. Decision Analysis and Behavioral Research. Cambridge University Press, New York.

Wadsworth, R. and J. Treweek. 1999. Geographical Information Systems for Ecology: An Introduction. Addison – Wesley – Longman Limited, Edinburgh Gate, England.

Walker, C.H. 2001. Principles of Ecotoxicology. CRC Press, Boca Raton, FL.

Walter, C.B., O'Neill, E. and R. Kirby. 1986. 'ELISA' as an aid in the identification of fish and molluscan prey of birds in marine ecosystems. J. Exp. Mar. Biol. Ecol. 96:97–102.

Walters, C.J. and S.J.D. Martell. 2004. Fisheries Ecology and Managements. Princeton University Press, Princeton.

Waples, R. (1998). Separating the wheat from the chaff: patterns of genetic differentiation in high gene flow species. Heredity 98:438–450.

Ward, R.,M. Woodwark, M. and D.O.F. Skibinski. 1994. A comparison of genetic diversity levels in marine, freshwater and anadromous fish. J. Fish. Biol. 44:213–232.

Ward, R. D. 2000. Genetics in fisheries management. Hydrobiologia 420:191–201.

Warde, W. and J.W. Petranka. 1981. A correction factor table for missing point-center quarter data. Ecology 62:491–494.

Wartenberg, D., S. Ferson, and F.J. Rohlf. 1987. Putting things in order: A critique of detrended correspondence analysis. Am. Nat. 129:434–448.

Warwick, R.M. and K. R. Clarke. 1991. A comparison of methods for analysing changes in benthic community structure. J. Mar. Biol. Assoc. U.K. 71:225–244.

Warwick, R.M. and K.R. Clark. 1994. Relearning the ABC: Taxonomic changes and abundance/biomass relationships in disturbed benthic communities. Mar. Biol. 118:739–744.

Wastler, T.A. 1969. Fourier Analysis: Applications in Water Pollution Control. U.S. Dept. of the interior.

Watson, J. 2004. HoloMar: A Holographic Camera for Subsea Imaging of Plankton. Sea Technol. 45(12):53–55. December.

Weibull, J.W. 1996. Evolutionary Game Theory. Labrinth Books, Millwood, New York.

Weinberg, S. 1981. A comparison of coral reef survey methods. Bijdragen tot de Dierkunde 51(2):199–218.

Weisberg, H.F. and B.D. Bowen. 1977. An Introduction to Survey Research and Data Analysis. Freeman, San Francisco.

Weiss, G.H. 1983. Random walks and their applications. Amer. Sci. 71:65–71.

Welch, P.S. 1948. Limnological Methods. McGraw-Hill, New York.

Wells, S.M. (ed.). 1988. Coral Reefs of the World. 3 Vols. IUCN, Cambridge, U.K.

Welsh, A.H., Peterson, A.T. and S.A. Altmann. 1988. The fallacy of averages. Amer. Nat. 132:277–288.

West, B. J. 1985. An Essay on the Importance of Being NonLinear. Springer-Verlag, New York.

Westwacott, S, Teleki, K., Wells, S. and J. West. 2000. Management of bleached and severely damaged coral reefs. IUCN, Cambridge.

Wetzel, R.G. and G.E. Likens. 2000. Limnological Analysis. Springer-Verlag, New York.

Wheeler, M. 1976. Lies, Damn Lies, and Statistics: The Manipulation of Public Opinion in America. Liveright, New York.

White, A.T., Hale, L.Z., Renard, Y. and L. Cortesi. 1994. Collaborative and Community-based Management of Coral Reefs. Kumarian Press, Inc., Bloomfield, CT.

White, G.C. 1996. NOREMARK: Population estimation from mark-resighting surveys. Wildlife Society Bulletin 24:50–52.

Whittaker, R.H. 1956. Vegetation of the Great Smokey Mountains. Ecol. Monogr. 26(1):1–80.

Whittaker, R.H. 1967. Gradient analysis of vegetation. Biol. Rev. 42:207–264.

Whorff, J.S. and L. Griffing. 1992. A video recording and analysis system used to sample intertidal communities. J. Exp. Mar. Biol. Ecol. 160:1–12.

Widder, E., Frey, L. and J. Bowers. 2005. Improved Bioluminescence Measurement Instrument. Sea Technol. 46(2):10–15. February.

Wiegert, R.G. 1962. The selection of an optimum quadrat size for sampling the standing crop of grasses and forbs. Ecology 43:125–129.

Wilcox, T.E. and B. Fletcher. 2004. High-frequency sonar for target re-acquisition and identification. Sea Technol. 45(6):41–45.

Williams, B.K., Nichols, J.D. and M.J. Conroy. 2002. Analysis and Management of Populations. Academic Press, New York.

Williams, G.P. 1997. Chaos Theory Tamed. Joseph Henry Press, Washington, D.C.

Williams, W.T., Lambert, J.M. and G.N. Lance. 1966. Multivariate methods in plant ecology. V. Similarity analyses and information-analysis. J. Ecol. 54:427–446.

Winer, B.J. 1971. Statistical Principles in Experimental Design. McGraw-Hill, New York.

Wood, S., Nulph, A., and B. Howell. 2004. Application of Autonomous Underwater Vehicles. Sea Technol. 45 (12):10–14. December.

Woodin, S.A. 1986. Experimental field manipulations of marine macroinfaunal assemblages: benefits and pitfalls. In: Biology of Benthic Marine Organisms, (eds.) Thompson, M.-F., Sarojini, R. and R. Nagabhushanam, Oxford & IBH Publ. Co. Pvt. Ltd., New Delhi, pp. 185–192.

Wright, M., Bakus, G.J., Ortiz, A., Ormsby, B. and D. Barnes. 1991. Computer image processing and automatic counting and measuring of fouling organisms. Comp. Biol. Med. 21:173–180.

Wright, S. 1951. "The genetical structure of populations." Ann. of Eugenics 15: 323–353.

Yates. F. 1981. Sampling Methods for Censuses and Surveys. Macmillan, New York, 458 pp.

Yong, M.Y. and R.A. Skillman. 1975. A computer program for analysis of polymodal frequency distributions

(ENORMSEP), FORTRAN IV. Fish. Bull. 73: 681.

Young, L.J. and J.H. Young. 1998. Statistical Ecology – A Population Perspective. Springer, New York.

Zar, J. H. 1999. Biostatistical Analysis. Pentice-Hall, London.

Zeitzschel, B., Diekmann, P. and L. Uhlmann. 1978. A new multi-sample sediment trap. Mar. Biol. (Berlin) 45: 285–288.

Zolman, J.F. 1993. Biostatistics: Experimental Design and Statistical Inference. Oxford University Press, New York.

Index

Abalone, DNA sequence analysis of, 203

Abundance, concordance of dominance and, 160

Abundance-based coverage estimators (ACEs), 147

Accidental sampling, 19–20. *See also* Haphazard sampling

Accumulation curves, 142

Acetate to ergosterol method, 324

Acoustic turbidity measurement, 305

Adaptive cluster sampling, 30

Adaptive sampling, 25–26

Additive utility models, in decision analysis, 199

Adequate measurements, 62

Adjusted Similarity index, 155

Aerial photography, in coral sampling, 313

Aerial surveys, large scale sampling via, 56

Age classes, in populations, 132–133

Age specific life tables, 134–136

Agglomerative clustering, 218

Aggregated distribution, 123, 124, 125. *See also* Statistical distributions
in sampling design, 12

Aggregation, 125. *See also* Aggregated distribution

Airborne remote sensing, satellite remote sensing versus, 58

Alcohol, 333

Algebraic equations, analytical methods for solving, 287

Allomone collector, 314, 316

Allozymes, 204, 205

Alpha diversity, 142

Alpha (α) probabilities, in descriptive statistics, 80–81

ALVIN submersible, 308, 309

Analysis. *See also* Community analysis; Cluster analysis; Frequency analysis; Power analysis; Principal Component Analysis (PCA); Ranking analysis; University of Washington Fisheries Sciences Spatial Analysis Laboratory
genetic, 203–206
of population genetic structure, 202

Analysis of covariance (ANCOVA), 103–104, 252, 253. *See also* Multiple Analysis of Covariance (MANCOVA)

Analysis of molecular variance, 202

Analysis of Similarities (ANOSIM) procedure, 243, 254
in community analysis, 229, 231–233

Analysis of variance (ANOVA), 88–96, 252, 254. *See also* Multivariate ANOVA (MANOVA); Non-Parametric Multiple ANOVA (NPMANOVA)
with Mantel test, 229
regression and, 109

Analytical models
for ecosystems, 285
of population growth, 138

Angle of inclination, of underwater slopes, 2–3

Animal counts, in image processing, 185

Animals, food selectivity of, 131–132. *See also* Birds; Mammals; Reptiles

Animal tags, 312

Animation, in multimedia development, 191

Aquarius project, 308

Aquatic ecotoxicology, 197

Aquatic genetics, DNA sequence analysis in, 204–205

Quantitative Analysis of Marine Biology Communities: Field Biology and Environment
by Gerald J. Bakus
Copyright © 2007 John Wiley & Sons, Inc.